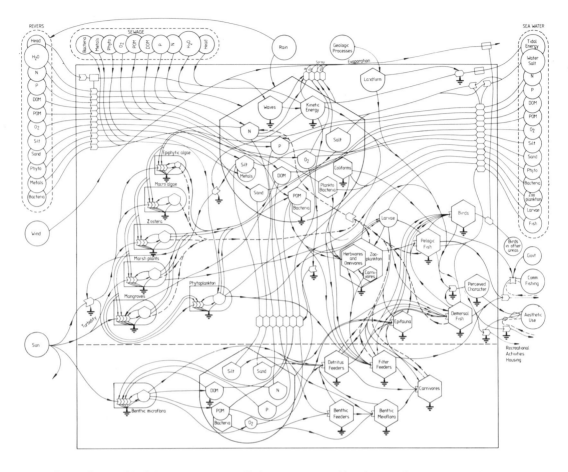

Energy-flow model of the Upper Waitemata Harbour, a mangrove fringed estuary in northern New Zealand.

Estuarine Ecosystems: A Systems Approach

Volume I

Author

George A. Knox
Professor of Zoology
University of Canterbury
Christchurch, New Zealand

CRC Press, Inc.
Boca Raton, Florida

Library of Congress Cataloging-in-Publication Data
Main entry under title:

Knox, G. A.
   Estuarine ecosystems.

   Bibliography: v. 1, p.
   Includes index.
   1. Estuarine ecology. I. Title.
QH541.5.E8K55   1986   574.5′26365   85-29983
ISBN 0-8493-6996-7 (v. 1)
ISBN 0-8493-6997-5 (v. 2)

This book represents information obtained from authentic and highly regarded sources. Reprinted material is quoted with permission, and sources are indicated. A wide variety of references are listed. Every reasonable effort has been made to give reliable data and information, but the author and the publisher cannot assume responsibility for the validity of all materials or for the consequences of their use.

Direct all inquiries to CRC Press, Inc., 2000 Corporate Blvd., N.W., Boca Raton, Florida, 33431.

© 1986 by CRC Press, Inc.

International Standard Book Number 0-8493-6996-7 (v. 1)
International Standard Book Number 0-8493-6997-5 (v.2)

Library of Congress Card Number 85-29983
Printed in the United States

To Dorothea
my wife, whose love, devotion, and understanding have supported and sustained me throughout my professional career.

PREFACE

Over the last 2 decades there has been an increasing interest in estuaries both as a focus for scientific research and from the public at large. Estuaries throughout the world have been preferred sites of human settlement. As a result they have often been subjected to the most intensive use applied to any marine area. In addition they receive the impact of many human activities throughout the entire watershed of the rivers that flow into them. In many parts of the world stopbanking, channelization, reclamation, development of ports and marinas, and waste discharges have irreversibly altered the nature of many estuarine systems. It is only gradually, and comparatively recently, that we have come to know the multipurpose values that they serve, their high productivity, and the key role that they play in the economy of coastal waters.

Earlier work on estuaries tended to concentrate on the problems of life in estuaries, but in more recent years the focus of much of the research on these fascinating ecosystems has been on an understanding of the processes which control their productivity and dynamic functioning. This emphasis on processes such as primary production, grazing, predation, secondary production, decomposition, detritus formation, role of dissolved organic matter, decomposition and the role of microorganisms, food webs, energy flow, nutrient cycling, and so on are the topics that are dealt with in this book. As a consequence it has not been possible to deal with the considerable literature on the biology of estuarine organisms, especially salinity adaptations, species distribution patterns, and community structure.

Thus the approach used in this book is somewhat different from that used in many texts in marine ecology in that the emphasis is almost exclusively on ecological processes and the use of systems approaches towards an understanding of such processes. As H. T. Odum has stated in his excellent text *Systems Ecology,* ''If the bewildering complexity of human knowledge developed in the twentieth century is to be retained and well used unifying concepts are needed to consolidate the understanding of systems of many kinds and to simplify the teaching of general principles.'' This book is an attempt to bring some order to the understanding of one of the most complex ecosystem types, estuarine ecosystems. Wherever possible the energy circuit language of symbols and diagrams developed by H. T. Odum has been used as a basis for understanding. Other systems languages and approaches are also discussed, and in some cases translated into energy circuit language in order to assist in the development of general principles.

As will be seen by the reference list at the end of this book, there is considerable recent literature on estuarine ecosystems although the list includes only a fraction of the published work. Because of this growing volume of published research I have had to be selective in the material included in these 2 volumes. Thus, of necessity I have concentrated on the more recent work published since the late 1960s, and in particular during the last decade. Regretfully, except in a few instances, it has not been possible to develop the history of the concepts being discussed. Examples have been carefully chosen from the pool of published research to illustrate these concepts. There are doubtless other examples that could have been equally useful and I apologize to authors whose work has not been included. Readers will find a bias towards work carried out in the U.S. since it is there that the bulk of recent process-orientated work has been carried out. However, I have attempted to balance this with examples from a range of geographic areas and to include some of the Southern Hemisphere work which is perhaps not so widely known.

I hope that this book will prove useful to advanced undergraduates, graduate students, and to professionals engaged in research on estuarine ecosystems, as well as those interested in problems of estuarine management. It will also have been worthwhile if it stimulates others to work on some of the fascinating aspects of research on estuarine ecosystems that have been discussed.

George A. Knox

THE AUTHOR

George A. Knox, M.B.E., F.R.S.N.Z., was Head of the Department of Zoology, University of Canterbury, Christchurch, New Zealand from 1959 to 1978. He is now Professor-Emeritus in Zoology.

Professor Knox was born in New Zealand and received his education at the University of Canterbury where he was appointed as staff member in 1948. He has been a Visiting Fellow at the East-West Center, Honolulu, and a Visiting Professor at the Department of Oceanography, Texas A & M University and the Department of Environmental Engineering Sciences, University of Florida, Gainesville. He has visited and worked in laboratories in the U.S., Canada, South America, Japan, Australia, Western Europe, the U.S.S.R., and China.

Professor Knox's major research interests have been the study of estuarine ecosystems with special reference to energy flow and systems analysis, problems of estuarine and coastal zone management, and studies of the pelagic and benthic ecosystems beneath the sea-ice in McMurdo Sound, Antarctica. He has published over 100 scientific papers, written three books, and was editor and co-author of two further volumes. He has received a number of awards and fellowships for his contributions to science: Fellow of the Royal Society of New Zealand (F.R.S.N.Z.), 1963; Hutton Medal, Royal Society of New Zealand, 1978; Conservation Trophy, New Zealand Antarctic Society, 1980; Honorary Member, Scientific Committee for Antarctic Research, 1982; Member of the Most Excellent Order of the British Empire (M.B.E.), 1985; New Zealand Marine Sciences Society Award for Outstanding Contribution to Marine Science in New Zealand, 1985; and New Zealand Association of Scientists' Sir Ernest Marsden Medal for Service to Science, 1985.

Throughout his career Professor Knox has been active in international scientific organizations. He has been a member of the Scientific Committee on Oceanic Research (SCOR) and the Special Committee for the International Biological Programme (SCIBP). He has been a member of the Scientific Committee for Antarctic Research (SCAR) since 1969, serving 4-year terms as Secretary and President, and a member of the Governing Board of the International Association for Ecology (INTECOL) since 1965, also serving 4-year terms as Secretary-General and President.

ACKNOWLEDGMENTS

I would like to express my appreciation to the late Edward Percival, who by his enthusiasm and teaching skills started me on my career as a marine biologist; to Howard Odum who inspired my interest in the energy analysis approach to ecosystem modeling; and to my colleagues in New Zealand and in many parts of the world and my students with whom I have discussed many of the ideas in this book.

My thanks are also due to all those who gave permission for me to reproduce original figures. I am indebted to Margaret Hawke and Lyn Tomkies for the typing of the manuscript and to John Black and Paul Cook for assistance with the drafting of the figures. Finally, I am indebted to Sandy Pearlman and the staff of CRC Press for their patience and support during the preparation of this volume and for seeing the project through to the completion of such a high quality publication. Any errors that remain are the author's responsibility and I would appreciate it if any reader cares to point them out.

ESTUARINE ECOSYSTEMS: A SYSTEMS APPROACH

Volume I

Volume II

TABLE OF CONTENTS

Chapter 1

THE APPROACH

I. WHAT IS AN ESTUARY?

If a marine biologist, a geographer, and a physical oceanographer each were asked to define an estuary, the three definitions might differ considerably. The biologist would emphasize that estuaries provide living organisms with widely varying salinities, and he would draw attention to their high productivity. The geographer would stress the fluvial, marine, and sedimentary processes at work in a drowned river mouth. The physical oceanographer's attention would be drawn to the interplay of river flow, tidal fluctuations, and density distributions.

Consequently, there are a variety of definitions that have been used in the literature. At the First International Conference on Estuaries held in Georgia in 1964, a confusing array of definitions were proposed, and these are to be found in the proceedings of the conference edited by Lauff[500] under the title *Estuaries*. To most people, an estuary is a place where rivers meet the sea and come under the influence of the tides forming a transition zone between fresh water and sea water. However, many other types of coastal water bodies are estuarine in character, and a broader concept is generally adopted by workers on estuarine problems. There is wide agreement that variable salinity is an essential feature of all estuarine systems, and the definition proposed by Pritchard[743] has generally been accepted in the literature. According to Pritchard: "an estuary is a semi-enclosed coastal body of water which has a free connection with the sea and within which sea water is measurably diluted with fresh water derived from land drainage." However, as Day[178] points out, such a definition excludes saline lakes whose waters contain salts with a different chemical composition from that of the sea. It also excludes marine inlets and lagoons without fresh inflow, and inlets on arid coasts whose salinity is the same as that of the sea. Excluded also are fjords which exhibit many of the features of estuaries. According to Pritchard,[743] the Baltic Sea, whose salinity is effectively stable over wide areas, is not an estuary, but this is not accepted by many Scandinavian workers.

Day[174] also notes two further difficulties with Pritchard's definition. The phrase " . . . a free connection with the open sea . . . " would exclude those estuaries which are usually cut off from the sea in the dry season, while the phrase " . . . diluted with fresh water . . . " would exclude estuaries which, far from being diluted with sea water, become hypersaline when evaporation exceeds fresh water inflow. In southern Africa, the Gulf of Mexico, Western and Southern Australia, and other arid countries there are many blind estuaries which are closed by sand bars for shorter or longer periods. Most of them burst open to the sea during floods, but thereafter they become hypersaline. Thus, Day[176] has proposed an amended definition: "an estuary is a partially enclosed coastal body of water which is either permanently or periodically open to the sea, and within which there is a measurable variation of salinity due to the mixture of sea water with fresh water derived from land drainage."

Many coastal shallow water bodies that are permanently cut off from the sea, or only intermittently open are called lagoons. There are particularly extensive series of lagoons along the eastern and Gulf of Mexico coasts of the U.S., Mexico, Brazil, West Africa, Natal, the southern and eastern shores of the Indian Peninsula, southwest and southeast Australia, Alaska, Siberia, around the Mediterranean, and around the southern Baltic, Black, and Caspian Seas. Lagoons grade into other coastal systems such as semienclosed marine bays, fresh water lakes, and into estuaries. There are four main lagoonal environments: fresh water dominated, brackish, sea water dominated, and hypersaline. Any given lagoon may comprise one (as in many small- to medium-sized lagoons) to, rather exceptionally, three of these environments (as in several of the larger and more elaborate systems). Salinity patterns, shallow depths, soft sediments, the well-mixed nature of the water column through wind action, organic richness, and often a marked tendency for rapid environmental change are salient estuarine characteristics.[34]

II. SPECIAL FEATURES OF ESTUARINE ECOSYSTEMS

The salient feature of estuarine ecosystems is its high productivity.[476] This arises from the unique set of geomorphological, physical, chemical, and biological factors that are characteristic of estuarine systems. While it is a truism that no two estuaries are identical, nevertheless, they do share a number of essential features, and in recent years research has led to the formulation of a number of generalizations concerning their structure and function. Many of the characteristics of estuarine systems are shared with adjacent coastal systems; but it is the combination of characteristics that make them unique. Above all, it is the fluctuating salinity patterns that give rise to their unique biological characteristics.

A. Circulation Patterns and Salinity Distributions

Circulation patterns and salinity distributions are among the most distinctive characteristics of estuaries.[210] The combined influences of fresh water inflow wind, waves, and tidal action result in particular patterns of water movement. These patterns range from stratified two-layered systems, characteristic of drowned river valleys and fjords, to well-mixed vertically homogeneous systems (see Chapter 2, Section IV). Nonstratified estuarine circulation is common in shallow embayments where the water is constantly mixed by wind and tide. It is also common in bays lacking a good supply of fresh water from land drainage. Circulation of water transports nutrients, propels plankton, distributes suspended fish larvae and invertebrates, flushes away wastes from animals and plants, cleanses the system of pollutants, controls salinity patterns, shifts sediments about, mixes water masses, and generally does useful work.[96]

Salinity fluctuations are a very important feature of the estuarine environment. Superimposed on the broad salinity gradient (of high mean salinities at the mouth to low mean salinities at the head) is a series of salinity oscillations. These oscillations vary in duration and amplitude throughout the estuary. In addition, shallow water and proximity to land promote heat exchange and lead to much greater temperature ranges and fluctuations than those found in the open ocean.

B. Estuarine Sediments

Estuarine sediments are unique, having their own characteristics and complexity.[620] In general, there is an upstream-downstream gradient from mud to sand, owing to the interplay of different physical factors. There is also a gradient in substratum texture throughout the intertidal areas, from one of finer particles at low tide to coarser particles higher up the shore. The character of the sediments is profoundly modified by the activity of burrowing invertebrates, especially the deposit or sediment feeders which are continually turning over the substrate. It is also modified by the deposition of fecal pellets, especially from filter-feeding bivalves, and the input of organic matter from rivers and from the marsh plants that line the margins. In addition to storing materials by adsorption, the sediments are sites of intense microbial activity. Through this activity the sulfur, nitrogen, and phosphorus cycles continually decompose the complex organic compounds of plant and animal detritus, and make it available in usable forms such as ammonia, nitrate, or phosphate.[635,1033]

C. Estuarine Productivity

Estuaries include some of the most productive plant systems found on this planet. Primary production is complex involving in various combinations the following groups of plants:

1. Macrophytes (seagrasses, sedges, rushes, cordgrasses, mangroves, etc.)
2. Microepiphytic algae
3. Macroepiphytic algae
4. Benthic microalgae (diatoms, flagellates, blue green algae)
5. Benthic macroalgae
6. Phytoplankton

No single study has simultaneously measured all these components of organic matter function. They are, however, highly dynamic and productive, e.g., a dense seagrass meadow may be composed of more than 4000 plants m^{-2} and may have a standing stock of 1 to 2 kg dry wt m^{-2}. Associated with this is an increase in bottom area by 10 to 20 times, as a result of leaf surface. Productivity of the seagrasses alone reported ranges from at least 5 to 15 g C m^{-2} day^{-1}, and when other producers such as the epiphytic microalgae are included, daily production can be well over 20 g C m^{-2} day^{-1}.[570]

Westlake[986] reviewed plant productivity on a global scale and concluded that when agricultural systems were discounted, tropical rain forests appeared to be the most productive of all (5 to 8 kg m^{-2} organic dry weight per annum), but salt marshes, reed swamps, and submerged macrophytes were the next most productive (in the range 2.9 to 7.5 kg m^{-2} $year^{-1}$). Mean net primary production for estuaries as a whole is about 2 kg m^{-2} $year^{-1}$, as compared to means of 0.75 for the total land and 0.155 for the total ocean.[575,804]

D. Estuarine Food Webs

Direct grazing by herbivores in general consumes only a small proportion of the macrophyte and macroalgal production. The great bulk of the organic matter produced (something over 90%) is processed through the detrital system. A much greater proportion of the primary production of the phytoplankton and benthic microalgae is, however, consumed by zooplankton, planktivorous fishes, the interstitial micro- and meio-fauna, surface deposit feeding mollusks, fishes and polychaete worms, and filter-feeding invertebrates. Annual plant growth and decay provides continuing large quantities of organic detritus.[161,229,635,912] In addition, there is often a very considerable input of detritus from river inflow, especially during storm events.[616] This detritus becomes colonized by a host of bacteria, fungi, microalgae, and other microorganisms.[229] Detrital particles and their associated microorganisms provide a basic food source for primary consumers such as zooplankton, most benthic invertebrates, and some fishes.

Many of the estuarine consumers are selective or indiscriminate feeders on particles in suspension in the water column, or in the sediment that they ingest. Thus, most of the biota that inhabits estuaries are best described as particle producers (microalgae and detritus derived from plant growth) and particle consumers,[141] and it is difficult to relate these two groups to the traditional primary producer vs. primary consumer categories. Thus, the first trophic level in the estuarine ecosystem is best described as a mixed trophic level of particle consumers, which in varying degrees are herbivores, omnivores, or primary carnivores.

E. Factors that Enhance Estuarine Productivity

There are six processes occurring at the estuarine interface of land and sea that tend to enhance primary productivity.

1. Fresh Water Inflow

River inflow results in considerable inputs of nutrients and organic matter, both particulate and dissolved. It also plays a major role in the nutrient trap effect detailed below.

2. Nutrient Trap Effect

Sediment characteristics and circulation and salinity patterns result in estuaries acting as nutrient traps or sinks, causing essential elements to be recycled over and over.[203,815] This nutrient trap effect is caused by three factors:[659]

1. The nature of estuarine sediments, with their high percentage of clay minerals, having a great adsorptive capacity which produces sediments containing large quantities of adsorped nutrients, trace elements, and other materials.
2. The process of biodeposition, whereby filter feeders remove enormous quantities of suspended matter which are compacted and extruded as feces or pseudofeces, to be incorporated in the sediments.[494,495]
3.   Through a combination of horizontal ebb and flow of water masses of different salinities, there is a tendency for nutrients to become trapped.

Fresh water flowing from estuarine watersheds is relatively rich in nitrogen and phosphorus, and creates a gradient from high to low concentrations as one moves towards the sea.[142,734,1041] The primary producers utilize these nutrients and those in the sea water to produce plant biomass. Phytoplankton and detritus produced from the death and decay of the marsh macrophytes tend to settle towards the bottom as they are flushed down the estuary. They carry a large proportion of the nitrogen and phosphorus, which they have assimilated in the surface waters, along with them to the bottom. Typical estuaries are stratified with a surface flow of fresher water (Figure 1A) and a counterflow of sea water forming a "salt wedge" along the bottom.[210,744] These two layers are separated by density variations due both to salt concentrations and temperature differences. The sheer forces at the junction of the upper and lower layers gives rise to turbulent mixing. This "counter current" system is responsible for nutrient trapping or conservation in estuaries.[461] Both living and dead particulates which settle through the pycnocline, or zone of maximum vertical density differential into the counter current, are carried up the estuary along with their nutrient contents.

3. Tidal Mixing

The rise and fall of the tide (Figure 1B) in an estuary is another factor promoting vertical mixing of nutrient-rich water from the bottom.[576] When the volume of tidal exchange is large compared with river input to an estuary, vertical salinity gradients may be broken down so that the estuary is virtually of the same salinity from top to bottom. The most noticeable salinity gradient is then the horizontal one from the river to the open sea. Under such conditions, nutrients regenerated at the sediment surface are carried rapidly to the surface water and used in primary production. A well-documented example is Narragansett Bay, Rhode Island.[491]

4. Tidal Marsh Nutrient Modulation

At times when nutrients are high in the upper estuary surface waters they tend to be taken up rapidly in the tidal marshes, mud flats, and bottom sediments.[142] At times of low nutrient concentrations in surface waters, a net release of nutrients occurs.[279] From their study of nutrient fluxes across the sediment-water interface in the turbid portion of the Patuxent Estuary in Chesapeake Bay, Boynton et al.[64] concluded that: "In general it appears that nutrient fluxes across the sediment water interface represent an important source to the water column in summer when photosynthetic demand is high and water column stocks are low and, conversely, serve as a sink in winter when demand is low and water column stocks high, thereby serving as a "buffering" function between supply and demand." Overall, in the long term the reservoir of nutrients in the sediment remains relatively constant; in the short term, however, they act as nutrient filters or modulators.[24,45] The marshes also tend to trap particulate nitrogen and phosphorus and convert them into orthophosphate, ammonia,

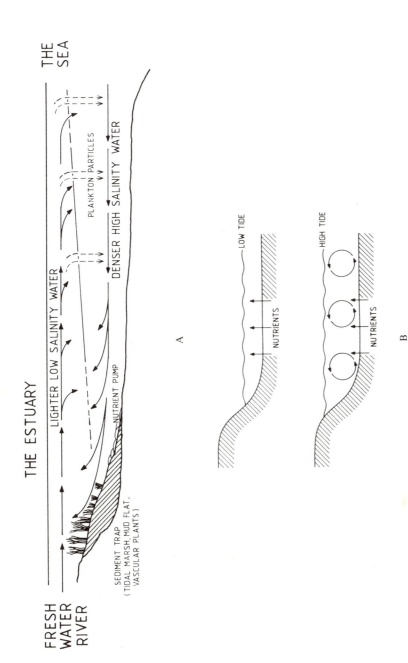

FIGURE 1(A). Schematic diagram of the nutrient conserving and modulating mechanisms in estuaries, including the two-layered salt wedge; plankton circulation pattern; sediment trap; the tidal marsh, vascular plant "nutrient pump", and deep bottom sediment modulators. (After Correll, D. L., *Bioscience*, 28(10), 649, 1978. With permission. Copyright c 1978 by the American Institute of Biological Sciences.) (B) Process of tidal mixing of nutrients in the shallow water estuaries (From Mann, K. H., *The Ecology of Coastal Waters. A Systems Approach*, Blackwell Scientific, Oxford, England, 1982, 3. With permission.)

and dissolved organic phosphorus and nitrogen, which are then exported back to the open waters of the estuary.[24,189]

5. Sediment Trapping

Rivers deliver to estuaries large quantities of mineral particulates derived from land erosion, e.g., the Rhode River estuary, U.S., receives about 1.2 t ha^{-1} of estuary per year from land runoff.[143] When a fresh water river flows into an estuary, the current velocity drops, the pH and ionic composition of the water are altered, and all but the fine clay fraction of the mineral particulates are deposited in a rather short distance. This depositional zone is termed the sediment trap (Figure 1A). In the Rhode River, Correll et al.[143] found that sediments were deposited in this zone at an average rate of about 11 tons ha^{-1} year^{-1}. Over time this process can sequentially produce tidal mud flats, low tidal marshes, high tidal marshes, and finally, terrestrial land. These sediments are rich in nutrients and organic matter, and at the midtidal-mud flat stage of sedimentation, support large populations of seagrasses and marsh plants.

6. Vascular Plant "Nutrient Pump"

Eelgrasses and marsh plants are believed to have the capability of acting as "nutrient pumps" between surface water and bottom sediments.[141] Thus, on the one hand, they can take up nutrients from the sediments, and lose them to the water via death and decomposition, leaching from leaves, herbivorous activity, or perhaps direct excretion. Gallagher et al.[270] showed that living *Spartina* loses a few percent of the photosynthate it produces when it is washed by the rising tide. On the other hand, their leaves can take up nutrients directly from the water, at least under some conditions, and translocate them to their roots. Dense eelgrass (*Zostera marina*) beds in the Izembek Lagoon of Alaska, take up phosphorus from bottom sediments at the rate of 166 mg P m^{-2} day^{-1}, and excrete it into tidal waters as orthophosphate at the rate of 62 mg P m^{-2} day^{-1}.[566,567] This eelgrass nutrient pump activity resulted in significant diel fluctuations in overlying surface water and sediment interstitial phosphorus concentrations.

F. Secondary Production in Estuaries

Secondary production in estuaries is also high. They are sheltered areas of high localized productivity, and in addition are important nursery areas for a wide variety of fish and crustaceans, including many species of commercial importance. Secondary production of benthic macrofauna is high compared to that of other ecosystems, possibly because of a high level of food, easy availability of food, and a preponderance of opportunistic species characterized by high growth rates and a rapid turnover.[1030]

Increasingly, in recent years we are recognizing the importance of estuaries to coastal fisheries. A few statistics highlight this. Korringa[487] showed how the inlets and estuaries of the coasts of Holland serve as nursery grounds for skate (*Solea*) and plaice *(Pleuronectes)*. Young fish congregate there in densities of up to 1 m^{-2}, feeding on the rich assortment of benthic organisms living in a mosaic of sediment types. In a report on South Florida's mangrove-bordered estuaries[945] it was estimated that in 1968 the commercial landing of species linked with the mangrove-food web included 14.5 million kg of prawns worth $15.7 million, 1.7 million kg of spotted sea trout worth $1.0 million, and 6.8 million kg of blue crabs worth $1.2 million. Sykes[907] has stressed the economic importance of estuarine dependent species to the U.S.: "Biological investigations have proven that commercial fisheries in the continental shelf of the United States are largely for species which spend a portion of their lives in estuaries. Nearly two-thirds of the total catch of fish and shellfish from waters off the east coast of the United States and well over half of the entire U.S. commercial catch is made up of estuarine dependent species." This conclusion was confirmed by

McHugh,[552] who recorded that of the 2925 $\times$ 10[6] kg per annum of fish landed in the U.S., about 2025 $\times$ 10[6] kg were species dependent on estuaries as nursery areas.

Pollard[727] has assessed the importance of New South Wales estuaries to commercial and sport fisheries. Over the 10-year period 1962—1963 to 1972—1973, the average annual catch of fish fluctuated around 14 million kg, and the combined fish, crustacean, and molluskan catch around 22.5 million kg. Total value at 1972—1973 wholesale prices was $27 million (Australian dollars). The proportion of this catch taken directly from estuaries was around 47% and the proportion of its total value about 50%. The estuarine dependent portion of the total catch was estimated to be about 66% by weight and over 70% by value. He also estimated that recreational fisheries caught at least the equivalent of 25% of the commercial catch.

Ritchie[793] has documented the importance of mangrove-dominated estuaries in northern New Zealand to local fisheries. For the Arapaoa River estuary, a many-branched and mangrove-fringed arm of Kaipara Harbour with an area of 4750 ha, he found that the annual flounder catch supported six full-time and two part-time fishermen. In addition, there were seven active oyster farms with three farmers gaining a full-time living from their farms. The area also produced moderate incidental catches of other fish species including kahawai, parore, mullet, shark, and an extensive amateur fishing for snapper. Thus, this small area supports the equivalent of 12 family units full-time, and many more in part, with an annual income in excess of $200,000 (New Zealand dollars).

III. ESTUARINE VALUES AND VULNERABILITY

From man's viewpoint, estuaries have multiple values that have too often in the past not been recognized by developers and planners. Their values lie in the biological resources of fishes and shellfish which are of prime economic importance; their function as a fundamental link in the development of many species of fish and crustaceans (including economically important species such as flatfish, mullet, and prawns), and in the migration of important species such as salmon; their provision of feeding and breeding sites and stopover points on migration routes for many species of ducks, geese, swans, and a great variety of wading birds; their mineral resources of sand, gravel, and sometimes oil; their provision of harbors and transportation routes for commerce; their provision of locations for housing and industrial plants; the recreational opportunities they provide for hunting, fishing, boating, swimming, and aesthetic enjoyment; and the opportunities they provide for scientific investigation.[6]

Estuaries not only have multipurpose values, but they are also extremely vulnerable to impact from man's activities. This vulnerability arises from some of the important characteristics of the estuarine environment as discussed above. These have been listed by Odum[657] as:

- **Estuarine productivity and nutrient trap effect**—As discussed above, the nutrient trap effect plays an important role in maintaining the high productivity of estuarine ecosystems. Unfortunately it also causes estuaries to become pollution sinks with the deposition and concentration of pollutants from petroleum by-products, persistent pesticides, and radioactive chemicals.[89,663,1042]

- **The unique structure of estuarine food webs**—The role of vascular plant production in estuarine food webs has been discussed above. Often, a single plant species such as the cordgrass *Spartina* is outstandingly important as a primary producer in the system. Because of the low diversity of plant species and the importance of plant detritus in the energy-flow pathways, the flow of energy is particularly susceptible to human impact through the destruction of plant producing areas by stopbanking, channelization, and reclamation.

- **Vulnerability of estuarine organisms**—Many estuarine organisms are living at or near the limits of their tolerance range, and consequently they are sensitive to alterations such as increased water temperatures from thermal power plants, decreased oxygen concentrations resulting from oxygen depletion following heavy organic pollution or eutrophication, or from the introduction of additional stress from low levels of pollutants.
- **Sedimentary control of the estuarine environment**—The important role played by the sediments in estuarine ecosystems has been discussed above. Man's activities can profoundly modify the nature of these sediments.
- **Importance of fresh water inflow**—The salient characteristic of estuaries is the gradient from fresh water to marine conditions, with the tidal rhythm in salinity gradients. Alterations to fresh water inflows can seriously interrupt and disturb the salinity regime to which the organisms are adapted. This salinity regime is of the utmost importance to the nursery function of the estuary in the life cycles of many species.

Estuaries throughout the world have always been the preferred sites of human settlement. As a result they have often been subjected to the most intensive use applied to any marine area. In addition, they receive the impact of many human activities throughout the entire watershed of the rivers draining into them.

IV. CURRENT ESTUARINE RESEARCH DIRECTIONS

Much recent research has focused on gaining an understanding of the mechanism whereby the high productivity of estuaries is maintained, and in determining the relationships between estuaries and adjacent marine waters. The following questions have been addressed:

1. Why are estuaries highly productive systems?
2. What are the components of primary production within various estuarine ecosystems, and what are their relative importance?
3. What are the mechanisms that maintain the high productivity?
4. What are the pathways and the rates of energy (carbon) flow within estuarine ecosystems, and what are their relative importance?
5. What are the roles of microorganisms within estuarine ecosystems, and what are the relative importance of aerobic and anaerobic processes?
6. What are the pathways and rates of chemical fluxes in estuarine ecosystems?
7. What are the rates and magnitudes of the flows of organic carbon and nutrients between estuarine marshes and coastal waters (or, are estuarine marshes and water bodies net importers or exporters of organic carbon and nutrients)?

V. THE SYSTEMS APPROACH

In marine science, as data have accumulated, so too have hypotheses about the structure, function, and patterns of estuarine ecosystems. However, it is only over the last 20 years or so that significant advances have been made in the form of testable models used to clarify our understanding of everything from water circulation patterns, to individual species population dynamics, to the behavior of an estuary undergoing pollution stress. Earlier work was devoted largely to autecology, physiological ecology, population ecology, and community ecology. More recently, attention has been directed towards the interactions between organisms, rather than the organisms themselves. Studies have focused on the flow of energy and matter through food webs and the cycling of elements. This has led to a recognition that in the diverse systems studied, there are recognizable patterns that are the properties of interacting components or ecosystems.

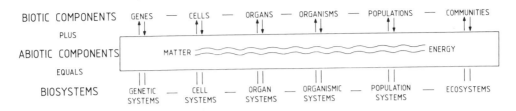

FIGURE 2. Levels of organization spectrum or hierarchy. Ecology focuses on the right-hand portion of the spectrum, that is, the levels of organization from organisms to ecosystem. (From Odum, E. P., *Basic Ecology*, Saunders College Publishing, Philadelphia, 1983, 4. With permission.)

It is appropriate at this point to make mention of the concept of "levels of organization."[638] Figure 2 depicts the levels of organization spectrum or hierarchy. Interaction with the physical environment (energy and matter) at each level produces characteristic functional systems. A *system* consists of "regularly interacting and interdependent components forming a unified whole" *(Webster's Collegiate Dictionary)*, or, from a different point of view, "a set of mutual relationships constituting an identifiable entity, real or postulational."[499] Hall and Day[350] consider that "Any phenomenon, either structural or functional, having at least two separable components and some interaction between these components may be considered a *system*."

An important consequence of hierarchical organization is that as components, or subsets, are combined to produce larger functional wholes, new properties emerge that were not present at the level below. Accordingly, an *emergent property* of an ecological level or unit cannot be predicted from the study of the components of that level or unit. In this book the emphasis is on the ecosystem level. Studies of whole ecosystems are often technically and logistically difficult. Nevertheless, as Mann[576] points out "the scientific reasons for working at the ecosystem level are compelling. Once we can see that ecosystem processes are occurring, and are influencing events at the level of organisms and populations, it becomes essential to our understanding of events at any of these levels to know about the system processes."

A. Systems Analysis

Mathematical symbols provide a useful shorthand for describing complex ecological systems, and mathematical equations permit formal statements of how ecosystem components are likely to interact. "The process of translating physical or biological concepts about any system into a set of mathematical relationships, and the manipulation of the mathematical system thus derived is called systems analysis."[971] The mathematical system is called a model and is an imperfect and abstract representation of the real world.

Jeffers[427] has pointed out that contrary to the belief of many ecologists, systems analysis is not a mathematical technique, or even a group of techniques. Rather, it is a broad research strategy that certainly includes the use of mathematical techniques and concepts, but in a systematic, scientific approach to the solution of complex problems. Basically it is a tool for understanding.

The application of systems analysis procedures to ecology has come to be known as *systems ecology*. The approach is a holistic one, holism being defined as "the philosophy of studying the total behaviour (or other total attributes) of some complicated system."[350] Odum[635] states that: "As the formalised approach to holism, systems ecology is becoming a major science in its own right for two reasons: (1) extremely powerful new formal tools are now available in terms of mathematical theory, cybernetics, electronic data processing, and so forth, and (2) formal simplification of complex ecosystems provide the best hope for solutions to man's environmental problems that can no longer be trusted to trial-and-error, or one-problem-one-solution procedures that have chiefly been relied on in the past."

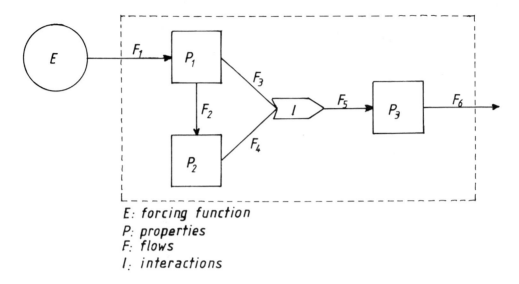

E: forcing function
P: properties
F: flows
I: interactions

FIGURE 3.   Compartment diagram illustrating the basic elements in a systems model. (From Odum, E. P., *Basic Ecology*, Saunders College Publishing, Philadelphia, 1983, 10. With permission.)

VI. MODELS AND MODELING

A model is any abstraction or simplification of a system; it is a simplified version of the real world. In their simplest form models may be verbal or graphic (informal). By providing an abstract and simplified description of some system, they may be used simply to guide research efforts or outline a problem for more detailed study. Ultimately, however, models must be statistical or mathematical (formal) if they are to be used for prediction of dynamic change.

Since ecological systems comprise many components that are highly interactive, reduction of the number of components is a necessary part of the modeling procedure. The complexity of the real world is usually simplified by the aggregation of processes and components that are similar into functional groups, such as trophic (i.e., feeding) levels, particle size, functional guilds, and so forth. For any particular model we must decide how simple or how complex it should be. If it is too simple it may not describe the system; if too complex we may be swamped by detail and ultimately develop a model no less complicated than the system being modeled.

Models can be evaluated in terms of three basic properties or goals: realism, precision, and generality.[971] Realism refers to the degree to which the mathematical statements of the model, when translated into words, correspond to the biological concepts that they are intended to represent. Precision is the ability of the model to predict numerical change and mimic the data upon which it is built. Generality refers to the breadth of applicability of the model (the number of different situations in which it can be applied).

Modeling begins with the construction of a diagram or "graphic model", as illustrated in Figure 3. In this model there are four basic elements: (1) *state variables* (or system variables), or sets of numbers (S_1 to S_3) which are used to represent the state, or condition of the system at any one time (these are the components or compartments within the system, e.g., trophic levels); (2) *flows* (or transfer functions) representing the interactions or functional relationships between the state variables; (3) *forcing functions* or inputs, e.g., energy source which drives the system; (4) *interaction functions* where forces and state variables interact to modify, amplify, control flows, or create new "emergent" state variables. In addition, the constants of the mathematical equations are termed parameters or coefficients. Finally, there is the output for the system as a whole.

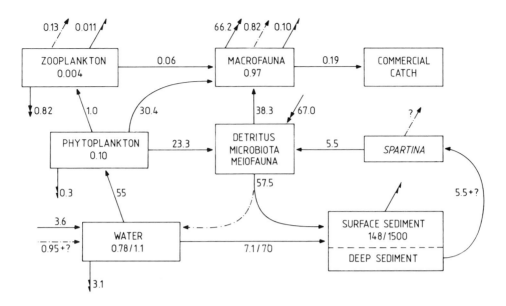

FIGURE 4. Annual budget of zinc in the Newport River estuary. Values inside compartments are mg Zn m^{-2} for the 31 km^2 estuary at half tide depth of 1.3 m. Values on arrows represent annual fluxes between compartments, mg Zn m^{-2} $^{-1}$. Arrows are defined as follows: $\longrightarrow$ = Flushing (inner volume only) plus emigration; $-\cdot\rightarrow$ = Excretion of soluble metals to water; $\rightarrow\rightarrow$ = Deposition of unassimilated materials to detritus. (From Wolfe, D. A., *Estuarine Research,* Vol. 1, Academic Press, New York, 1975, 660. With permission.)

Models may be either *stochastic* or *deterministic*. Stochastic models are ones in which predicted values depend on probability distributions (i.e., an attempt is made to include the effects of random variability in forcing functions and parameters). Deterministic models are ones in which predicted values may be determined exactly (i.e., chance variation is ignored). Stochastic models are mathematically difficult to deal with, and so deterministic models are mostly used.

There is a number of mathematical tools that can be used in systems analysis. The three major ones are set theory and transformations, matrix algebra, and difference and differential equations. For the purposes of computer simulation, the third of these is generally used. Computer-simulated models permit one to predict the probable outcomes as the parameters in the model are changed, as new parameters are added, or as old ones are removed.

Contrary to the feeling of many who are skeptical about modeling-complex ecosystems, information about only a relatively small number of components is often sufficient for the construction of an effective model, because key factors and emergent and other integrative properties dominate or control a large part of the system being modeled.

VII. APPROACHES TO SYSTEMS ECOLOGY

A. Compartmental Systems Approach

The compartmental approach to ecological modeling is described by Patten[699] and Van Dyne.[954] Workers using this approach have usually been interested in the gross dynamics of whole ecosystems as energy processing or nutrient-cycling units. Ecosystems are seen as consisting of compartments (or pools) of energy or nutrients. Each pool may represent a species population or a trophic level comprising the aggregated species populations at that level. For the purposes of the model, the complicated processes associated with the populations making up each pool are assumed to counterbalance one another, resulting in the simple behavior of the pool as a whole. Figure 4 shows such a compartmental diagram.

For any particular model we must decide how simple or how complex it should be. If it is too simple it may not accurately describe the system, if too complex it often becomes difficult to see the significant relationships. The degree to which a model is aggregated depends not only on what we know about the system but also what we want to know, and what kinds of questions we wish to ask.

B. Energy Analysis

Organizing and understanding the complexity of estuarine systems can best be carried out by the construction of simulation models. Such models, as we have seen, are based on the relations between forcing functions and state variables, relations among state variables, and the behavior of state variables. While there are a number of approaches to modeling and simulation, the principal approach adopted in this book is that of energy analysis.[638]

Energy-circuit diagraming is a visual mathematics, which uses energy as a common denominator for combining the models of chemical, physical, biologic, geologic, and economic subsystems. The first step in energy analysis is the construction of an overview of the system using energy-language diagrams. Symbols representing units and processes within the system are connected by pathways representing the flows of energy from sources outside the selected boundaries of the system, through the web of the system, and finally out as degraded used energy. The Frontispiece is an example of a comprehensive energy-flow model for the Upper Waitemata Harbour, a mangrove-fringed estuary in northern New Zealand. The symbolic language that is used is presented in detail in Odum,[643,644,646-648,650] Odum and Odum,[654] and Hall and Day.[350] Readers are referred to these publications for details. Figure 5 presents the basic symbols that are used with a little explanation.

Fundamental to the use of the various symbols are: (1) the requirement of conservation of matter and energy and the degradation of some energy into waste heat during useful work processes, (2) the importance of feedbacks and interactions in the function of complex systems, and (3) the use of some fundamental physical and biotic principles. Although the symbols were originally invented for the diagraming of energy-flows, they are equally useful for material-flow pathways. As a general rule it is permissible to use flow diagrams of energy, materials, or a combination of the two, as long as it is explicitly stated (and remembered) what is being diagramed, and as long as the energy relations that drive the processes are kept in mind. Two fundamental concepts in energy analysis are *energy quality* and the *maximum power principle*.

1. Energy Quality

In order to be able to use energy-flow diagrams for simulation or empirical modeling, the energy values of the components and flows must be calculated. These values are initially calculated in energy-flow units of different types (energy of sunlight, wind, tide, etc.) and cannot be strictly equated. However, they may be converted to equivalents of one type of energy according to the concept of embodied energy.[649] The energy transformation ratio (ETR), or quality factor, is the energy of one type required through processing to develop flow of another type, for example, the sunlight required to develop plant matter is the embodied energy of that plant matter in solar equivalents (joules of sunlight required).

Along a food chain, low-quality energy in the form of sunlight is transformed and concentrated through successive stages to high-quality energy in the top carnivores. It takes many joules or calories of dilute energy to form one joule or calorie of concentrated energy such as a top carnivore. Ecological literature has conventionally used the kcal (Calorie) as the unit for expressing energy-flows. However, the demands of international convention has led to the increasing use of the kilojoule (kJ) as the unit of energy (1 kcal = 4.187 kJ). Both systems will be used according to that used by the author of the research under discussion.

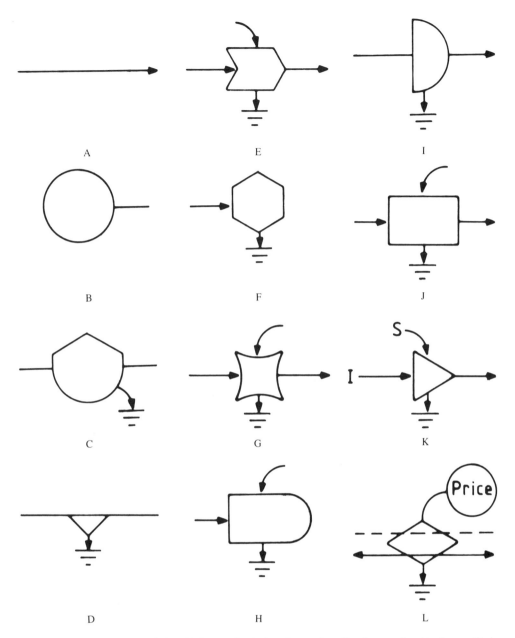

A

B

C

D

E

F

G

H

I

J

K

L

FIGURE 5. Basic energy symbols. (A) *Energy circuit.* A pathway whose flow is proportional to the quantity in the storage or source upstream. (B) *Source.* Outside source of energy delivering forces according to a program controlled from outside; a forcing function. (C) *Tank.* A compartment of energy storage within the system storing a quantity as the balance of inflows and outflows: a state variable. (D) *Heat sink.* Dispersion of potential energy into heat that accompanies all real transformation processes and storages; loss of potential energy from further use by the system. (E) *Interaction.* Interactive intersection of two pathways coupled to produce an outflow in proportion to a function of both; control action of one flow on another; limiting factor action; work gate. (F) *Consumer.* Unit that transforms energy quality, stores it, and feeds it back autocatalytically to improve inflow. (G) *Switching action.* A symbol that indicates one or more switching actions. (H) *Producer.* Unit that collects and transforms low-quality energy under control interactions of high-quality flows. (I) *Self-limiting energy receiver.* A unit that has a self-limiting output when input drives are high because there is a limiting constant quantity of material reacting on a circular pathway within. (J) *Box.* Miscellaneous symbol to use for whatever unit or function is labeled. (K) *Constant-gain amplifier.* A unit that delivers an output in proportion to the input I but changed by a constant factor as long as the energy source S is sufficient. (L) *Transaction.* A unit that indicates a sale of goods or services (solid line) in exchange for payment of money (dashed). Price is shown as an external source. (Redrawn with permission from Odum, H. T., *Systems Ecology: An Introduction,* 1983; copyright 1983, John Wiley & Sons.)

2. Maximum Power Principle

According to the principle of maximum power,[655] systems that develop structures to interact high-quality energy with low-quality energy utilize the high quality to amplify the low-quality flows so that more work is done by both. Systems that obtain the most energy and use it most effectively are the ones that survive. The maximum power principle can be stated as follows: Those systems that survive in competition among alternative choices are those that develop more power inflow and use it best to meet the needs of survival. They do this by: (1) developing storages of high-quality energy; (2) feeding back work from storage to increase flows; (3) recycling materials as needed; (4) organizing control mechanisms that keep the system adapted and stable; and (5) setting up exchange with other systems to supply special energy needs.

C. Flow Analysis

Flow analysis[249] applies the tools of input-output economics[504] to the flow of nutrients, energy, or other conserved materials in an ecosystem. It developed from an analogy between the transactions table of the input-output economist and the matrix of nutrient or energy flows between the compartments that the ecologist uses in constructing a flow model. The economist's transactions table resembles this in that it summarizes the flows of cash among sectors in a regional economy over some specified time period. A series of elementary calculations enables the economist to utilize the "Leontief inverse matrix", which provides a set of multipliers measuring the relative dependencies of sectors in the economy on one another. Under appropriate assumptions, this same computation can lead the ecosystem analysist to a set of measures of the relative dependencies of the total flows through the various compartments in the model on one another.[250]

The first ecological application of this technique was that by Hannon[355] in the use of energy-flow models as a means of making structural comparisons among ecosystems. Patten et al.[700] and Finn[248,249] developed this application further devising quantitative measures of the cycling efficiency of both nutrients and energy in ecosystem models. Richey et al.[789] applied Finn's definition of cycling index to the comparison of carbon flow structure in four-lake ecosystems. The application of flow analysis will be discussed with examples in Volume II, Chapter 1, Section VII.

Chapter 2

THE ESTUARINE ENVIRONMENT

I. INTRODUCTION

In the previous section the characteristics and special features of estuaries were outlined. Here we shall examine the geomorphology, sedimentology, hydrology, and chemistry of estuaries in more detail. However, an exhaustive treatment would be inappropriate in this book, and only sufficient detail will be provided to give the necessary setting for the sections that follow.

Although a number of geological events have contributed to the formation of estuaries, the majority are drowned river valleys due to eustatic changes in sea level. As Gorsline[313] remarks, estuaries are transient features in the geological time scale. They are constantly being altered by the deposition and erosion of sediments, and suffer extreme modification during small changes in mean sea level. Many estuaries are probably no more than 3000 years old, and within recent years erosion in the drainage basin and consequent sedimentation in estuaries has been accelerated by forest clearing and agricultural practices.[177]

In the understanding of estuarine processes it is necessary to realize that extreme short-term conditions are more important than long-term ones.[23] For example, sudden flooding can cause a river to discharge more material into an estuary in a matter of days than occurs in years under normal conditions.

There are various classifications of estuaries based on parameters such as topography, salinity structure, patterns of stratification, and circulation. Some of these will be discussed below.

II. GEOMORPHOLOGICAL FEATURES

From a geomorphological point of view, Pritchard[743] recognizes four primary subdivisions of estuaries: (1) drowned river valleys, (2) fjord-type estuaries, (3) bar-built estuaries, and (4) estuaries produced by tectonic processes.

1. *Drowned river valleys (coastal plain estuaries)* — These were formed during the Flandrian transgression which ended about 3000 BC. In this type of estuary sedimentation has not kept pace with inundation and the topography remains similar to that of a river valley. Sediments grade from muds in the upper reaches, to coarse sands at the mouth. Drowned river valleys are usually restricted to, and are common, in temperate latitudes where the amount of sediment discharged by rivers is often relatively small. Examples of this type are the Thames and the Mersey in the U.K., the Chesapeake Bay system in the eastern U.S.,[741] the Knysna in South Africa,[174] and the Fitzroy River in Western Australia.[429]

2. *Fjords* — Fjords have been formed in areas covered by Pleistocene ice sheets which deepened and widened existing river valleys to a typical U-shape and left rock bars or sills of glacial deposits at their mouths. They normally have rocky floors with a thin covering of sediment. Fjords are well developed in Norway,[807] British Columbia,[759] Chile, and the Fjordland Region of the South Island, New Zealand.[36]

3. *Bar-built estuaries* — These are usually shallow estuarine basins separated from the sea by barrier sand islands and sand spits, broken by one or more inlets. In these estuaries recent sedimentation has kept pace with inundation, and they have a characteristic bar across the mouth. They are often only a few meters deep and have

extensive lagoons and shallow waterways inside their mouths. Such estuaries are especially common in tropical regions, or in areas where there is active coastal deposition of sediments. Classic examples of such estuaries are the extensive network of marine bays of Texas[644] and the Gippsland Lakes in Victoria, Australia.[429] Other examples are the Avon-Heathcote Estuary, New Zealand[485] and estuaries along the Atlantic coast of the U.S.[377]

4.   *Estuaries produced by tectonic processes* — This is a catchall classification for estuaries not clearly included in the other divisions, although it must be emphasized that many estuarine systems are formed by a combination of geomorphological processes. Coastal indentations formed by faulting or by land subsidence, such as San Francisco Bay, are included in this category.

These estuarine topographic characteristics are important because they can influence the chemical composition of, and the processes occurring in the waters and bottom sediments. The work of Jennings and Bird[429] has shown that there are six dynamic factors which modify the characteristics of Australian estuaries, and Day[175] notes that the same factors are operative in South Africa. The first of these is rainfall and runoff from the drainage basin. It is not only the total distribution that is important, but also its distribution throughout the year; both droughts and flash floods can have marked effects on the estuarine biota. The second and third dynamic factors which affect estuaries are the prevalence of high-energy waves and the tidal range. The ocean swell and tidal range have complementary effects on estuaries. The swell rapidly builds up bars at the mouths of estuaries, whereas the tides entering and leaving the estuary have a scouring action. Tidal amplitude has an important effect on the amount of exposure of tidal flats.

The remaining factors are the influence of biotic factors on sedimentation, the mineralogy of the beach sands, and the lithology of derived dune formations and neotectonic effects. Some of these factors will be discussed in more detail in the succeeding sections.

III. ESTUARINE SEDIMENTS

In terms of sedimentation, estuaries are very complex environments, or strictly a series of environments. One principal reason for this is that the sediments themselves can originate from a number of sources; these include sediments of terrestrial origin transported by rivers (fluvial sediments) and sediments from the sea (marine sediments). On high-energy coasts where littoral drift is heavy, a flood-tide delta of marine sand may largely occlude the inlet, and if the tides have a high amplitude the marine sand is carried rapidly to the head of the estuary. Conversely, on low-energy coasts where littoral drift is minimal, little marine sand enters the estuary. The estuary basin may remain deep, or if the drainage basin of the estuary is prone to erosion and the estuary is river-dominated, the estuary basin can become largely filled with fluvial sediments. These are extreme cases and generally there is an upstream-downstream gradient of fine silts and clays of fluvial origin to medium or coarse sand of marine origin at the mouth.[180] The sedimentological properties of estuaries have been reviewed by Postma,[737] Dyer,[209,210] and Day.[180]

A classification scheme (Figure 1) is generally used to describe differences in sediment texture by reference to the proportion of silt, sand, and clay. The commonly used size ranges of particles in each group are given in Table 1. Such classifications are essentially arbitrary and many such gradings are to be found in the engineering and geological literature. One commonly used scale is the Wentworth scale of units since the grain sizes are grouped in a geometric series with a ratio of two, and because the sizes correspond closely to the mesh openings of sieves in common use.

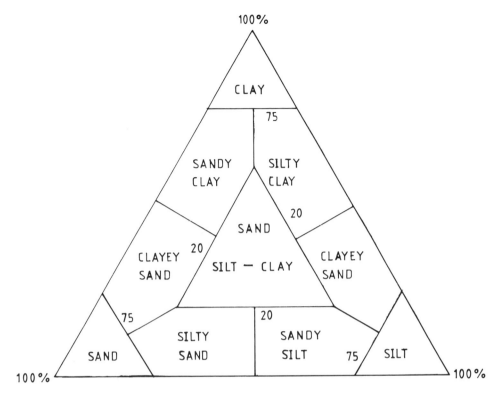

FIGURE 1. Classification scheme for sediment texture according to percentage composition of silt, clay, and sand. (From Parsons, T. R., Takahashi, M., and Hargrave, B. T., *Biological Oceanographic Processes,* 2nd ed., copyright 1977, Pergamon Press. With permission.)

Table 1
COMMONLY USED
SEDIMENT PARTICLE
SIZE NOMENCLATURE

Sediment	Size (mm)
Gravel	
Boulders	>500
Cobbles	25—500
Pebbles	10—25
Fine gravel	2—10
Sand	
Very coarse sand	1—2
Coarse sand	0.5—1.0
Medium sand	0.250—0.500
Fine sand	0.100—0.250
Very fine sand	0.050—0.100
Silt	
Coarse silt	0.200—0.050
Medium silt	0.005—0.020
Fine silt	0.002—0.005
Clay	<0.002

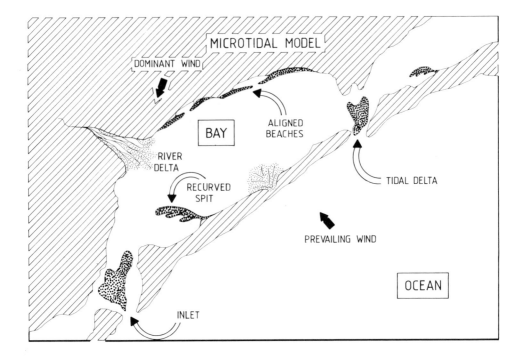

FIGURE 2.   Microtidal estuary model. Sand bodies in this type of estuary are mostly storm- or river-generated (e.g., washover fans; river deltas) or wave-generated (e.g., aligned beaches; recurved spits). Tidal deltas are usually small. (From Hayes, M. O., *Estuarine Research,* Vol. 2, Cronin, L. E., Ed., Academic Press, New York, 1975, 7. With permission.)

A. Tidal Range and Coastal Plain Estuaries

Much of the material published on estuaries is concerned with large coastal plain estuaries where the tidal range may be considerably less than half the mean depth, and where the water flowing into the estuary usually has a residence time of several days. Such estuaries are commonly either partially mixed or highly stratified.[664] Examples are the Chesapeake Bay and the New England estuaries of the U.S. Atlantic coast, the Thames, Tay, and Severn estuaries of the U.K., and the Gironde of France.

The morphology of the sediments in coastal plain and bar-built estuaries is determined by the interaction of a number of process variables, including: (1) tidal range, (2) tidal currents, (3) wave conditions, and (4) storm action. Of these four, variations in tidal range have the broadest effect in determining large-scale differences in the morphology of sedimentation.[377] Based on tidal range, Davies[164] proposed a classification which was further elaborated by Hayes.[377] Three types were recognized: microtidal estuaries, mesotidal estuaries, and macrotidal estuaries.

1. Microtidal Estuaries: Tidal Range 0 to 2 m

Characteristic deposits occurring in microtidal estuaries (Figure 2) include: (a) washover fans deposited during storm-surge floods of major storms; (b) wave-built features such as aligned beaches, recurved spits, cusputate spits, and bay mouth bars; (c) river deltas; and (d) small flood-tidal deltas. The processes that dominate in microtidal estuaries are created by wind and wave effects.

2. Mesotidal Estuaries: Tidal Range 2 to 4 m

These estuaries (Figure 3) differ from the microtidal estuaries in that sediments deposited by tidal currents begin to predominate. The barrier islands are short and stubby, and the

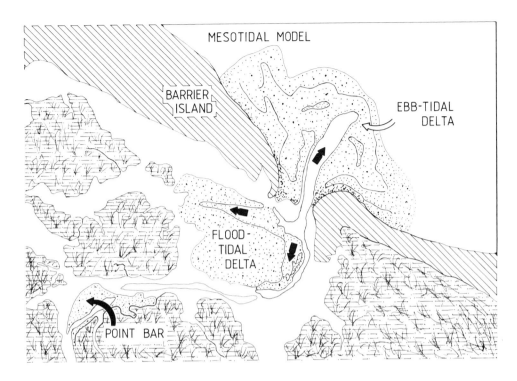

FIGURE 3. Mesotial estuary model. Tidal deltas and point-bar deposits are the principal sand bodies occurring in this type of estuary. (From Hayes, M. O., *Estuarine Research,* Vol. 2, Cronin, L. E., Ed., Academic Press, New York, 1975, 11. With permission.)

tidal deltas are large and conspicuous. Meandering tidal channels occur behind the barriers; point-bar deposits containing bedforms generated by tidal currents usually predominate in these channels. The principal sand deposits are the tidal deltas.

a. Ebb-Tide Deltas

The components of a typical ebb-tide delta (Figure 4) include a *main ebb channel* which usually shows a slight-to-strong dominance of ebb-tidal currents over flood-tidal currents. The main ebb channel is flanked on either side by *channel-margin linear bars,* which are levee-like deposits formed by the interaction of ebb- and flood-tidal currents with wave-generated currents. At the end of the main channel is a relatively steep, seaward-sloping lobe of sand called the *terminal lobe.* Broad sheets of sand, called *swash platforms,* flank both sides of the main channel. Usually isolated *swash bars,* built by swash action of waves, occur on the swash platform. Marginal tidal channels dominated by flood-tidal currents, called *marginal-flood-channels,* usually occur between the swash platform and the adjacent updrift and downdrift beaches.

b. Flood-Tidal Deltas

An example of a typical flood-tidal delta morphology as shown in Figure 5 consists of:

1. *Flood ramp* — Seaward-facing slope on the sand body over which the main force of the flood current is directed. It is always covered with flood-orientated sand waves.
2. *Flood channels* — Channels dominated by flood currents that bifurcate off the flood ramp.
3. *Ebb shields* — Topographically high rims or margins around the tidal delta that protect portions of it from modification by ebb currents.

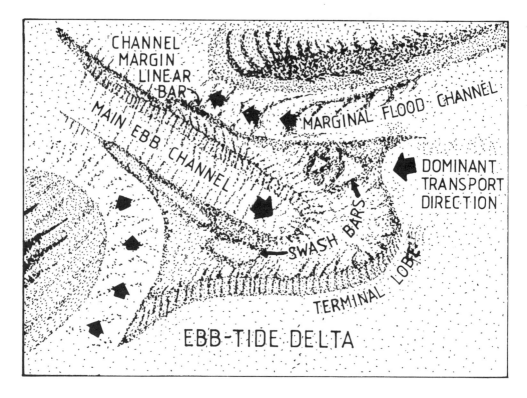

FIGURE 4.   Model of the morphology of an ebb-tide delta. Arrows indicate dominant direction of tidal currents. (From Hayes, M. O., *Estuarine Research,* Vol. 2, Cronin, L. E., Ed., Academic Press, New York, 1975, 13. With permission.)

4.   *Ebb spits* — Spits formed by ebb-tidal currents.
5.   *Spill-over lobes* — Lobate banks of sediment formed by unidirectional currents.

3. Macrotidal Estuaries: Tidal Range > 4 m

The most prominent feature of this type of estuary (Figure 6) is the overwhelming dominance of tidal currents. Such estuaries are usually broad-mouthed and funnel-shaped. Sand deposition is generally concentrated in the center of the estuary away from the shore, which is usually dominated by broad, muddy tidal flats. The sand bodies are long linear features orientated parallel with the tidal currents.

The morphologies described above are important in providing a range of habitats for estuarine organisms. Due to their coarse deposits and shifting topography, the deltas and sand bars have a low faunal diversity and the populations that do occur undergo large fluctuations in numbers.

B. Deposition, Erosion, Flocculation, and Turbidity

Sand, silt, clay, and colloidal particulates such as humus are carried in suspension in river water. The physicochemical changes that occur when fresh and salt water mix have been reviewed by Postma,[737] Dyer,[209] and Burton.[87] Coarse particles entering an estuary are deposited according to hydrodynamic factors. Their deposition within the estuary is controlled by the speed of the currents and the particle size of the sediments. The relationship between current speed and the erosion, transportation, and deposition of sediments is shown in Figure 7. From the Figure it can be seen that for pebbles 10^4 μm (1 cm) in diameter, erosion of the sediments will take place at current speeds over 150 cm sec^{-1}. At current speeds between 150 and 90 cm sec^{-1} the pebbles will be transported by the current, while at speeds less

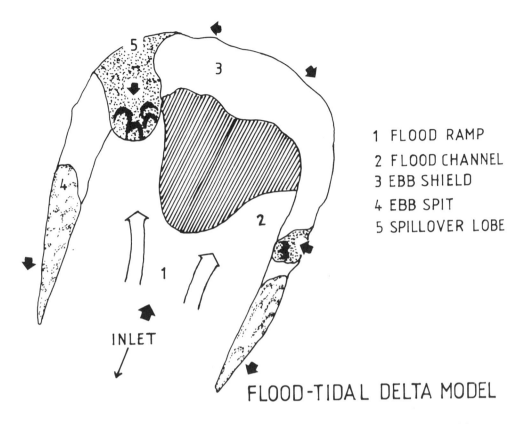

1 FLOOD RAMP
2 FLOOD CHANNEL
3 EBB SHIELD
4 EBB SPIT
5 SPILLOVER LOBE

INLET

FLOOD-TIDAL DELTA MODEL

FIGURE 5. Model of the morphology of a flood-tide delta. Arrows indicate dominant direction of tidal currents. (From Hayes, M. O., *Estuarine Research,* Vol. 2, Cronin, L. E., Ed., Academic Press, New York, 1975, 17. With permission.)

than 90 cm sec^{-1} they will be deposited. Similarly, for a fine sand of 10^2 μm (0.1 mm) diameter, erosion will occur at speeds greater than 30 cm sec^{-1} and deposition will occur at speeds less than 15 cm sec^{-1}. For silts and clays a similar relationship exists. However, erosion velocities are affected by the degree of consolidation which is a function of the water content of the sediment. The greater the degree of consolidation the higher the erosion velocity. The consequences of these relationships are that in estuaries with fast flowing rivers and strong tidal currents at either end, all sizes of sedimentary particles will be eroded and transported. As the currents slacken the larger-sized particles will be deposited first and only in the calmer middle reaches of an estuary and especially in the slack water at high tide overlying the intertidal flats will the currents be slow enough for the finer particles (silts and clays) to be deposited.

The rate of deposition, or settling velocity of sediments, is related to the particle diameter. These relationships are shown in Table 2. From this table it can be seen that sands and coarser materials settle rapidly, and any sediment coarser than 15 μm will settle within one tidal cycle. For finer particles the settling velocities are much lower and the clay and silt particles will be unable to fall and settle in one tidal cycle. Consequently, the waters of estuaries tend to be turbid as the silt and clay particles are carried about until they eventually settle on the mud flats. Carricker[96] states that: "The characteristically turbid nature of estuarine water is the product of the interplay of: (1) the particulate matter from the watershed and off-inlet shores and bottoms, reworking and scouring of estuarine bottoms by tidal currents and waves, loosening of bottom sediments by burrowing animals, and decomposition of pelagic and benthic estuarine organisms; (2) the net two-layered opposing estuarine

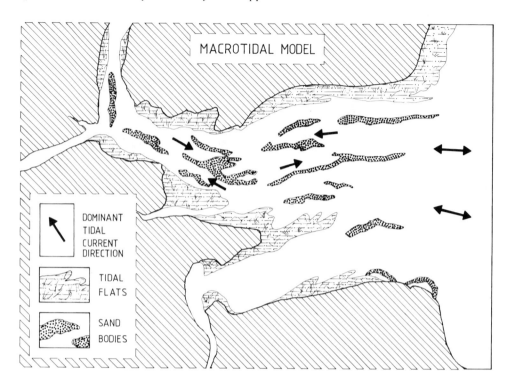

FIGURE 6.   Macrotidal estuary model. The principal sand bodies are linear sand bars located in the central portions of the estuary that are built by tidal currents. (From Hayes, M. O., *Estuarine Research,* Vol. 2, Cronin, L. E., Ed., Academic Press, New York, 1975, 21. With permission.)

circulation; (3) the mixing of fresh and sea water and consequent flocculation of finer particles; and (4) the presence of relatively quiet sedimentation areas provided by semi-enclosures and widening of the estuarine basin.'' To these can be added the particulate organic matter (detritus) derived from the death and decay of the seagrasses, marsh plants, and the mangrove-litter fall.

As mentioned previously, there is throughout the intertidal areas a gradient in substratum texture of one of finer particles at low tide level, to coarse particles higher up the shore. Morgans[608] postulates that this gradient, which is the reverse of what would be expected on the basis of velocity alone, is due to a combination of two processes. First, during high tide, the wind, which is usually well-developed, creates a water chop that lifts the finer sediment from the shallows and keeps it in suspension long enough for most to be removed by the full tide. Sediments at the lower levels of the shore are less disturbed by wind chop. The second process is active when the wind drops; fine particles take an appreciable time to fall and they are given less time to settle on shallow banks than in the deeper parts, simply because of tidal rise and fall. Figure 8 illustrates these processes on a sand beach at Howick in the Upper Waitemata Harbour, New Zealand. It is interesting to note the finer deposits on the eelgrass *(Zostera)* flat due to the slowing down of water velocity by the leaves of the plants.

For cohesive suspended particles, mainly in the clay (<2 μm) and colloidal ranges, their behavior is modified by processes causing coagulation or flocculation and, in some cases, by disaggregation of the flocculated material. The surface charges on naturally occurring particulate material has been investigated by several workers using electrokinetic measurements.[22,619,740] Silt and clay particles bear negative surface charges due to the adsorption of anions, particularly OH^-, cation substitution in the crystal lattice and broken bonds at the

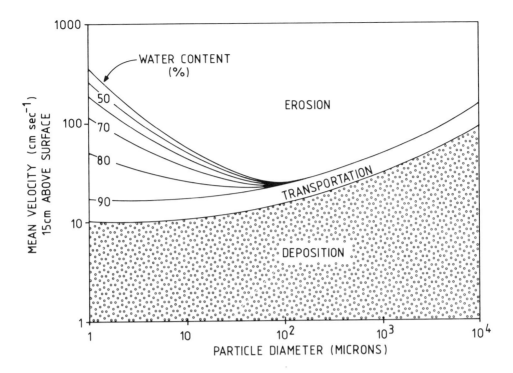

FIGURE 7. Erosion, transportation, and depositional velocities for different grain particle sizes. Also illustrated is the effect of the water content of the sediment on the degree of consolidation which in turn modifies the erosion velocities. (From Postma, H., *Estuaries*, Lauff, G., Ed., Publication No. 83, American Association for the Advancement of Science, Washington, D.C., 1967, 158. With permission; copyright 1967 by AAAS.)

Table 2
SETTLING VELOCITIES OF SEDIMENTS

Material	Median diameter (μm)	Settling velocity (m day^{-1})
Fine sand	125—250	1040
Very fine sand	62—125	301
Silt	31.2	75.2
	15.6	18.8
	7.8	4.7
	3.9	1.2
Clay	1.95	0.3
	0.98	0.074
	0.49	0.018
	0.25	0.004
	0.12	0.001

After King, C. M., *Introduction to Marine Geology and Geomorphology*, Edward Arnold, London, 1975, 196. With permission.

edge of the particles. These negative charges are balanced by a double layer of hydrated cations. The thickness of this double layer depends mainly on the ionic concentration of the water in which the particles are suspended. River water usually has a low electrolyte content and the charges on the particles repel one another. Estuarine water has a high electrolyte content so that the repulsive charges diminish, and when the particles collide they unite to form a large spongy network or floccule. The flocculation of silt particles, most types of

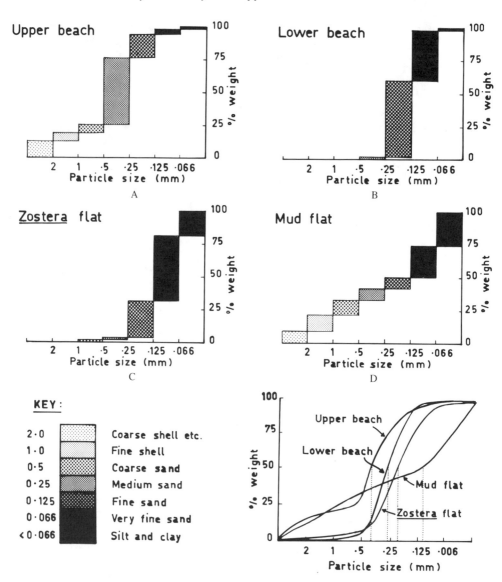

FIGURE 8.   Distribution of sediment particles analyzed according to the Wentworth scale for three types of beach (A to C) at Howick, Upper Waitemata Harbour, New Zealand (upper beach, lower beach, and *Zostera* flat). In D the same information is presented as cumulative curves with the addition of the curve for a low tidal mud flat in Lyttelton Harbour, New Zealand. (From Morton, J. E. and Miller, M. C., *The New Zealand Seashore,* Collins, London, 1968, 41. With permission.)

humus, and clay minerals *illite* and *kaolinite* mainly occurs at salinities of 1 to 4‰, but montmorillonite flocculates slowly as the salinity increases to full sea water. Sewage which contains large carbohydrate and protein molecules as well as polyvalent metallic ions, promotes flocculation, while bacteria and other organic particles act as binding agents. As more and more particles are added to the floccule it grows until its diameter may exceed 0.5 mm. Over 90% of the spongy floccule is water, and it sinks in still water at about 0.4 cm sec^{-1}.

Flocculation starts at the head of the estuary and as the floccules grow they drift downstream with the water becoming increasingly turbid. In many estuaries this produces a so-called "turbidity maximum". The presence and magnitude of this turbidity maximum is controlled

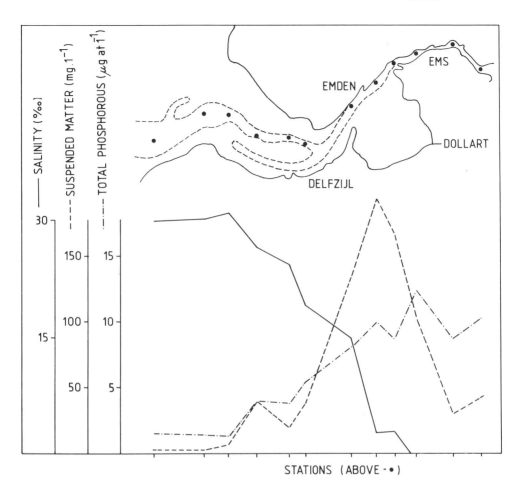

FIGURE 9.   Turbidity maximum in the Elms Estuary, The Netherlands. Suspended matter — mg l^{-1}; salinity — % NaCl; total phosphorus — μg-atom l^{-1}. Observations were taken at the surface of the estuary at the stations indicated on the map. (From Postma, H., *Estuaries,* Lauff, G., Ed., Publication No. 83, American Association for the Advancement of Science, Washington, D.C., 1967, 172. With permission; copyright 1967 by AAAS.)

by a number of factors, including the amount of suspended material in the sea water, the estuarine circulation patterns, and the settling velocity of the available material.[737] In the Elms Estuary in the Netherlands (Figure 9) the turbidity maximum is located at the meeting point of river and tidal currents, a region where sedimentary material accumulates.

The magnitude and location of the turbidity maximum depends on both the particle settling velocity of the suspended sediment, and the contribution of suspended sediment from the seaward end of the estuary.[247] This is due to the inward flow of sea water and its contained sediment along the bottom of the estuary, until it reaches a stagnation point where inflow ceases. At this stagnation point the sea water and its sediment rise to mix with the fresher surface water, and it is here that the turbidity maximum occurs. The position of the stagnation point and the turbidity maximum varies with the strength of the river flow, moving upstream with low river flow, and downstream with high river flow.

High turbidity cuts down light penetration and consequently influences primary production. The significance of this will be discussed later, as will the roles of organic matter in the sediments.

C. Sediment Properties

An important property is the degree of sorting; that is the proportion of particles which are similar in diameter to that of the median particle size. The Upper Beach sample in Figure 8 is a poorly sorted one, containing much shell and coarse sand, while the Lower Beach sample is well sorted and has 97% of the sediment in the fine sand and very fine sand grades. The Mud Flat sample, although containing a high proportion of fine sediments which are well sorted, nevertheless contains appreciable quantities of the coarser fractions.

In general, exposed shores and the upper parts of shores in moderate exposure tend to have ill-sorted deposits, while sheltered shores and the lower parts of shores in moderate exposure tend to have well-sorted ones. Changes in particle size and the degree of sorting result in important changes in the physicochemical properties of the sediment which are reflected in the diversity and kinds of animals and plants characterizing the deposits. Among the most important of these are the interstitial pore space, water content, mobility, and depth to which the deposits are disturbed by wave action, the salinity and temperature of the interstitial water, the oxygen content, the organic content, and the depth of the reducing layer. The following considers each of these in turn.

1. Interstitial Space and Water Content

The interstitial water of a beach is either retained in the interstices between the sand grains as the tide falls, or is replenished from below by capillary action. The amount of water that can be retained is a function of the available pore space which is dependent on the degree of packing and the degree of sorting of the sediment. In poorly sorted sediments the smaller grains pack into the interstices between the larger particles and thus reduce the percentage pore space. Coarse ill-sorted sandy beaches have a relatively low porosity (approximately 20%) whereas in more sheltered situations where the deposits are well-sorted the water retention may approach 45%.

The rate of replacement of water lost by evaporation from the surface is dependent upon the diameter of the channels between the sand grains. These channels decrease in size with a decrease in grain size so that the capillary rise is greatest in the fine deposits. Thus, in fine deposits where the slope of the shore is low and the water retention (porosity) is high, the sediment is permanently damp; whereas in coarse ill-sorted deposits where the slope is steep and the water retention low, the sediment contains less water and dries out quickly.

Related to the above are the properties of thixotropy and dilatancy which affect the ease with which animals can burrow into the substratum.[104] All visitors to the seaside have noticed the whitening of wet sand under the foot. This is due to the water being driven from the interstices by the pressure applied until the sand becomes hard packed and dry. This property is called dilatancy and such sands are called dilatant. They are difficult to penetrate because the application of pressure causes them to harden. Dilatant sands usually have a water content of less than 22% by weight. When the water content of the sands is greater than 25% the sands become thixotropic and are softer and easier to penetrate. Thixotropy describes a system which becomes less vicious upon agitation and thixotropic sands show a reduction in resistance with increased rate of shear in contrast to dilatant sands which show an increase in resistance. The most notorious examples of thixotropic sands are quicksands which liquify when pressure is applied. In burrowing worms, especially in experiments with *Arenicola*, it has been demonstrated that the speed of burrowing is dependent upon the moisture content of the sediments and their resistance to shear.

2. Temperature

The temperature within the sediments is determined by insolation, evaporation, wind, rain, tidal inundation, and the amount of pore water. There is a gradient of temperature across the intertidal zone with maximum and minimum values occurring at high-water mark

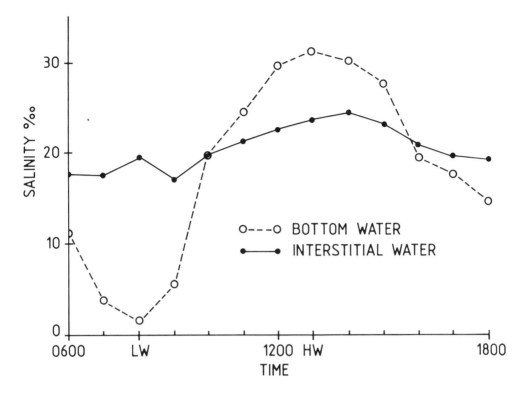

FIGURE 10.   Salinity of bottom water and interstitial water at a low tide station in the upper part of the Avon-Heathcote Estuary, New Zealand.[961]

and low-water mark, respectively. Marked vertical temperature gradients can develop within the upper 10 cm below which the temperature is fairly uniform approaching that of the sea water. The vertical gradient is much steeper in the summer in temperate regions than in the winter. Thus, animals living in the sediment are buffered against the temperature extremes which occur at the surface when the tide is out.

3. Salinity

Fundamentally, the interstitial salinity of estuarine sediments represents an equilibrium between that of the overlying sea water and the fresh water seeping out from the land. In general there is a horizontal salinity gradient from low water to high water. The nature of this gradient depends on the pattern of estuarine circulation and salinity stratification. There may be a considerable difference between the interstitial salinities and those of the overlying water (see Figure 10 for some data from the Avon-Heathcote Estuary, New Zealand). It can be seen that the interstitial salinity is considerably dampened when compared to that of the overlying water. Tube-building invertebrates which irrigate their burrows can play a significant role in maintaining the interstitial water salinity and other chemical properties at approximately that of the overlying water.[9,602] Many species cease irrigation when the salinity of the overlying water falls below a certain level.

Interstitial salinity variations are greatest on intertidal flats. During exposure the salinities of surface sediments are subject to dilution by rain and concentration by evaporation. In a 2-month study, September to October, of salinity in a *Salicornia-Spartina* marsh at Mission Bay, San Diego, California, the water retained on the marsh had a higher salinity than that of the bay (ca. 34‰) for 75% of the time, exceeding 40‰ for 37% of the time, exceeding 45‰ for 10% of the time and had recorded a maximum value of 50‰.[67,721] The movement of such water off a marsh can produce "slugs" of high salinity water in the main body of the estuary.

4. Oxygen Content

The oxygen content of the interstitial water depends to a large extent on the drainage through the sediment. Porosity and drainage time increases sharply when there is 20% or more of fine sand in a deposit and in a like manner the oxygen concentration of the interstitial water varies with the percentage concentration of fine sands. On the whole, coarse sands have more oxygen than do fine sands and muds. In poorly drained mud flats there is a pronounced vertical gradient in oxygen concentration.[68] In one study the values varied from saturation at the surface (due to the photosynthetic activity of the diatoms) to 1.4 mℓ O_2 ℓ^{-1} at 2 cm and 0.3 mℓ ℓ^{-1} at 5-cm depth.

As discussed below, the vertical gradient of oxygen content is correlated with the amount of organic matter in the deposit and the depth of the reducing layer. Animals which live in the deeper layers where oxygen levels are low must either be tolerant of anaerobic conditions or retain a connection with the oxygenated surface layers and maintain a flow of water to and from the body. Burrowing polychaete worms either live in U-shaped burrows through which water can be circulated *(Abarenicola)* by a pumping action of the body, or they maintain a circulation through their simple burrows or tubes by ciliary and/or muscular action. Bivalves living in anaerobic sediments communicate with the surface via their siphons and thus maintain a circulation of oxygenated water across their gills.

5. Organic Content

Sediments are profoundly modified by the input of organic matter which becomes incorporated in the deposits as detritus. This detritus is derived from a variety of sources (see Section IV). Since organic particles are light and only settle out where the water is quiet there is an inverse relationship between the organic content of a deposit and the turbulence of the water and hence grain size. Organic detritus tends to clog the interstices between the grains and bind them together to give the characteristic muddy sand and sand deposits.

The most typical estuarine sediments are those that consist of clays, silts, and organic materials, and are found principally in the upper reaches and quiet lateral parts of estuaries. These sediments have a characteristic vertical layering (Figure 11) in bands of color[228,245] due to the one-way supply of light and oxygen and the biological activity of the burrowing animals, the meiofauna, and the microflora and fauna. At the sediment-water interface there is a layer of often semifluid yellow or brown (due to the presence of ferric iron) oxidized sediment which is swept into suspension by the stronger currents. In this layer the redox-potential as measured by the Eh is around $+400$ mV close to the surface, and around $+200$ mV deeper in the sediment. Below this layer is the "grey zone" or redox-potential-discontinuity (RPD) layer, a layer where oxidizing processes become replaced by reducing processes. According to Fenchel and Reidl:[245] "Food availability higher than oxygen input sufficient for the oxidation of food causes anaerobic conditions; hence the steepness and depth of the RPD depend basically on the equilibrium 'food: oxygen flow into the interstices.' " The depth of the RPD layer depends upon the organic content and grain size composition (mean size, sorting, clay fraction) of the sediment; an increase of clay and organic matter sharpens and raises the RPD towards the sediment surface while an increase of mean grain size and sorting causes the RPD to sink deeper into the mud. In this RPD layer, oxygen as well as reduced compounds such as hydrogen sulfide are present in small amounts, while Eh decreases quickly from positive to negative values. The third and deepest layer is the "black zone" or "sulfide zone". This layer is totally anaerobic and H_2S occurs in substantial amounts, up to 700 mg ℓ^{-1} in the interstitial water of muddy sediments, while values nearing 300 mg ℓ^{-1} are common.[245] Redox-potential of the interstitial water varies between -100 and -250 mV. Considerable amounts of H_2S are found as ferrous sulfides,

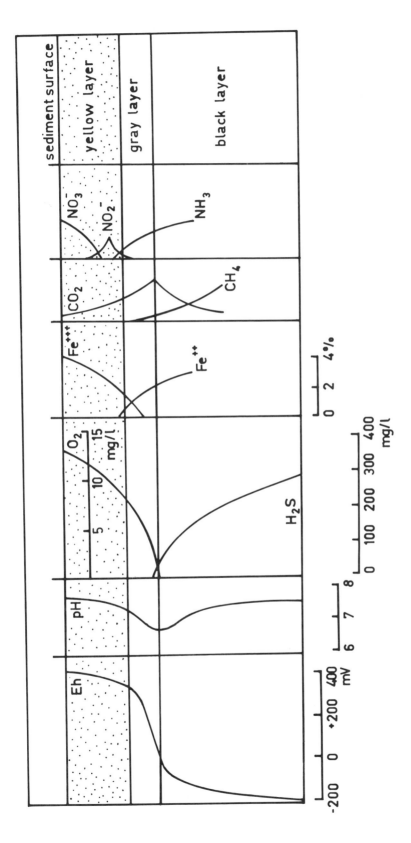

FIGURE 11. Schematic representation of Eh and pH profiles and the vertical distribution of some compounds and ions in estuarine sediments. Fully oxidized layer dotted. (From Fenchel, T., *Ophelia*, 6, 61, 1969. With permission.)

giving the sediment its characteristic black color. These layers undergo vertical migrations that are correlated with the following changes in the external parameters:[228,245,310,669] (1) increase of protection from water movement reduces permeability and brings the RPD closer to the surface area in long-term range; (2) higher temperatures cause a higher position of the RPD, thus giving rise to seasonal changes in level; and (3) input of organic matter causes the RPD level to rise; even a circadian rhythm is observed where sunlight, due to the oxidizing activity of phototrophic bacteria and algae, keeps the RPD down but it rises to the sediment surface for a major part of the night. The biochemical processes, mediated by the abundant microorganisms that occur within these layers and the "sulfide layer" in particular are of profound importance for the functioning of the estuarine ecosystem and will be considered in more detail later. Discussion of the impact of the macrofauna on sediment structure and properties will be dealt with in Chapter 5, Section IV.C.3.

IV. ESTUARINE CIRCULATION AND SALINITY PATTERNS

Current patterns within estuaries are complex due to the intermingling of fresh and salt water of different densities and the configuration of the estuarine basin. They are further complicated by the action of the tides. As river water is less dense than sea water, it tends to flow seawards as a surface current while the sea water tends to flow up-estuary as a bottom current. If there is no mixing between the two layers, the sea water forms a *salt wedge* extending towards the head of the estuary. As the tide rises and falls, the tip of the salt wedge advances and retreats. The volume of water between high and low water levels is known as the *tidal prism* and as it increases from neap to spring tides so does the velocity of the tidal current.

Usually there is some degree of eddy diffusion and turbulent mixing at the interface between the surface layer of fresh water and the sea water of the salt wedge. Depending on the degree of mixing, the water column may become vertically stratified. Mixing may proceed until there may be a gradual increase of salinity with depth, or if there is a high degree of turbulent mixing (as may occur in a shallow estuary due to wind and waves) the salinity becomes homogenous from the surface to the bottom.

A classification of estuaries based on the current system and the distribution of salinity has been developed by Pritchard,[742,744] Cameron and Pritchard,[93] and others. The major factors which affect the current system and the distribution of salinity are the ratio of river discharge to tidal flow, the degree of turbulence, and the configuration of the estuarine basin, especially width and depth. Estuaries are generally classified in three groupings.

A. Positive or Normal Estuaries

Most estuaries are of this type in the sense that there is an increase in salinity from the head where the river enters, to the sea at the mouth. Further, there is a net flow seaward over a tidal cycle. Positive estuaries may be subdivided according to the degree of stratification or mixing of salt and fresh water.

1. Salt Wedge Estuaries

These are characterized by a dominant fresh water inflow, a small tidal range, and a large depth to width ratio (Figure 12, A and B). A salt wedge of sea water on the bottom penetrates up-estuary, to a distance which is inversely dependent on river flow. Salt water mixes into the outward flowing fresh water by advection, but the mixing of fresh water downwards into the salt water is minimal. The degree of mixing across the boundary between the two water masses is largely dependent on the volume of fresh water flow. As the velocity of the surface layer of river water increases, friction at the interface with the lower salt water layer will increase and internal waves will develop leading to the trapping or *entrainment*

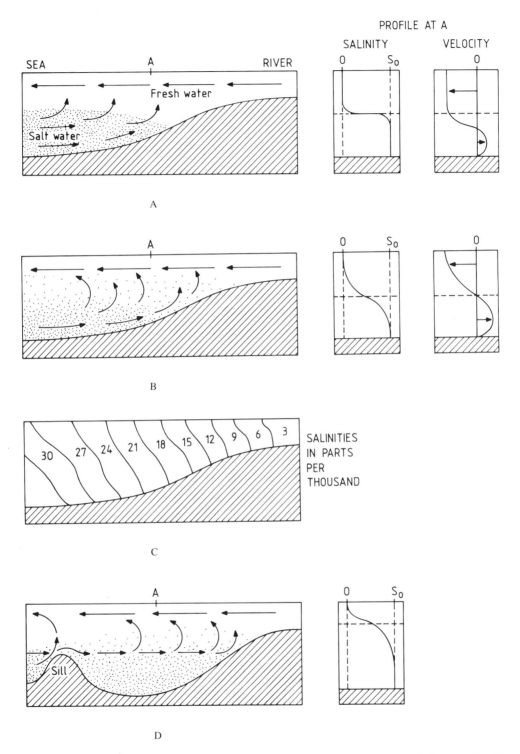

FIGURE 12. Diagrams illustrating the main types of estuaries as seen in longitudinal vertical section. (A) Salt wedge estuary; salt water is stippled. (B) Partially mixed estuary; salt and fresh water partially mixed by tidal movements and internal waves. (C) Vertically homogeneous estuary; isohalines for salinity are shown. (D) Fiord; salinine water trapped by sill.[743]

of salt water in the upper layer. A layer of mixed water of varying depth develops between the fresh water and the sea water with marked haloclines at the boundaries. Such estuaries are often referred to as highly stratified estuaries. There is little tidal mixing in a salt wedge estuary. The estuary of the Mississippi River belongs to this type as do some fjord systems.

If the width of the estuary exceeds 0.5 km, Corioli's force becomes significant and the interface between the two layers will be tilted laterally. In the Northern Hemisphere the depth of the surface layer will be greatest nearest the right bank looking downstream and the thickness of the bottom layer greatest near the left bank. The reverse is found in the Southern Hemisphere.

2. Partially Mixed Estuaries

These are estuaries in which varying degrees of mixing between the outward-flowing surface fresh water and the inward-flowing bottom sea water occur such that the distinct boundaries that are found in salt wedge estuaries do not occur. Turbulent mixing between the two layers may be increased by different factors; it increases when the ratio between tidal flow and river flow approaches 1:1 and when there are irregularities in the channel bed. It is accentuated in shallow estuaries where the volume of the tidal prism is large compared with the total volume of the estuary basin.

The addition of saline bottom water to the upper layer not only increases its salinity but also its volume until it may be many times the volume of river water that entered the estuary.[744] Pritchard estimated that in the estuary of the James River the net seaward flow in the upper layer was 20 times the discharge of the river and the net landward flow in the lower layer was 19 times the river flow. Superimposed on these flows is the much larger volume transported up and down the estuary by the flood and ebb tides.

The velocity of the upper layer is greatest at the surface and decreases with depth until the interface with the lower layer is reached. At this point the velocity is zero and below this the velocity of the lower layer increases until it is retarded by friction with the bed of the estuary.

The salinity of the upper layer increases down-estuary and the salinity of the lower layer decreases up-estuary to the tip of the salt wedge. There is thus a gradual increase in the water salinity from the surface to the sea (Figure 12B). Examples of this type of partially mixed estuary are the James River, the Mersey, the Thames, and the Hawkesbury River.

3. Vertically Homogeneous Estuaries

In these estuaries (Figure 12C) salinity decreases from the mouth towards the head without a vertical gradient in salinity. This is a result of turbulent mixing and is characteristic of shallow estuaries with a large tidal range where the ratio of tidal inflow to river flow is of the order of 10:1. Estuaries of this type are the Solway Firth in the U.K. and Netarts Bay and Coos Bay, Oregon.

In some estuaries a lateral gradient of salinity develops with a near vertical boundary between the more saline water flowing upstream and the less saline water flowing downstream, on the left and right sides of the estuary, respectively. This type of estuary occurs in the relatively wide lower reaches of the Delaware and Raritan estuaries.

B. Hypersaline or Negative Estuaries

These estuaries have a reversed or "negative" salinity gradient with the salinity increasing from sea water values at the mouth to hypersaline values in the upper reaches where the water level is below sea level, so that the net flow is landward. Such conditions are found in regions subject to periodic drought. The classic example of a hypersaline estuary is the Lagoon Madre in Texas described by Hedgpeth.[382]

C. Periodically Closed Estuaries

These are bodies of coastal water referred to by Day[172] as *blind estuaries* and termed *estuarine lagoons* by Jennings and Bard.[429] They may be open to the sea at frequent intervals or they may be closed for periods of a year or more. When they are closed there is no tidal rise and fall, and consequently no tidal currents. Fresh water enters from the river and circulation is dependent on the residual river current and wind stress on the surface. According to the ratio between evaporation and seepage through the bar on the one hand, and fresh water inflow plus precipitation on the other, the salinity will vary.[178] The estuary may become hypersaline; it may retain its normal salinity when the mouth is closed or it may become hyposaline. There are many such estuaries in southern Africa such as Uhe Umgababa estuary,[665] in western Australia, and in other arid areas.

V. DISTRIBUTION PATTERNS OF ESTUARINE ORGANISMS

There is vast literature on the distribution patterns of estuarine plants and animals. Details will be found in many reviews including Lauff,[500] Wolff,[1027,1029] Perkins,[710] and Day.[177] Here we will concentrate primarily on the distribution patterns of macrobenthic animals, as they have been the subject of the bulk of the studies. From these studies there emerges a picture of the animals responding in an often complex manner to gradients of environmental factors such as sediment grain size and organic content, salinity distributions, and tidal height. The biotic change along the estuarine complex-gradient is perhaps best referred to as a "coenocline", or community gradient, and the estuarine ecosystem gradient as an "ecocline".[999]

Early estuarine benthic ecologists found it useful to classify segments of the estuarine ecocline into zones of similar biotic composition and relate these zones to the distribution of salinity. A variety of classification schemes were developed, especially by researchers working in large homiohaline brackish systems such as the Baltic and the Zuiderzee (reviewed by Segerstraale[823] and Remane[776]). The proliferation of classification schemes resulted in a symposium in 1958 to consider brackish water classification. The consensus classification adopted was the "Venice System".[905] Carricker[96] further expanded the Venice System by incorporating Day's[173] physiographic subdivisions and biotic distributional classes based on his studies of South African estuaries. This classification is shown in Table 3.

The Venice classification has been widely used by benthic ecologists working in large homiohaline brackish water systems in northwestern Europe and has also been applied to North American estuaries.[57,59] However, its application to estuaries in the Dutch Delta area has been criticized by Wolff.[1027] More recently, Boesch[60] has investigated the zonation of the macrobenthos along a homiohaline estuarine gradient in the Chesapeake Bay-York River estuary, and a seasonally poikilohaline estuarine-gradient in the Brisbane River estuary, Australia, by assessing assemblage similarity and patterns of species distributions.

Boesch found interesting differences and similarities between the Chesapeake Bay-York River estuary and the Brisbane River estuary. The differences were mainly attributable to the relative homiohalinity of the former and the poikilohalinity of the latter. The macrobenthic coenocline in the Chesapeake system was gradual and relatively constant with zones of somewhat accelerated change occurring in the 20 to 15‰ range and in the 8 to 3‰ range. In the Brisbane estuary, on the other hand, the coenocline change was much more abrupt and appeared to be controlled by the boundary condition of seasonally minimum salinity rather than by average conditions. Benthic habitats throughout most of the estuary were inhabited by estuarine endemic forms. Thus, the zonation patterns along the estuarine gradient did not relate in any meaningful way to existing estuarine classification schemes such as the Venice scheme. Boesch concluded that designation of portions of the estuary as polyhaline, mesohaline, oligohaline, etc. was futile.

Table 3

CLASSIFICATION OF ESTUARINE ZONES RELATING THE VENICE SYSTEM CLASSIFICATION TO DISTRIBUTIONAL CLASSES OF ORGANISMS

Venice system			Ecological classification
Divisions of the estuary	**Salinity ranges ‰**	**Zones**	**Types of organisms and approximate ranges of distribution in estuary, relative to divisions and salinities**
River	0.5	Limnetic	
Head			
Upper reaches	0.5—5	Oligohaline	
Middle reaches	5—18	Mesohaline	
	18—25	Polyhaline	
Lower reaches	20—30	Polyhaline	
Mouth			
Lower reaches	20—30	Polyhaline	
Mouth	30—40	Euhaline	

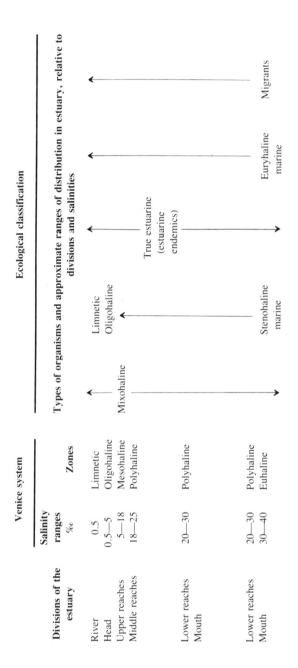

(Ecological classification labels: Mixohaline; Limnetic; Oligohaline; True estuarine (estuarine endemics); Stenohaline marine; Euryhaline marine; Migrants)

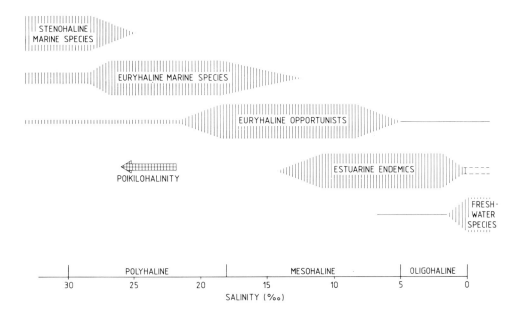

FIGURE 13. Distributional classes of species in a homeohaline estuary. (From Boesch, D. F., *Ecology of the Marine Benthos,* Coull, B. C., Ed., University of South Carolina Press, Columbia, 1977, 260. With permission.)

Similarities, however, were evident between the two estuaries in terms of classes of species distribution patterns. Both estuaries had what can be termed euryhaline marine species, euryhaline opportunists, and estuarine endemics. A general model based on these groupings is shown in Figure 13. In a homiohaline estuary the dominant and characteristic species of the macrobenthos progress from stenohaline marine species on the continental shelf to a diverse assemblage of euryhaline marine species in the lower reaches of the estuary. Many of the euryhaline species, although they are known fully from marine habitats, are much more common and abundant in the lower estuary. The coenocline further grades into domination by opportunistic euryhaline species and estuarine endemics broadly around a salinity of 18‰. The euryhaline opportunists are mostly annelids and include several well-known opportunistic species.[324] (See Chapter 5, Section VI for further discussion of such species.) Euryhaline opportunists decline in importance and number up-estuary as their salinity tolerance limits are reached and give way, broadly at a salinity of 5‰, to virtually complete dominance by estuarine endemics. The estuarine endemics decline in importance and numbers entering fresh tidal water. The effect of poikilohalinity is to displace the euryhaline species down-estuary, allowing the domination of much of the estuary by estuarine endemic forms.

In general, the estuary limits of marine and estuarine organisms are set by tolerance of low salinity. Similarly, the down-estuary limits of fresh water species are generally set by tolerance of high salinities. In addition to salinity tolerances of adult individuals, larval and juvenile tolerances and the effect of salinity on reproduction may also determine distributional limits along the estuarine coenocline.[469,816] The factors affecting the down-estuary distribution of euryhaline marine and estuarine endemic species is not well understood. Boesch[60] points out that estuarine endemics are very rarely found in salinities which remain above 15% in homiohaline brackish waters although they are known to be able to survive for long periods in high salinity. The work of Cain[92] on reproduction and development of the bivalve *Rangia cuneata* which is generally not found in salinities above 15‰ provides some insight into the problem. Cain found that a change of salinity either up from 0‰, or down from 10 to 15‰, was necessary to induce spawning in *Rangia,* and that embryos and early larvae survived only in salinities between 2 and 10‰. It is apparent that the distributional limits of estuarine endemic species are maintained through behavioral and physiological adaptations, coupled with exclusion by the pressures of competition and predation down-estuary.

There is no simple explanation of distribution patterns within estuarine ecosystems. While salinity gradients are perhaps the dominating factor, important differences between different estuaries exist in geomorphology, sedimentology, hydrography, anthropogenic modification, and biogeography. The latter can be important, for example in the Avon-Heathcote Estuary, New Zealand,[485] there are only six species of crabs, whereas the Brisbane River estuary has 23 species.[860] Increased intraspecific competition in the latter will be an important factor influencing distribution patterns.

Chapter 3

PRIMARY PRODUCTION

I. INTRODUCTION

As mentioned in the introductory section, estuarine primary production is a complex process involving, in various combinations, seed plants (both emergent and submerged), macroalgae, epiphytic microalgae, benthic microalgae, and phytoplankton. The mix of the various groups and their degree of dominance varies with latitude, geographic position, estuarine geomorphology, sedimentation patterns, fresh water inflow, and a host of other factors. The various components of the primary production mix have generally been studied by specialists in the groups concerned, and there are only a limited number of integrated studies attempting to measure simultaneously the production of the major producers.

In this section the various producer groups will be considered separately from the viewpoint of their geographic distribution, standing crop (biomass), productivity, and the factors that influence production. Finally, the relative contributions of the various groups in representative systems will be compared.

II. SALT MARSHES

A. Development and Geographic Distribution of Salt Marshes

Salt marshes are intertidal ecosystems developed in sheltered situations where silt and mud can accumulate. Geologically they are a relatively recent landform. Most appear to have originated about 3000 years ago when the sea level rise slowed sufficiently to favor marsh development. Tidal marshes are plastic coastal features shaped by the interaction of fresh water, sea water, sediments, and vegetation. For stability they require protection from high-energy waves and they therefore usually develop in a lagoon, or inlet behind barrier islands,[915] or in the protection of an estuary. In such situations the slowing of water currents permits the deposition of fine sediment and the building up of extensive, gently sloping beaches. Colonization of such sites leads to a further slowing down of tidal currents and accelerating deposition of fine sediments. Growth of the plants is often able to keep pace with sediment deposition so that over long time periods considerable accumulations of sediment and peat can occur. In places where the sea level is rising relative to the land, accumulations of sediment several meters deep may occur. Occasionally along a low-energy coastline, salt marshes may front the open sea such as the marshes of the north coast of the Gulf of Mexico in Western Florida[908] and parts of Louisiana, the north Norfolk coast in England,[875] and the coast of the Netherlands.[39] Chapman[106] give a good account of salt marsh formation.

The vegetation pattern of marshes is influenced by a complexity of environmental factors, including frequency and range of tides, salinity, micro-relief, substrate, ice-scouring, and storms. In addition, historical factors, and more recent anthropic factors including fires, cutting, dyking, grazing, and ditching can profoundly influence the distribution of species on the marshes.

One of the characteristics of a mature salt marsh is the presence of creeks and drainage channels which follow a very characteristic pattern. They form a network over the marsh becoming progressively narrower and shallower as they subdivide in a more or less regular pattern (Figure 1). It is through this network that the tidal waters enter and leave the marsh.

In general, salt marshes are confined to the temperate regions of the globe and replaced in tropical regions by mangrove vegetation. On the landward side the transition zone between

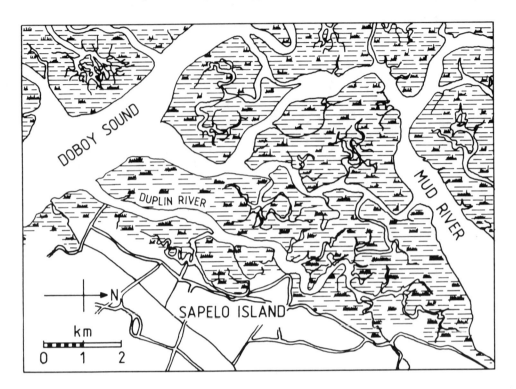

FIGURE 1.   Diagram of the marshes of Sapelo Island, Georgia. The classic work on *Spartina* marshes was done in the tidal watershed of the Duplin River, a brackish tidal stream which is tributary to Doboy Sound. (From Wiegert, R. G., Christian, R. R., and Wetzel, R. L., *The Ecology of a Salt Marsh,* Pomeroy, L. R. and Wiegert, R. G., Eds., Springer-Verlag, Berlin, 1981, 9. With permission).

the marsh proper and the adjacent zone dominated by terrestrial vegetation may be abrupt. In this case the ecotone is dominated by shrubby plants. In other cases the transition zone may be extensive, and in this case generally low-growing plants less tolerant of salinity take over. Up-estuary the transition to fresh water swamp is commonly characterized by species of *Typha, Scirpus, Cyperus,* and *Phragmites.* One feature is the widespread occurrence of dominant genera such as *Salicornia, Spartina, Juncus, Arthrocnemum,* and *Plantago.* The genus *Salicornia* (saltwort) is cosmopolitan occurring in both hemispheres, as does the genus *Suaeda.* Marsh plants have been extensively studied by botanists who recognize a wide variety of plant community types.[112] Chapman[108] also identifies nine geographic regions within most of which a number of subgroups are identified.

In most marshes a characteristic banding or zoning pattern occurs. Some examples of typical patterns are shown in Figure 2. On the Atlantic coast of North America which has over 600,000 ha of salt marshes, the smooth cordgrass *(Spartina alterniflora)* dominates the area between mean sea level and mean high water. Just above this zone are found one or more species of needle rush (mainly *Juncus roemerianus)* together with the saltgrass *(Distichlis spicata), Spartina patens,* and various species of glasswort *(Salicornia).* On the Pacific coast the cordgrass is absent and the region above mean high water is dominated by the arrowgrass *(Triglochin maritima)* with *Distichlis spicata* and the tufted hairgrass *(Deschampsia caespitosa)* above. On the European side of the North Atlantic the flora is much more diverse and heterogeneous; for example, in the U.K., there are marked differences between the marshes bordering the North Sea, the English Channel, and the Atlantic. The North Sea marshes frequently have as co-dominants the sea pink *(Ameria),* sea lavender *(Limonium),* sea plantago *(Plantago maritima),* and species of *Spergularia* and *Triglochin.* The Atlantic

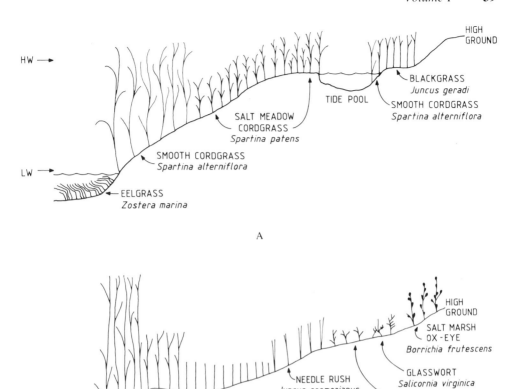

FIGURE 2.   Some characteristic salt marsh vegetation patterns. (A) Northern U.S. Atlantic coast; (B) Southern U.S. Atlantic coast; (C) Northern U.S. Pacific coast; (D) Ashmead, Knysna Estuary, South Africa. (A, B, and C from Gallagher, J. T., *Coastal Ecosystem Management,* Clark, J. R., Ed., John Wiley & Sons, New York, 1977, 752. With permission; D from Day, J. H., *Estuarine Ecology with Special Reference to Southern Africa,* Day, J. H., Ed., A. A. Balkema, Rotterdam, 1981f, 77. With permission.)

marshes tend to be used for cattle and sheep grazing and are dominated by the grasses *Puccinellia* and *Festuca.* On the south coast of England, the cordgrasses *Spartina townsendii* and *S. angelica* have been spreading rapidly and replacing the original more diverse flora. In South Africa the ricegrass, *Spartina maritima,* dominates in a band above mean water. Then follows a band of glasswort *Sarcocornia,* with mixed plants of *Chenolea diffusa* and *Limonium linifolium.* This in turn on the higher shore is replaced by a band of *Sporobolus virginicus* with the rush *Juncus kraussii* and occasional plants of *Disphyma crassifolium.*

B. *Spartina*-Dominated Marshes

In salt marshes world-wide one of the most frequent dominants is the cordgrass *Spartina* which unlike most other salt marsh species, often forms pure stands of a single species. In North America, large expanses of intertidal marsh (particularly along the southeastern coast) are dominated by the smooth cordgrass *Spartina alterniflora* Loisel. The Mississippi delta and the coastal plain to the west of it contain some 500,000 ha of salt marsh. A marsh of approximately equal area lies along the east coast of North America from Virginia to

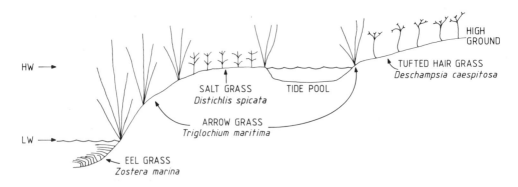

FIGURE 2C

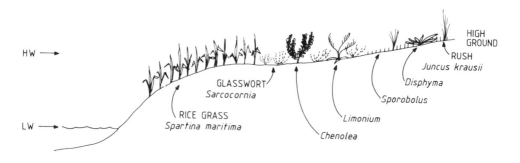

FIGURE 2D

Georgia.[111] Extensive *Spartina* marshes also occur along the Atlantic coast of South America.[985] Salt marsh formations in tropical latitudes are usually thought to be limited,[165] but they have now been shown to be much more extensive. *Spartina alterniflora* along the Atlantic coast from Florida south, and *S. braziliensis* in Brazil are colonizing species in front of mangrove woodlands. Also along tropical coasts with a well-defined dry season, *Spartina* marshes are to be found on the landward side of the mangroves.

Until the 19th century introduction of *Spartina alterniflora* into Europe from North America, only *S. maritima,* a small species rarely exceeding 0.2 m in height, was found in northwest Europe. However, *Spartina* has only beomce a major component of salt marsh vegetation in northwest Europe since the appearance of the *S. maritima — S. alterniflora* hybrid, *S. x. townsendii* H. & J. Groves, and the fertile amphidiploid derived from it, *S. angelica* Hubbard.[532] Since the hybrid and the amphidiploid are morphologically very similar, and often indistinguishable. they are often referred to collectively as *S. townsendii (sensu lato).* Since the first recorded occurrence of *S. townsendii* in 1870,[577] the species has spread both as a result of deliberate plantings and natural dispersal to occupy an estimated total area of about 25,000 ha in northwest Europe.[755] Both *S. maritima* and *S. x. townsendii* have been introduced into New Zealand.[35] While the former is restricted in its distribution, the latter is now widespread and the fertile amphidiploid appears to have also arisen in New Zealand. In parts of New Zealand, for example the Pelorus Sound region,[653] the species is spreading actively.

A characteristic feature of *Spartina* marshes is the development of natural levees along the banks of the tidal creek (Figure 3). Due to this development there are extensive areas that are reached only by the fortnightly spring tides. In these areas increased salinity due to evaporation from the marsh surface profoundly influences the size and distribution of the dominant plants. The development of the levees also influences the tidal movement of water

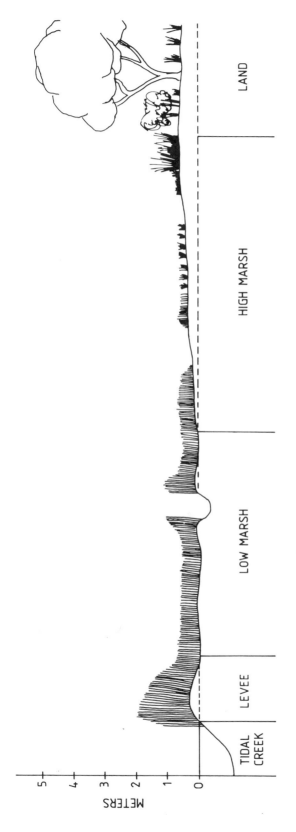

FIGURE 3. Schematic cross-section through the Georgia salt marsh, showing levee and low marsh with tall *Spartina*, and high marsh with short *Spartina*. The level reference line is approximately at mean high water. (From Wiegert, R. G., Christian, R. R., and Wetzel, R. L., *The Ecology of a Salt Marsh*, Pomeroy, L. R. and Weigert, R. G., Eds., Springer-Verlag, Berlin, 1981, 11. With permission.)

over the marsh. In a fully developed system of natural levees, water does not flow laterally from the tidal streams into and out of the marsh, but is instead guided by the levees up the tidal creeks to central distribution points in the interior of the marsh. A perched water table behind the levees holds interstitial water that is more saline than the tidal water and largely anaerobic.[621,622] The course of microbial processes and the success of the dominant plants is influenced greatly by this permanently anaerobic ground-water.

The Duplin River marsh, Sapelo Island, is typical of the marshes of the southeastern coast of North America. There *Spartina*, which occupies 95% of the marsh,[766] can be divided into at least four rather distinct types of stands.[732] The stands adjacent to the creeks are tall (> 2 m), robust, and have a low density (30 to 50 stems m^{-2}). Creek banks, which occupy 8% of the marsh surface and comprise the zone most frequently inundated, have sediment salinities of 25 to 28%.[275] Broad, low-density stands of spindly plants are found at the heads of the creeks where the drainage is poor. At a slightly higher elevation are broad, flat swards of short *S. alterniflora* which contribute approximately 45% of the *S. alterniflora* coverage.[768] Density of the short (25 cm) *Spartina* may exceed 300 plants m^{-2}, and the sediment salinities average 35 to 40‰. In addition, robust plants, with a density of 80 m^{-2}, 0.5 to 1.0 m tall, form patches totaling 20% of the area.[274]

The distribution of the various growth forms and their associated fauna and microflora appears to be a function of the edaphic features created by the drainage patterns. The remaining marsh macrophytes occupy the fringe of the salt marsh flats and the interface with the high ground. Where the slope is gradual *Fimbristylis castanea* and *S. patens* are found. On the edges of the high salinity salt flats, typically two species of glasswort, *Salicornia virginica* and *S. bigelovii* are mixed with *Batis maritima*, *Sporoblus virginicus*, very short *Borrichia frutescens*, dwarf rushes, *Juncus roemerianus*, and *Limonium nashii*.

C. Primary Production

As Pomeroy et al.[732] point out, the distribution and productivity of salt marsh vascular plants results from the interaction of environmental gradients with the physiological capabilities and versatility of the plant species. Even within a single species such as *Spartina alterniflora*, there are large differences in growth form and production at different locations in an environmental gradient.[452,940] Plants with C_4 photosynthesis such as *Spartina* are generally characterized by high rates of net photosynthesis with high temperature optima, high light saturation, and lowered transpiration rates.[732] On the other hand, the rush *Juncus roemerianus* which occupies regions of lower salinity and the succulent species living on the high salinity flats and fringing areas of the marsh are C_3 species.[18]

Pomeroy et al.[732] have described a series of experiments investigating the physiological ecology of *S. alterniflora*, a C_4 plant and *J. roemerianus*, a C_3 plant. Experiments were carried out using a mobile laboratory to carry out gas-exchange experiments on intact, *in situ* plants enclosed in temperature- and humidity-controlled chambers.[294,295] The problem for salt marsh plants is to obtain CO_2 without the loss of water vapor through transpiration. There is a close analogy between salt marsh plants and desert plants, which experience an absolute water shortage, and the physiological strategies for coping with the problem are similar. Although water as such is never a problem in a salt marsh, it must be desalted at significant physiological cost.

In their studies on Sapelo Island, Pomeroy et al.[732] found that ambient gas-exchange tracking experiments gave results for net production somewhat higher than harvest estimates. Blum et al.[55] also noted consistently higher productivity estimates from gas-exchange data for the same species in a North Carolina marsh. While estimates of net primary production from gas exchange in the Sapelo Island study agreed closely with those from harvest data for short *S. alterniflora*, they exceeded those of harvest data for *J. roemerianus* and tall *S. alterniflora* by factors of 1.4 and 2.7 respectively. These higher values may be in part an artifact of the experimental situation.

1. Methods of Estimation

The net primary production (P_n) of the vegetation covering a unit area of ground is the mass, or energy, incorporated by photosynthesis (gross primary production or P_g) less that respired (R) during a given time interval.[599] That is

$$P_n = P_g - R$$

As detailed above both P_g and R can be determined by the measurement of CO_2 fluxes. However, there are technical difficulties in doing this in a salt marsh.

Net above-ground primary productivity (P_nAG) in salt marshes has traditionally been estimated by clear-cut harvest techniques.[468,599,767,848,950] Reviews of primary productivity estimates for salt marsh plants by Keefe[452] and Turner[940] reveal the variability of P_nAG estimates. Differences in P_nAG methodologies have been investigated by Kirby and Gosselink,[468] Linthurst and Reimold,[524] Shew et al.,[832] and Hardisky.[362] The harvest techniques are known as destructive techniques. In contrast, a variety of nondestructive techniques have also been used and a number of studies have made comparisons between the two techniques.[362]

In the harvest techniques, a number of quadrats selected at random, generally 0.25 or 0.5 m^{-2} in area, are clipped each month from the study site. The clipped material from each quadrat is separated into live and dead material which is then oven dried to a constant weight. Estimates of primary production have been based on peak live-standing crop biomass, but as Kirby and Gosselink[468] have shown, such estimates are too low. They suggested that accounting for production losses due to mortality and disappearance during the growing season, and changes in standing crop between sampling periods yields a better estimate. Various methods of arriving at production estimates taking such matters into account have been detailed by Smalley,[849] Milner and Hughes,[599] and Wiegert and Evans.[1007]

Among the nondestructive techniques Hardisky[362] describes a tagging technique in which live and dead culms are individually tagged and at intervals measured for height, and the number of live and dead leaves counted. Differences in height and number of live and dead leaves between successive intervals were analyzed by the method described by Hardisky and Reimold[1076] to give an estimate of annual production. Using this technique on a marsh near Brunswick, Georgia, Hardisky[362] estimated P_nAG at 635 g dry wt m^{-2} year^{-1} compared to a destructively estimated Pl_nAG of 931 g dry weight m^{-2} year^{-1}. The difference was attributed to errors associated with the harvest technique.

Long and Woolhouse[532] combined field measurements of leaf photosynthesis, using a ^{14}C technique, coupled with measurements of increase in leaf area per unit of ground over the growing season, to compute photosynthetic rate. Extrapolating from these estimates of stand photosynthetic rate, and assuming that the photosynthetic rate is zero when the canopy is submerged at high tide, gross production can be estimated.

Long and Mason[531] have recently reviewed the varous methodologies for the estimation of P_n using destructive techniques. As noted, such techniques determine P_n from the sum of the changes in plant biomass (ΔW) and all the losses over a given time interval. They redefine net primary production as:

$$P_n = \Delta W + L + G + E \qquad (1)$$

where ΔW = change in biomass, L = losses by death or shedding, G = loss to grazers, and E = loss through root exudation.

Losses of shoot material may be determined using nondestructive techniques or determined by measuring the change in the amount of dead material and accounting for the rates of loss of this material. Long and Mason[531] point out that an added complication is tidal input and export of dead vegetation:

Table 1
METHODS USED TO ESTIMATE P_n IN SALT MARSHES BY EXTRAPOLATION FROM BIOMASS MEASUREMENTS

1. Maximum live dry weight

$$P_n = W_{l(max)}$$

Assumptions:
   No carry-over of biomass from one year to the next
   No death occurs before the maximum biomass is gained

2. Maximum standing crop

$$P_n = W_{l(max)} + W_d$$

Assumptions:
   No carry-over of either biomass or dead material from one year to the next
   Dead material does not decompose before the maximum biomass is obtained

3. Maximum-minimum

$$P_n = W_{l(max)} - W_{l(min)}$$

Assumptions:
   As for method 1, but accounts for any carry-over of material between years

4. International Biological Programme[599]

$$P_n = \Sigma(\Delta W_l) \text{ (negative } \Delta W_l \text{ is taken as zero)}$$

Assumptions:
   Death and growth do not occur simultaneously
   P_n is never negative

5. Smalley's method[524]
P_n^i for any given interval is determined according to the following conditions:

IF $W_l > 0$ and $W_d > 0$	THEN $P_n^i = W_l + W_d$
IF $W_l < 0$ and $(W_l + W_d) > 0$	THEN $P_n^i = W_l + W_d$
IF $W_l > 0$ and $W_d < 0$	THEN $P_n^i = W_l$
IF $(W_l + W_d) < 0$	THEN $P_n^i = 0$

$$P_n = \Sigma P_n^i$$

Note: $W_{l(max)}$ = Maximum biomass recorded during the year, $W_{l(min)}$ = Minimum biomass recorded during the year, W_d = Mass of dead plant material, Δ = Net change in a quantity between two sampling dates, P_n^i = Primary production between two sampling dates, P_n = Annual primary production.

From Long, S. P. and Mason, C. F., *Saltmarsh Ecology,* Blackie and Son Limited, Glasgow, 1983, 98. With permission.

$$L = \Delta W_d + D + T_e + T_i$$

where ΔW_d = changes in dead vegetation, D = loss due to decomposition, T_e = tidal export of material, and T_i = tidal import of material.

   The vast majority of P_n estimates in salt marshes have been based on extrapolation from maximum biomass, i.e., the sum of biomass and weight of dead material. These are the methods 1 to 3 listed in Table 1 and they assume that no material is lost before maximum biomass is achieved and that no growth occurs after the maximum has been reached. This clearly is not the case.[204] Methods 4 and 5 have been developed to alleviate these difficulties. Table 2 compares P_n determined by actual measurements of the components in Equation 1 above with estimates calculated by extrapolation methods. It is clear that all the extrapolation methods underestimate P_n and this should be borne in mind in the succeeding discussion.

Table 2
ESTIMATES OF NET ANNUAL PRIMARY PRODUCTION[a]

Species location	*Spartina anglica* E. England[204]	*Puccinellia maritima* E. England[420]	*Spartina patens* Georgia[524]	*Juncus gerardii* Maine[524]
(Method)				
Max live dry weight	341	310	942	244
Max standing crop	506	800	—	—
Max—min	283	254	—	—
IBP method	283	—	705	244
Smalley's method	271	420	1674	562
Eqn. $L = \Delta W_d + D + T_e + T_i$[b]	702[c]	807	3925	616

[a] According to the extrapolation methods listed in Table 1 compared with evaluations according to the equation $L = \Delta W_d + D + T_e + T_i$.
[b] None of these studies takes account of rood exudation.
[c] Losses measured by direct observation of permanent quadrats.

From Long, S. P. and Mason, C. F., *Saltmarsh Ecology*, Blackie and Son Limited, Glasgow, 1983, 99. With permission.

Table 3
GROSS PRIMARY PRODUCTION (Pg), RESPIRATION (R), AND NET PRIMARY PRODUCTION (Pn) BASED ON GAS EXCHANGE MEASUREMENTS

Species	Location	Pg	R/Pg	R	Pn	Ref.
Spartina alterniflora	New York	4490	0.60	2700	1790	409
Spartina patens	Maryland	1760	0.58	1020	740	198
Spartina angelica	E. England	4500	0.69	3120	702	204

Note: All figures represent organic dry weights.

From Long, S. P. and Mason, C. F., *Saltmarsh Ecology*, Blackie and Son Limited, Glasgow, 1983, 100. With permission.

Long and Mason[531] summarize the measurements of salt marsh gross primary production (P_g) each carried out by a different CO_2 exchange method (Table 3). They all suggest that respiratory losses account for 60 to 70% of P_g. Long and Mason suggest that this respiratory loss is not surprising when the costs of maintaining a very high proportion of belowground tissue are considered (see Section II.C.4.) and when the considerable energy expenditure that is involved in salt exclusion and secretion, soil reoxidation, and symbiotic nitrogen fixation[63] is taken into account.

Estimation of belowground biomass and production is much more difficult than that of the aboveground component,[307] and as a consequence the majority of studies have measured aboveground production only. Typically some kind of coring device is used and the material must be extensively washed before being separated into live and dead fractions. Separation is difficult although staining techniques are a help. Many investigators have found live-dead separation techniques to be inadequate and report belowground biomass as a combination of live and dead material.[188,306,854] The change in biomass technique, originally used for prairie vegetation[149] is often used to estimate productivity.

2. Aboveground Primary Production
As Mann[576] points out, at least 90% of the work on systems processes in salt marshes

Table 4

ORGANIC PRODUCTION ESTIMATES FOR ESTUARINE
EMERGENT MACROPHYTES (g C m^{-2} year^{-1})

Locality	Species	Production	Ref.
North Carolina	*Juncus roemerianus*	280—612	255, 892, 967, 1021
Florida	*Juncus roemerianus*	425	380
East and Gulf coasts, U.S.	*Spartina alterniflora*	220—2000	467, 607, 640, 849, 1020
Long Island	*Spartina patens*	500	368
North Carolina	*Spartina patens*	650	967
	Spartina patens with *Scirpus robustus*	670	967
California	*Spartina foliosa*	108—276	581
Sapelo Island	*Spartina alterniflora*		
	Short	520	273, 525
	Tall	1480	
	Distichis spicata	4214	
	Spartina cynosuroides	5996	
	Spartina patens	3824	
	Sporobolus virginicus	1372	
England	*Spartina townsendii*	900	532
New Zealand	*Spartina townsendii*	959	653
Oregon	Mixed communities *(Carex, Juncus, Scirpis)*	92—1120	217
California	*Salicornia virginica*	3844	581

has been carried out in North America, so consequently this and the succeeding sections relating to marsh primary production will report mainly on work carried out on marshes dominated by the cordgrass *Spartina alterniflora*. Table 4 lists organic production estimates for a range of estuarine emergent macrophyte species from North America and elsewhere. From this Table it can be seen that not only are there wide variations in productivity between species, but also there is variation within individual species. Species with the highest productivities are *Distichlis spicata* (4214 g C m^{-2} year^{-1}), *Spartina cynosuroides* (5996 g C m^{-2} year^{-1}), and *Spartina patens* (3824 g C m^{-2} year^{-1}), at Sapelo Island, Georgia. However, in terms of total contribution to overall marsh production along the east coast of North America, these species make a minor contribution. Overwhelmingly on an aerial basis the *Spartina alterniflora* marshes dominate, extending from the Gulf of St. Lawrence to Texas.

Hatcher and Mann[375] have compared the production in various salt marshes along the Atlantic and Gulf coasts of the U.S. (Figure 4). They record a net production of 289 g C m^{-2} year^{-1} in Nova Scotia, a minimum of 133 g C m^{-2} year^{-1} in New Jersey, and a maximum of 1153 g C m^{-2} year^{-1} in Georgia. In spite of local variations, possibly due to nutrient variations, it is apparent that net production increases towards the tropics.

Long and Woolhouse[532] and Knox[480] give data for the production of *Spartina townsendii*. The value for this species in the Humber estuary, England, was 900 g C m^{-2} year^{-1} and the same species in Pelorus Sound, New Zealand, had a production of 959 g C m^{-2} year^{-1}. Data for production in the Knysna estuary, South Africa[337] are comparable to those from marshes in the Northern Hemisphere.

Comparing the productivity of estuarine emergent macrophytes in one marsh with those in another marsh depends upon the spatial scale in which it is framed. Production varies widely over a given marsh depending on the duration of tidal inundation, drainage, substratum elevation gradient, age of the marsh, sediment-nutrient status, and salinity. Variation in

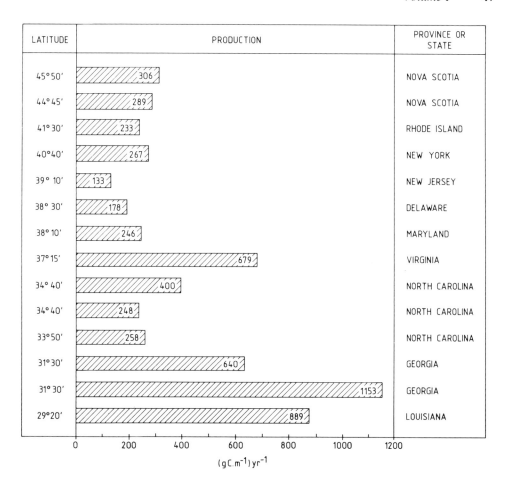

FIGURE 4.   Productivity of *S. alterniflora* at various sites on the Atlantic and Gulf coasts of North America. Notes in left hand column: (1) tall form under optimal conditions; (2) average of range of ecotypes, weighted according to percentage occurrence; (3) arithmetic mean of production in different ecotypes; (A) accounting for losses during the growing season; (B) losses during the growing season assumed to be 25% of final biomass. (From Hatcher, B. G. and Mann, K. H., *J. Fish. Res. Board Can.*, 32(1), 86, 1975. With permission.)

production and biomass of *Spartina* marshes is particularly large. On every marsh there is a tendency for *Spartina* to be taller and more productive along the edges of the creeks, and shorter and less productive as distance from the creek increases (Table 5). Gallagher et al.[273] found almost a 2.5-fold difference in production between tall and short *S. alterniflora* (often referred to as ecophenes). The tall *Spartina* had a mean biomass of 1966 g m^{-2} (dry weight) while the shortest stands back from the creeks contained 397 g m^{-2}. In the Pelorus marsh, New Zealand, tall *S. townsendii* had a biomass ranging from 932 to 1368 g m^{-2} whereas the short *S. townsendii* had a biomass ranging from 215 to 766 g m^{-2}. The proportion of the various types of stands within a marsh becomes an important factor.

The annual net aboveground primary production figures in Table 4 do not include organic matter leached by tidal water or removed by grazing herbivores. The latter losses are small but the former may be significant. Gallagher et al.[270] found that leaching losses from *S. alterniflora* were about 6.1 g C m^{-2} year^{-1}, a small amount compared to shoot production. The organic matter lost, however, may be especially important because it becomes immediately available for heterotrophic metabolism. Turner,[941] in another study of the loss of soluble photosynthate, reported release of 400 mg C m^{-2} year^{-1}. The disparity between these two studies has not been resolved.

Table 5
**ANNUAL NET PRODUCTION, OR PEAK BIOMASS,
OF MARSH PLANTS IN 'LOW MARSH' (CREEK
BANK) AND 'HIGH MARSH' SITUATIONS (g dry
wt. m^{-2} year^{-1})**

Species	Low marsh	High marsh	Ref.
Spartina alterniflora (P_n)	3300	2200	640
Spartina alterniflora (P_n)	3700	1350	273
Spartina alterniflora (PB)	1966	397	273
Juncus gerardi (P_n)	3500	485	520
Spartina townsendii (PB)	1294—2082	327—933	480

Note: P_n = net production and PB = peak biomass.

3. Belowground Production

Studies of belowground production and decomposition in marsh ecosystems have lagged much behind aboveground studies for technical reasons. Sampling is difficult, especially in species where much of the belowground material consists of large irregularly distributed components such as tubers and rhizomes. Good et al.[307] have recently reviewed the growing body of literature on belowground production and decompositon.

Table 6 lists belowground productivity estimates for ten marsh species along the east coast of North America. The greatest number of estimates are for *Spartina alterniflora*. Estimates are variable ranging from 0.22 to 3.5 kg dry wt m^{-2} year^{-1} for the tall ecophene, to 0.45 to 6.2 for the short ecophene. However, despite the large differences that have been recorded in aboveground productivity for *S. alterniflora* ecophenes, belowground productivity for any given locality seem to be very similar and indicate that the total productivity of the short ecophene is not nearly as low as previously judged from aboveground data only. Estimates for the other species are less extensive but in general are of the same order of magnitude. The occurrence of peak belowground biomass appears to vary somewhat among the species. For *S. alterniflora* (all height forms), peak belowground biomass occurs in late spring to summer (May to August) with sites at lower latitudes generally peaking earlier.[306,346,854,949,991]

Pomeroy et al.[732] list estimates of the underground production of seven marsh plants on the Sapelo Island marsh (Table 7). It will be noted that while the belowground production was the same for both short and tall *S. alterniflora*, the belowground production of the former was 59% of the total production compared to only 34% for the latter. For the other species, apart from *Juncus roemerianus*, the belowground production was small compared to that of the aboveground.

Table 8 lists peak aboveground and belowground biomass estimates for three marsh species in New Zealand. The belowground biomass estimates for short *S. townsendii* and *Juncus* and *Leptocarpus* are among the highest recorded in any marsh species.

4. Root-Shoot Ratios

Root-shoot ratios in marsh species reflect differences in species and habitats, while the amount of photosynthate allocated to underground parts, range widely both within and between species. Table 9 gives root-shoot ratios (biomass) for four species of salt-water marsh plants along the western Atlantic coast along with some data from other localities. Root to shoot ratios as high as 50 have been reported for the short *Spartina alterniflora* ecophene,[266] although most of the estimates are considerably lower, ranging from 1.2 to 5.4. For the tall form, values range from 0.5 to 8.25. Generally in the same marsh, the ratios are generally much higher for the short ecophene than for the tall ecophene. Values for the rush *Juncus roemerianus* range from 0.80 to 8.7. The ratios in a New Zealand marsh

Table 6
**BELOWGROUND PRODUCTIVITY ESTIMATES FOR TEN
MARSH EMERGENT MACROPHYTES FOR SEVERAL
LOCATIONS ALONG THE WESTERN ATLANTIC COAST**

Salt-water species	Productivity (kg m^{-2} year^{-1})	Location	Ref.
Borrichia frutescens	0.82	Georgia	271
Distichlis spicata	1.07	Georgia	271
	3.40	Delaware	271
	2.78	New Jersey	306
Juncus gerardii	4.29	Delaware	271
	1.62	Maine	271
Juncus roemerianus	1.36	Mississippi	188
	4.4—7.6	Alabama	890
	3.36	Georgia	271
Phragmites communis	3.65	Delaware	271
(australis)	2.81	New Jersey	308
Salicornia virginica	0.43	Georgia	271
	1.43	Delaware	271
Spartina alterniflora	2.1	Georgia	271
Tall form	0.5	North Carolina	891
	2.9	New Jersey	305
	3.3	New Jersey	306
	2.4	New Jersey	308
	3.5	Massachusetts	949
	0.22	Maine	271
Short form	2.02	Georgia	271
	0.56	North Carolina	857
	0.46	North Carolina	891
	3.2	New Jersey	305
	2.4	New Jersey	306
	2.3	New Jersey	854
	3.6—6.2	Alabama	890
Spartina cynosuroides	2.2	Mississippi	188
	3.56	Georgia	271
Spartina patens	0.31	Georgia	271
	0.47	Delaware	271
	2.5	Massachusetts	949
	0.54	Maine	271
Sporobolus virginicus	0.58	Georgia	271

for the introduced *Spartina townsendii* for the short and tall ecophenes, were 1.7 and 11.6,[653] respectively, while the ratios for the rush *Juncus maritimus* and the sedge *Leptocarpus simplex* were 4.1 and 3.4, respectively.

High root-shoot ratios have generally been considered to be indicative of adaptive mechanisms, with unfavorable sediment conditions requiring greater root surface for each unit of abovegound material.[827] Early studies in Europe[757] revealed the common occurrence of high root-shoot ratios in a variety of halophyte species.

5. Factors Affecting Productivity

Within salt marshes there is considerable spatial and temporal environmental heterogeneity, much of which is related to the frequency and duration of tidal cover. Odum[634] has described salt marshes as pulse-stabilized systems due to the tidal regime which sets limits to the type of vegetation at particular tidal levels. Jeffries et al.[428] point out that because tidal cover in any marsh throughout the year is predictable, seasonal trends in the salinity, water potential, and osmolarity of the sediments in different parts of the marsh can be predicted. However,

Table 7
ESTIMATES OF PRIMARY PRODUCTION OF
MARSH PLANTS NEAR SAPELO ISLAND,
GEORGIA[271,273,274,522]

| Species | Annual production (g C m^{-2} year^{-1}) | |
	Aboveground	Belowground
Spartina alterniflora		
Tall	1482	770
Short	520	770
Spartina cynosuroides	5996	—
Spartina patens	3824	120
Sporobolus virginicus	1372	220
Distichlis spicata	4214	420
Juncus roemerianus	880	1.340
Salicornia virginica	—	140
Borrichia frutescens	—	320

Table 8
PEAK BIOMASS ESTIMATES FOR THE PRINCIPAL
EMERGENT MACROPHYTE SPECIES IN THE KAITUNA
MARSH, PELORUS SOUND, NEW ZEALAND[480,653]
(g dry weight m^{-2})

Species	Aboveground biomass	Belowground biomass
Spartina townsendii		
Tall form	1294—2082	2414—3442
Short form	327—933	6582—8072
Leptocarpus simplex	2353	8041
Juncus maritimus var australiensis	1782	7382

they also note that salt marshes display cyclical stability, as defined by Orians,[671] in as much as the salinity, water potential, and osmolarity of the sediments all fluctuate around the corresponding values for sea water which change little during the year. To this extent salt marshes are highly predictive, stable environments.

Levins[506] has stated that a general theory of the responses of organisms to a fluctuating environment must include the predictability and regularity of occurrence of a particular set of environmental conditions, the length of time that these conditions exist relative to the life cycle of an organism and their effects on its fitness, and the total investment required for reproduction. A number of recent studies including those of Jeffries et al.,[428] Linthurst,[523] and Good et al.[307] have investigated some of these factors in relation to the growth strategies of marsh plants. The major factors that have been considered are salinity, drainage, aeration, nitrogen, and pH. Most studies have concentrated on the aboveground component, but increasing attention is being paid to the responses of the belowground one.

a. Salinity

There is considerable literature on the salinity relationships and physiological responses of halophytes and I do not propose to discuss it here. Readers are referred to Waisel[965] and Chapman[108] for detailed discussions. The effect of salinity on the growth of *S. alterniflora* is one of the most widely studied factors affecting the distribution and zonation of this

Table 9
ROOT-SHOOT RATIOS FOR 6 SPECIES OF MARSH
EMERGENT MACROPHYTES ALONG THE EAST
COAST OF THE U.S.[307] AND 3 FROM NEW
ZEALAND[480,653]

Species	Ratio	Location	Ref.
Distichlis spicata	7.2	Georgia	266
	4.5	New Jersey	306
Juncus roemerianus	0.80	Mississippi	188
	3.26	Alabama	890
	3.7—8.7	Florida	493
	8.2	Georgia	266
Spartina alterniflora			
Tall form	1.43	Georgia	266
	0.3—0.4	North Carolina	891
	4.53	New Jersey	306
	8.25	Massachusetts	949
Short form	3.72	Alabama	890
	48.9	Georgia	266
	1.2—1.3	North Carolina	891
	4.7	New Jersey	854
	5.24	New Jersey	306
Spartina patens	5.58	New Jersey	306
	4.0	Massachusetts	949
Spartina townsendii			
Tall form	1.7—1.9	New Zealand	480
Short form	8.6—20.0	New Zealand	480
Juncus maritimus	4.1	New Zealand	480
Leptocarpus simplex	3.4	New Zealand	480

species.[348,606,621,693,850,851,1035] Mooring et al.[606] found that *S. alterniflora* grows best at 10‰. Similar results were observed by Haines and Dunn.[348] Nestler[621] found that the interstitial water in the sediments of the Sapelo Island marshes formed salinity clines across the marsh, with the lowest average values in the creek beds and increasing values with increasing distance from them (Figure 5A). Biomass, height, and leaf area (but not shoot number) were all negatively correlated with salinity (Figure 5B). Phleger[720] showed that *Spartina foliosa* in California, grown in nutrient solutions, made the best growth at zero salinity. Linthurst,[522] while working in North Carolina, found that salinity increases of 15‰ decreased biomass, density, and mean height of *S. alterniflora* and sediment MOM (macro-organic matter, i.e., > 2 mm) an average of 42, 32, 22, and 37%, respectively. Reductions in all the biological variable measurements were greater when salinity was raised from 30 to 45‰ than when it was raised from 15 to 30‰. Increase of salinity from 15 to 45‰ decreased biomass, density, mean height, and MOM 66, 53, 38 and 61%, respectively.

At high salinities, osmotic stress, or cell-membrane damage, is likely to be the primary growth regulating factor. Osmotic stress could result in reduced water uptake.[517] Membrane permeability changes could reduce the influx of necessary nutrients and/or cause leakage of nutrients from the roots to the surrounding substrate.

In a number of studies, belowground growth was also depressed with increasing salinity. Under laboratory conditions, Linthurst and Seneca[526] found similar salinity belowground growth responses in *S. alterniflora* to those reported by Linthurst[523] for the aboveground component. Similar results were reported by Parrondo et al.[693] Culture solutions of 0 to 16‰ had no effect on *S. alterniflora* growth, but a salinity of 32‰ significantly reduced root biomass. Haines and Dunn[348] studied the effect of ammonium nitrogen and salinity on

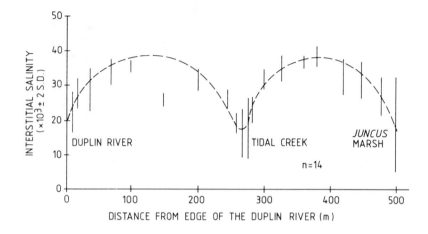

A

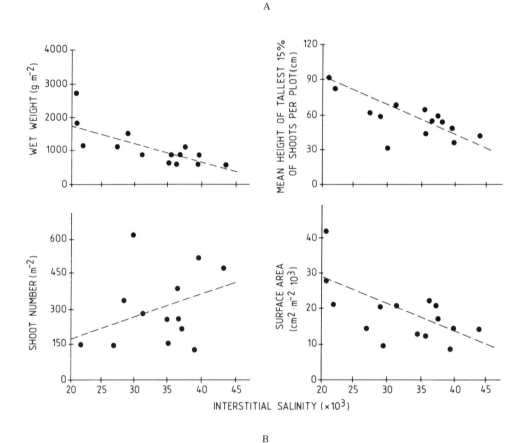

B

FIGURE 5.   (A) Distribution of salinity in a Georgia salt marsh; (B) correlation between salinity and *Spartina* production. (From Nestler, J., *Estuarine Coastal Mar. Sci.*, 5, 709, 1977a. With permission.)

S. alterniflora under controlled laboratory conditions. Salinity had a varying effect on root-rhizome growth depending on ammonium level. At low and intermediate N levels, there was no change in root weight and an increase in rhizome weight with increasing salinity. At high N level, root weight decreased with increasing salinity, while rhizome weight was greatest at intermediate salinity.

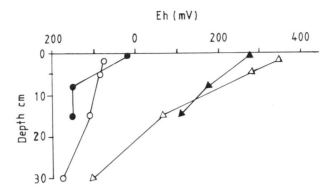

FIGURE 6. Relationship of Eh to depth of soil in tall and short *Spartina alterniflora*. ○, ● Short; △, ▲ Tall. ○, △ Data from Howes et al.[414] ●, ▲ Data from Linthurst and Seneca.[526] (From Chalmers, A. G., *Estuarine Comparisons*, Kennedy, V. S., Ed., Academic Press, New York, 1982, 237. With permission.)

Belowground biomass of *Spartina patens, S. foliosa, S. cynosuroides,* and *Sporoblus virginicus* also decreases with increased substrate or culture solution salinity.[269,317,693,850] In most cases, increased salinity reduced aboveground production more than belowground as reflected in an increased root-shoot ratio at high salinities.[348,693,851]

In addition to tidal inundation and subsurface water movement, plant growth itself can affect intertidal salinity by ion exclusion by plant roots.[851] A similar increase in sediment salinity was observed in two different fertilized plots of short *Spartina alterniflora*. In spite of the evidence for control of *S. alterniflora* growth and height by salinity, there are marshes in which both tall and short forms occur where salinity stress is not a factor. In the Great Sippewissett marsh in Massachusetts, there is no consistent pattern of salinity difference in tall and short grass areas, and no salinity greater than 33‰ has been measured.[414] In Louisiana, tall and short forms occur both on creek banks and inland areas, but there are no corresponding differences in interstitial salinity.[82]

Mendelssohn et al.[588] have found that *S. alterniflora* can respond to oxygen deficiency in its roots by switching from aerobic respiration to fermentation as a means of producing ATP. They speculated that reduced growth may be due to the decrease in carbon available for growth resulting from increased consumption of glucose during fermentation and subsequent diffusion of ethanol, an end-product of fermentation. Even though *S. alterniflora* can provide some oxygen to the roots via diffusion, conditions in the soil and rhizosphere are for the most part reducing[913] with high levels of hydrogen sulfide. Hydrogen sulfide can be toxic to plants due to inhibition of oxidase enzymes[725] and can inhibit nutrient uptake.

Oxygen diffusion from the roots of *S. alterniflora* can oxidize reduced sulfur components and raise the Eh of the sediment. Figure 6 shows that the Eh is higher in the root zone of tall *S. alterniflora* than in that of the short form. Howes et al.[414] found direct evidence that *S. alterniflora* itself can alter the Eh of the rhizosphere and that tall plants oxidize their rhizosphere more than the short plants. Eh was higher in plots which had been fertilized for several years than in adjacent plots. A positive feedback loop exists between rhizosphere oxidation and productivity. Oxidation of the root zone increases nutrient uptake capability and reduces the toxic effect of H_2S, thereby increasing productivity, which inturn increases rhizosphere oxidation.

b. Aeration

Aeration was found to have a positive effect on the growth of *S. alterniflora*.[522,523] Linthurst and Seneca[526] found that aeration of the root medium in combination with high nitrogen levels was more effective in overcoming the detrimental effect of high salinity than high nitrogen alone. Aeration alone also had a beneficial effect on plant growth at all salinities tested. They suggested that aeration enhanced the plants potential for nutrient uptake. The response of aboveground production to aeration was found to be greater than belowground production response. Linthurst and Seneca[526] noted that belowground production response was not equal for three *S. alterniflora* height forms, with the tall form showing the greatest increase in production under aerated conditions.

Figure 7 from Linthurst[521] depicts measurements of various physical and biological parameters associated with various aeration treatments under laboratory conditions. Substrate pH was affected by aeration (Figure 7a); reducing potential of the soil decreased with increased aeration and redox potentials adjusted to pH 7 were positive in the aerated substrate system (Figure 7b). There was no significant difference between redox potentials in the regularly flooded (RF) and stagnant aerated (SA) systems. According to Linthurst[523] this suggests that the ionic forms of the elements, as affected by redox potentials, might be similar in these two systems. Sulfide concentrations of the substrate were highest in the stagnant unaerated (SU) and stagnant aerated (SA) systems (Figure 7c).

Examination of the biological responses indicated that average culm height of the tallest plants per treatment was clearly representative of a height gradient (Figure 7a). Stem densities were not significantly different in the aerated systems, but they were lower in the unaerated greenhouse marsh (Figure 7e). Both aboveground and belowground biomass were enhanced in the aerated substrate system (Figures 7f and g). Root-shoot ratios were highest in the unaerated system.

It has been shown that the aerenchyma in the tissues of *S. alterniflora* are a primary pathway for oxygen maintenance of the roots through diffusion of oxygen to the root zone from the aboveground portion of the plants. Teal and Kaniwisher[914] suggest that this diffusion prevents internal annoxia. It has also been demonstrated that oxygen produced during photosynthesis improves the oxygen status of the submerged roots.[20]

Oxygen concentration in the base of *S. alterniflora* shoots declines during tidal submergence in the dark, and the shoot bases can become anoxic during nighttime at high tides.[296] Gleason and Zieman[296] also found that *S. alterniflora* shoot height was inversely proportional to the rate and extent of oxygen decline in the shoot bases during laboratory-simulated tidal submergence. Anoxia in the roots can cause cellular damage[296,526] and reduce the rate of ammonium uptake.[611]

c. Drainage

Several studies have monitored the effects of drainage on the growth of salt marsh species, especially *S. alterniflora*. Field studies have shown that reduced drainage conditions result in a decrease in the total biomass of *S. alterniflora*.[526,589] As noted with other environmental factors, aboveground production is affected more than belowground production. Decreased drainage reduced root-rhizome growth of three *S. alterniflora* height forms, although the reduction was not significant for the short form.[589]

In laboratory studies,[523] mean *in situ* pH, reducing potential, free sulfides, and salinity significantly decreased with increasing pot-surface elevation. Soluble salts, Na, P, K, Ca, and Mg concentrations also decreased with increasing elevation. Organic matter, $NH_4 - N$, sulfate Mn, Zn, and Cu were not significantly different between elevation treatments. Laboratory studies have also demonstrated that reduced drainage conditions generally result in increased *S. alterniflora* belowground biomass.[589,693] Plants grown under stagnant conditions (standing water over soil surface), however, had significantly reduced root biomass.[589] One problem is that the laboratory experiments do not accurately mimic field conditions.

Wiegert et al.[1005] have found that increasing sediment drainage in an intermediate-height *S. alterniflora* marsh increased plant biomass beyond that of an adjacent undisturbed plot,

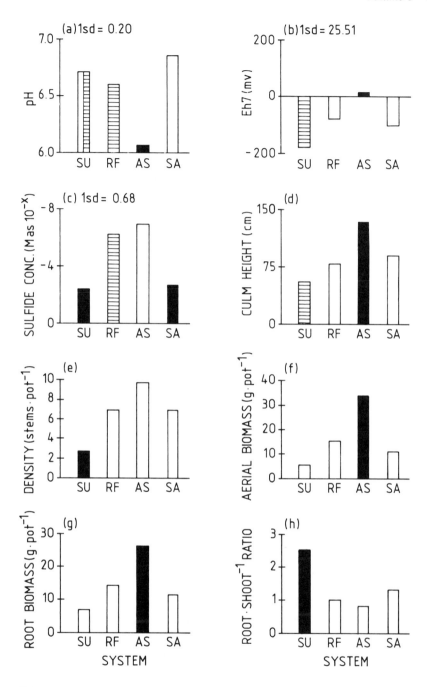

FIGURE 7. Measurements of the physical and biological parameters associated with various aeration treatments. SU — stagnant unaerated system; RF — regularly flooded system; AS — aerated substrate system; and SA — stagnant aerated water system. Bars of the same design are not significantly different. Eh adjusted to pH 7. Sulfide concentrations (molarity). (From Linthurst, R. A., *Am. J. Bot.*, 66, 685, 1979. With permission.)

and shifted a number of other characteristics towards those typical of tall *S. alterniflora* zones. King and Klug[465] have advanced the hypotheses that plant height and productivity are the result of the interaction between tidal-water movement, dissolved iron and sulfide concentrations are inversely related to soil-water movement and plant productivity and that the concentration of sulfide is inversely related to that of dissolved iron. The major input of iron to marsh sediments is via sedimentation and King and Klug[465] proposed that iron input will vary with distance from tidal creeks and frequency of flooding, and will therefore be greater in the tall *S. alterniflora* zone than in the short zone.

d. Tidal Inundation

Working in Long Island Sound, where the tidal range changes from 0.7 m near the mouth to 2.26 m near the head of the inlet, Steever et al.[876] found that marsh productivity was correlated with tidal range; the correlation coefficient being better than 0.96. They compared two sites in the same general area, one of which had the tidal range reduced by a gate, and found that there was a 26% reduction in productivity in the gated marsh. Finally, they showed that a variety of data from the Atlantic coasts of North America fitted the same trend (Figure 8). Odum[636] refers to tidal energy as an "energy subsidy," performing the work of mineral cycling, food transport, waste removal, and so on. He claims that, "It is clear that the energy subsidy provided by tidal flow more than compensates for the energy drain of osmoregulation required by a high salinity environment." He also suggests that a tidal range of about 2 to 2.5 m is optimum, and that greater ranges incur a stress. As Mann[576] points out, this energy subsidy theory is an attractive one, but that the evidence is mainly correlational and as we have seen, other factors are involved in determining marsh production.

e. pH

In laboratory experiments growth was optimal at pH 6 in comparison with that at pH 4 and 8.[523] Short *S. alterniflora* with its shallow root system[272] may be subjected to changing pH when periods of high temperature, low tides, and rainfall prevail. *S. alterniflora* growth was also inhibited at pH 8, a pH observed in substrates subject to dieback in a North Carolina marsh.[526] This pH increase could, in part, have contributed to the dieback phenomenon. In addition to substrate effects, pH has a direct effect on the root membrane and this can directly influence plant performance.

f. Nutrients

Numerous authors have investigated the effects of nitrogen on growth.[76,267,348,587,897,948] As pointed out by Pomeroy[730] salt marshes are nutrient sinks where there is normally a large excess of all nutrients, with the exception of nitrogen (nutrient cycles will be considered in detail in Volume II, Chapter 1, Section II). The response of many halophytes to the availability of a limited resource, nitrogen, is of particular interest because of the known role of soluble organic nitrogen compounds in osmoregulation in algae and higher plants.[391] The accumulation of amino acids (including proline, pipecolic acid, 5-hydroxy pipecolic acid, and methylated quaternary ammonium compounds such as glycine betaine and homobetaine) is an unusual feature of the nitrogen metabolism of many halophytes.[887] Stewart et al.[887] discuss the complex role of these substances in salt-tolerance.

S. alterniflora accumulates proline and glycine-betaine[99] to counteract the low osmotic potential in vacuoles in which the plant sequesters NaCl. Since each of these compounds is 12% nitrogen, accumulation of them by a plant experiencing salinity stress can significantly reduce the nitrogen available for growth. Cavalieri and Huang[99] found that fertilization of short *S. alterniflora* with ammonium nitrate resulted in increased growth, higher concentrations of proline and glycine-betaine, and lower-leaf water potentials, but that there was no effect with tall *S. alterniflora*. This implies that tall *S. alterniflora* growing under conditions of relatively low salinity stress have access to sufficient nitrogen for both plant growth and osmoregulation.

In laboratory experiments, an increase in nitrogen from a natural level to 168 kg ha^{-1} N increased biomass, density, and mean height 2.02, 1.46, and 1.26 times.[523,526] The high level of N and aeration together in comparison with the low level of these two factors,

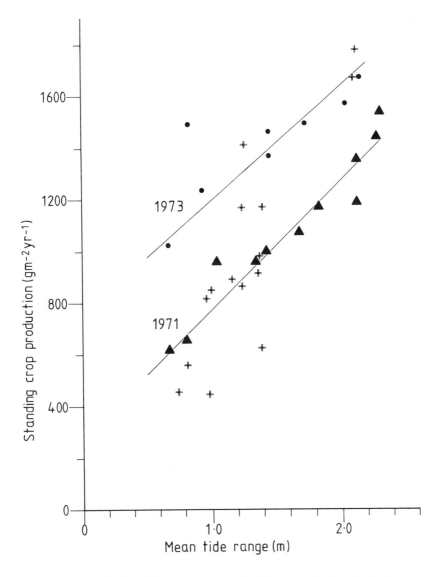

FIGURE 8.   Production of *Spartina* in relation to tidal range. Circles and triangles are author's own data in different years, with regression lines fitted by eye. Crosses are data from the literature. (From Steever, Z. E., Warren, R. S., and Niering, W. A., *Estuarine Coastal Mar. Sci.*, 4, 476, 1976. With permission.)

averaged over salinity, produced increases of 4.53, 2.71, 1.88, and 2.24 times in biomass, density, mean height, and macroorganic matter (MOM). Linthurst[523] postulates that growth enhancement by additions of $NH_4 - N$ is possibly due to (1) eliminating a portion of the $NH_4 - N$ vs. Na uptake competition; (2) creating a large N gradient, subsequently increasing the rate of diffusion to the root zone; and/or (3) increasing concentrations in a naturally nitrogen deficient system. Eleven times more biomass was observed in low salinity (15‰) nitrogen-enriched aerated treatments, than in the high salinity nitrogen-poor unaerated treatments. Linthurst[523] concluded that continued aeration is beneficial to the plant and significantly enhances the potential for nutrient uptake and for the general nitrogen response. Aeration may eliminate the need for energetically expensive metabolic adjustments by the plant at low-oxygen concentrations and/or minimize damage to root systems caused by

Table 10

EFFECTS OF NITROGEN FERTILIZATION ON GROWTH OF
SPARTINA ALTERNIFLORA **IN SALT MARSHES FROM**
MASSACHUSETTS TO LOUISIANA

Site	Fertilizer (rate of application)	Control (g/m²)	Fertilizer (g/m²)	Ref.
Massachusetts	Urea (290 g N m^{-2} year^{-1})	424	834	949
Delaware	Ammonium nitrate (240 g N m^{-2} year^{-1})	772	2104[a]	897
North Carolina	Ammonium sulfate (67.2 g N m^{-2} year^{-1})	450	1800	76
Georgia	Ammonium nitrate (20 g N m^{-2} year^{-1})	471	803	267
Georgia	Sewage sludge (100 g N m^{-2} year^{-1})	396	650	102
Louisiana	Ammonium sulfate (20 g N m^{-2} year^{-1})	1666	1916	697

[a]   g Fresh weight/m²; all others are g dry weight/m².

From Chalmers, A. G., *Estuarine Comparisons,* Kennedy, V. S., Ed., Academic Press, New York, 1982, 234. With permission.

extreme anaerobis. As a result, more energy is available for growth, and root integrity is maintained for normal uptake processes.

Table 10 presents the results of a number of investigations of the impact of nitrogen on *S. alterniflora* production. These studies have also shown that additions of phosphorus have no stimulatory effect. Gallagher[267] found that additions of ammonium nitrate stimulated growth of short *S. alterniflora,* but not of tall, leading to the commonly held view that growth of short *S. alterniflora,* but not tall, is nitrogen limited.

Grass in a short *S. alterniflora* marsh fertilized for 4 years with urea was almost, but not quite, transformed to the tall form.[951] Fertilization increased growth, greenness, height, and stem and leaf width. Some secondary factor apparently prevented complete convergence. The implication was that either nutrient availability is greater for tall *S. alterniflora* than for short, or that a secondary limiting factor is not active for the tall grass, or perhaps some combination of the two. Salinity, iron, and water availability have been suggested as possible limiting factors.[101]

Sediment nitrogen in tall and short *S. alterniflora* differs between marshes. In Georgia, it was found that while sediment nitrogen was the same in both tall and short stands, the annual accumulation of nitrogen in tall *S. alterniflora* stands was 2.5 times higher than in the short grass stand.[100] In this marsh, turnover rates of inorganic N either must be greater in tall zones than in the short *S. alterniflora* zone, or there must be some additional source of nitrogen for the tall grass. The potential for nitrogen fixation is greater in the tall zone than in the short,[357] tidally induced subsurface flow is greater in the tall zone,[765] and rates of NH_4^+ mineralization are also faster in the tall zone.[101] Maintenance of low salinities in creek bank sediments by tidal flushing helps prevent the salinity stress observed in fertilized high-marsh areas.[101]

Mendelssohn[587] found that short *S. alterniflora* was deficient in nitrogen status relative to the tall form, although interstitial NH_4^+ concentrations were six times higher in the short grass zone, supporting the idea that short *S. alterniflora* was nitrogen limited. Here the contention was again that some secondary factor was exerting its influence, in this case by

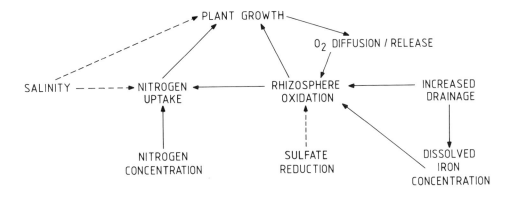

FIGURE 9. Relationship between factors directly and indirectly affecting growth of *Spartina alterniflora.* Solid lines are positive effects, dashed lines are negative effects. (From Chalmers, A. G., *Estuarine Comparisons,* Kennedy, V. S., Ed., Academic Press, New York, 1982, 239. With permission.)

preventing uptake and/or assimilation of the available NH_4^+. Morris[612] hypothesized that some edaphic factor was altering nitrogen-uptake kinetics and that this was responsible for the observed decrease in productivity with distance from the creeks. He suggested that this factor might be salinity, oxygen deficiency in the root zone, hydrogen sulfide, the exchange capacity of the sediments, or the diffusion rate in the soil.

g. Synthesis

Chalmers[101] presents a model (Figure 9) of the interactions between the factors which affect *S. alterniflora* growth, either directly or indirectly. The synthesis which follows is based on the conclusion in her paper. Edaphic factors control the heterogeneity in height, biomass, and productivity in *S. alterniflora.* Field and laboratory studies have shown that salinity is one factor which can influence *S. alterniflora* growth, but there are marshes where both tall and short forms occur in the absence of salinity gradients.

Fertilization experiments have demonstrated that growth of tall *S. alterniflora* is not nitrogen limited, but that the productivity of the short form can be increased by nitrogen additions. Other studies have shown that the apparent nitrogen limitation in the short form is not due to the shortage of available nitrogen, but to an attenuation in nitrogen uptake kinetics. Salinity stress-caused diversion of nitrogen to the production of osmotica can also reduce the amount of nitrogen available for growth.

High sulfide concentrations and consequent low oxidation-reduction potentials in the rhizosphere can affect nitrogen-uptake kinetics. Such conditions and anoxia can also cause structural damage or alterations in the roots which could affect nutrient uptake. Soil drainage, iron concentrations, oxygen diffusion from *S. alterniflora* roots, and plant productivity itself can all affect sulfide concentrations and redox potentials.

III. MANGROVE SYSTEMS

A. Introduction

The term "mangrove" refers to two different concepts. First it describes an ecological group of halophytic shrub and tree species belonging to some 12 genera in eight different families of plants.[965] In a second sense the term mangrove refers to the complex of plant communities fringing tropical sheltered shores. Chapman[112] prefers to refer to these communities or formations, as he terms them, as "Mangals",[110,561] reserving the term mangrove for the individual genera or species. Here the term mangrove system or ecosystem will be used. Mangrove ecosystems occur in tropical areas where they reach their greatest structural

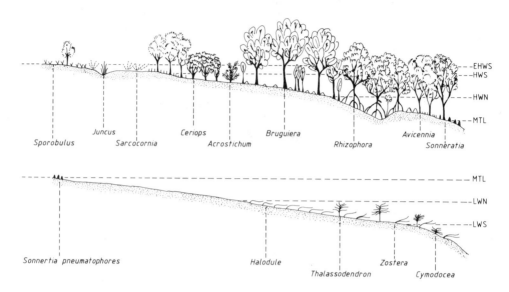

FIGURE 10.   Zonation on a mangrove shore in Morrumbene Estuary, Mozambique, east coast of southern Africa. (From Day, J. H., *Estuarine Ecology With Particular Reference to Southern Africa,* Day, J. H., Ed., A. A. Balkema, Rotterdam, 1981f, 90. With permission.)

and floristic diversity, and extend into subtropical areas in Asia, North America, subequatorial Africa, Australia, and New Zealand. Mangroves grow as far south as Okiwa Harbour (38° 01′ S) in New Zealand,[38] and Corner Inlet, Victoria (38° 55′ S) in Australia.[129] The most luxuriant and diverse mangrove forests occur in Southeast Asia where no fewer than 36 species of true mangroves occur. Biogeographers recognize two main vegetation groups. The Old World group with about 60 species and the New World group with only about 10 species.It is inferred that the adaptation of forest trees probably first occurred in the Indo-Malaysian area, and that mangroves spread and evolved from that center.

Mangrove ecosystems have attracted much curiosity and scientific attention and considerable literature has accumulated on the physiology of mangroves and the natural history of the communities in which they are the dominants. For recent reviews of this literature readers are referred to MacNae,[561] Lugo and Snedaker,[542] Walsh,[968] and Chapman.[110,111]

B. Distribution and Zonation

Mangrove forests range from the complex, species-rich assemblages such as those of the west coast of Malaysia, which may contain as many as 20 mangrove species (Figure 11), to ones composed of a single species such as the New Zealand *Avicennia marina* Forest. They grow from the highest level of spring tides down to almost mean sea level and on accreting shores. MacNae[561] and Saenger et al.[806] give many examples of vertical zonation in mangrove forests. On some gently sloping tropical estuarine shores, the mangrove swamp may extend up to 5 km from the main channel and may cover many thousands of hectares.

A typical zonation pattern can be illustrated by referring to a mangrove shore on the Morrumbene Estuary in Mozambique on the east coast of southern Africa (Figure 10).[175] From extreme high tide of springs (EHWS) to mean high water of springs (MHWS) there is a diverse and variable association of halophytes dominated by *Sporolobus virginicus,* the glasswort *Sarcocornia,* and the rush *Juncus kraussii.* The region from MHWS to MTL is covered with a mature forest of mangroves. This includes an upper fringe of *Avicennia marina,* thickets of *Ceriops tagal,* then *Bruguiera,* and *Rhizophora mucronata* in seepage channels and a lower fringe of *Avicennia* and *Sonneratia alba.* The genera *Bruguiera* and *Rhizophora* can be distinguished by their prop roots, especially in the latter species where the roots emerge from the trunk high above the ground and arch downwards. They form a dense, almost impenetrable tangle which traps sediments. The genus *Avicennia* has numerous breathing roots of pneumatophores growing vertically upwards from the underground root system. The pneumatophores of *Avicennia* in the lower fringe, the prop roots of *Rhizophora*, and the boles of these and other trees are covered with a moss-like association of algae including *Bostrychia*, *Calloglossa*, and other genera.

In the more diverse Malaysian forests (Figure 11) Watson[980] recognized five zones based on frequency of inundation. Beginning at the lowest level these are: (1) species growing on sediments *flooded at all times:* no species normally exists under these conditions, but *Rhizophora mucronata* may do so; (2) species on sediments *flooded by medium high tides:* species of *Avicennia*, *Sonneratia griffithii*, and bordering rivers, *Rhizophora mucronata;* (3) species on sediments *flooded by normal high tides:* most mangroves, but species of *Rhizophora* tend to be dominant; (4) species on sediments *flooded by spring tides only: Bruguiera gymnorhiza*, and *B. cylindrica;* and (5) species on land *flooded by equinoctial or other exceptional tides only: B. gymnorhiza* dominant, but *Rhizophora apiculata* and *Xylocarpus granatum* also coexist. The animals associated with these mangrove communites will be discussed in Chapter 5.

Most of the ecological literature on mangroves is concerned with the description of species composition and plant zonation. MacNae[561] has reviewed the zonation schemes proposed by various authors for the mangroves of the Indo-West Pacific. These schemes have been based on the frequency of inundation, salinity, or on dominant tree species. Some authors believe that this zonation represents a successional sequence leading via sediment accumulation to a terrestrial forest. Others consider that it represents a response to external forces (substratum, water flow, eustatic change) rather than a temporal sequence induced by the plants themselves.[930]

C. Environmental Factors

Literature on mangrove dynamics is heavily weighted towards the classic successional view.[107,110,968] This model emphasizes biotic processes inducing soil accumulation and plant community change from a pioneer through to a climax stage. In recent years an alternative model that seems to better fit the observed facts has been proposed.[930,932] This alternative model views mangroves as opportunistic organisms colonizing available substrates. In this context mangrove patterns are primarily seen as ecologic responses to external conditions of sedimentation, microtopography, estuarine hydrology, and geochemistry.[932]

According to Thom[932] there are three major components to the environmental setting of any locality in which mangroves occur: geophysical, geomorphic, and biologic. The first of these includes a variety of physical forces which operate from global to regional spatial scales such as sea level change, and climatic and tidal factors. The second component is essentially the product of the geophysical forces. It includes factors such as the depositional environment, the extent to which this is dominated by wave or river, or tidal processes[277] and the impact of the microtopography of the particular landforms (e.g., river levees, beach-ridge swales) on plant establishment, growth, and regeneration. Land surface elevation, drainage and stability, in combination with substrate or sediment properties (texture, composition, structure, etc.), nutrient inputs, and the salinity regime will produce environmental gradients within the coastal region. Different species, or different ecotypes within a species according to their physiological responses to the above factors, especially to moisture and/ or salinity stress conditions, will establish themselves where the combination of conditions are favorable. Growth patterns will be reflected in species and structural distributional patterns as well as gross and net productivity along environmental gradients. For example, Lugo et al.[540] reported differential responses in the *in situ* gas exchange characteristics of three mangrove species distributed along a salinity gradient in Florida. Such variations in phys-

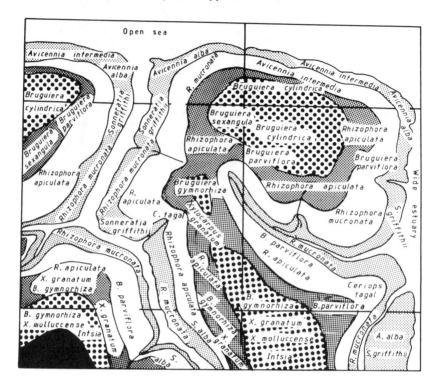

FIGURE 11.   A diagram showing the typical distribution of mangrove trees on the Malaysian west coast. (From MacNae, W., *Adv. Mar. Biol.*, 6, 90, 1968. With permission; after Reference 980.)

iological responses to an environmental variable are major factors affecting development of zonation patterns.[32]

Interspecific competition plays an important role in determining the species diversity and distribution patterns of a given area. In addition, the pool of species in the Atlantic, "New World", or southern Australia-New Zealand, is much more limited than in the tropical Indo-Pacific. As a consequence, *Avicennia* and *Rhizophora* occupy a broader range of habitats within any given are or depositional complex in the first three regions than is likely in the species diverse tropical Indo-Pacific.

Thom[932] illustrates five coastal settings in the form of generalized geometry and possible habitats for mangrove growth (Figure 12), while in Table 11 he relates these environmental settings to the geomorphic and ecological responses to the various geophysical processes. These examples clearly demonstrate how physical processes through geomorphic responses interact to produce distinctive habitat conditions. However, as Thom notes, there are patterns that do not fit nearly into the settings and it is therefore important, for each case, to evaluate the interaction of physical processes, landform-sediment products, and ecologic conditions. A major variable influencing the evolution of these landform-habitats is relative land-sea level movement. For example, for the maintenance over time of the well-zoned diverse mangrove forests of the equatorial hot-wet climate regions, the sea level must be relatively stable. More complex mosaic patterns of species will reflect rising or falling sea levels.

Thus, "mangrove distribution (zonation?) can be viewed as an opportunistic response of certain species to more or less favored changing environmental conditions whose characteristics within a region are primarily controlled by past and present geomorphic processes."[932]

Oliver[666] has discussed the role of environmental factors, with special reference to climate on the development of mangrove ecosystems. Figure 13 from Oliver summarizes the relationships of these environmental factors. It will be noted that the system is characterized by a number of feedback loops. The establishment of a mature, well-developed community has a modifying influence upon the very climate and other factors which caused the initial development. From the diagram it can be seen that climate has many indirect effects upon mangrove communities, particularly through its influence on the nature and scale of operation of the geomorphic processes outlined above, as well as in the complex relationships of soil biogeochemistry.

D. Adaptations

Mangroves are particularly interesting subjects for the study of adaptation. They live rooted in a saline anaerobic substrate; abundant salt is usually toxic to trees, and oxygen is necessary for root respiration, so the question is, how do these plants persist, grow, and reproduce successfully? In addition, the plants must cope with periodic fluctuations and extremes of the physicochemical parameters of their environment. Saenger[805] has recently reviewed morphological, anatomical, and reproductive adaptations, while Clough et al.[127] have done likewise for physiological processes. The following account is based largely on these two reviews.

1. Morphological, Anatomical, and Physiological Adaptations

Leaves of most mangroves exhibit a range of xeromorphic features,[874] such as a thick-walled epidermis, thick-waxy cuticles, a tomentum of variously shaped hairs, sunken stomata, and the distribution of cutinized and sclerenchymatous cells throughout the leaf. These are xeric characters which have probably developed in response to the physiological dryness of the environment.[833] Succulence is a common feature of mangrove leaves.

a. Salt Regulation

As a group, mangroves do not appear to be obligate halophytes, as many species grow

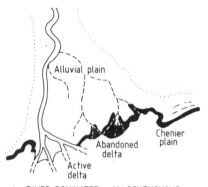

I. RIVER DOMINATED – ALLOCHTHONOUS

II. TIDE DOMINATED – ALLOCHTHONOUS

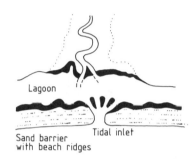

III. WAVE DOMINATED, BARRIER
LAGOON – AUTOCHTHONOUS

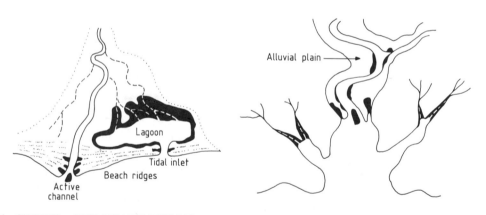

IV. COMPOSITE – RIVER AND WAVE DOMINATED   V. DROWNED BEDROCK VALLEY

FIGURE 12.   Generalized environmental settings for mangrove colonization and development (shaded). The five settings occur on coasts dominated by terrigenous deposition and reworking of sand, silt, and clay sediments. (From Thom, B. G., *Mangrove Ecosystems in Australia: Structure, Function, and Management,* Clough, B. F., Ed., Australian National University Press, Canberra, 1982, 9. With permission.)

well in fresh water.[560] However, most species seem to grow best at salinities which lie somewhere in the range between fresh water and sea water.[127] Mangroves growing in saline environments absorb sodium and chloride ions, and various physiological strategies have evolved to control uptake and concentration of these ions in metabolic tissues, although the various mechanisms involved are incompletely known.[968]

Table 11
GEOMORPHIC AND ECOLOGIC RESPONSES TO VARYING COMBINATIONS OF ENVIRONMENTAL PROCESS VARIABLES[a]

Setting		Processes			Geomorphic response		Ecologic response		Example
Tide	Rainfall	River discharge	Turbidity	Wave power	Landform diversity	Shoreline stability	Zonation diversity	Community stability	
I									
L	H	H	H	L	H	L	H	L	Mississippi
L	H	H	H	L	H	L	H	L	Orinoco
II									
H	H	H	H	L	M	M	M	M	Ganges
H	L	M	H	L	L	L	L	H	Ord
III									
M	M	L	L	H	L	H	L	M	El Salvador
M	L	L	L	H	L	M	L	M	Senegal
IV									
L	H	H	H	M	H	L	H	L	Grijalva
M	H	H	H	M	M	M	M	L	Burdekin
V									
M	M	M	M	H	L	H	L	M	Broken Bay

Note: Tide = H >4m, M 2—4 m, L <2m; Rainfall = H >1500 mm, M 700—1500 mm, L <700 mm; River discharge in m^3/sec = H >10,000, M 3000 to 1000, L <3000; and Wave power in $\times 10^7$ ergs/sec = H <100, M 10—100, L >10.

[a] Some attempt is made to quantify these variables for large river and open-ocean wave energy conditions. (H = high; M = moderate; L = low.)

From Thom, B. G., *Mangrove Ecosystems in Australia: Structure, Function, and Management,* Clough, B. F., Ed., Australian National University Press, Canberra, 1982, 12. With permission.

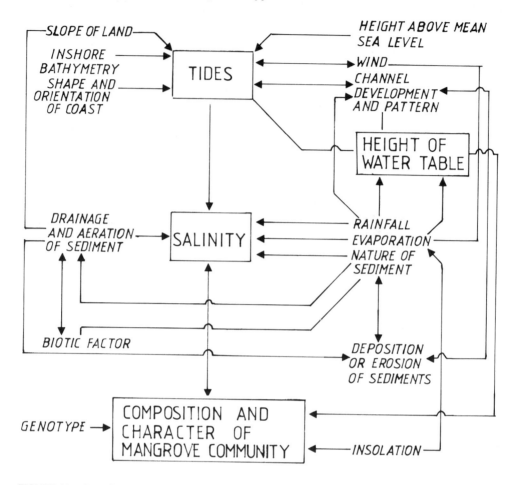

FIGURE 13.   Interrelationships of environmental factors in a mangrove ecosystem. (From Oliver, J., *Mangrove Ecosystems in Australia*, Clough, B. F., Ed., Australian National University Press, Canberra, 1982, 29. With permission.)

Various authors have suggested that mangroves regulate salt by three mechanisms: exclusion, extrusion, and accumulation. Scholander et al.[819] classified mangroves into salt-secreters and salt-excluders. However, the wide range of adaptations important in the maintenance of salt balance include the capacity of the roots to discriminate against NaCl,[110,818] the possession by some species of salt-secreting glands in the leaves,[127,805] the accumulation of salt in the leaves and bark,[110] and the loss of salt when leaves and other organs are shed. All of these adaptations, apart from the possession of salt-secreting glands by a few species, are found to varying degrees in all species of mangroves despite their wide generic origin. Thus, as Clough et al.[127] point out, it seems unwise to classify particular species as "salt-excluders" or "salt-accumulators" until the quantitative significance of the adaptations has been established clearly.

b. Sediment Interactions

Adaptations which aid mangroves in overcoming the problems of anaerobic sediments and in anchoring the plants in often semifluid sediments include the root systems with their diversity of form and function and the almost ubiquitous presence of aerenchyma and lenticels. Below the sediment surface, all mangroves possess a system of laterally-spreading cable roots with smaller descending anchor roots.[291] The root system is shallow (generally

Table 12
ROOT-SHOOT BIOMASS RATIOS OF
MANGROVE COMMUNITIES COMPARED
WITH NONMANGROVE VEGETATION TYPES

Mature mangroves	Root-shoot ratio	Ref.
Panama	0.68	301, 542
Puerto Rico	0.80	302
Australia (N.S.W.)	1.02	74
Nonmature mangroves		
Florida	1.73	542, 859
Australia (N.S.W.)	1.41	74
Dwarfed mangroves		
Australia (Vic.)	1.70	128
Nonmangrove		
Mean of six communities	0.20 ± 0.04	128, 944

less than 2m) in most species and tap roots have not been reported.[968] Despite the shallowness of the root system, the ratio of belowground biomass to aboveground biomass (root-shoot biomass ratio) is higher than that of other vegetation types (Table 12). Most species of mangroves possess in varying combinations an array of aboveground root types.[290,707] These include: (1) surface cable-roots: *(Aegialitis, Excoecaria);* (2) pneumatophores: negatively-geotrophic, unbranched *(Avicennia, Xylocarpus),* or branched *(Sonneratia)* roots arising from the cable root system; (3) knee-roots: modified sections of the cable roots with a period of negative-geotrophic followed by a period of positive-geotrophic growth *(Bruguiera, Ceriops, Lumnitzera);* and (4) stilt-roots: positively-geotrophic arching *(Rhizophora)* or straight *(Ceriops)* generally branched roots arising from the trunk and growing into the substrate. Evidence that these structures are adaptations to subterranian root aeration and to physical anchoring of the plant come from a variety of sources. The most apparent is that those mangroves growing at lower tidal levels and consequently the most frequently inundated, tend to possess the greatest array of aboveground root types, e.g., *Aviecennia, Rhizophora, Bruguiera,* and *Sonneratia.* This, together with the presence of parenchymatous tissue[290] and numerous lenticels in most of the aboveground roots[543] lend support to the root aeration hypothesis. The mechanisms of air uptake through the development of a negative gas pressure has been investigated by Scholander et al.[820] in *Rhizophora mangle* and *Avicennia germinans.* Direct evidence of the aeration function of stilt-roots in *Rhizophora* was obtained by Canoy,[1066] who noted an increase in the number of stilt-roots produced per square meter with increased temperature and consequently reduced dissolved oxygen concentrations in a thermally polluted environment. In *Rhizophora mangle* aerial roots normally have only 5% gas space before penetration into the substrate, compared with about 50% after penetration.[289] Lenticels are common in the periderm of the stems and roots of most mangroves.[543]

2. Reproductive Adaptations

Pollination in most mangroves occurs through the agency of winds, insects, and birds,[125] and most species possess small, nonsticky pollen grains.[1044] In several genera the fruits contain seeds that develop precociously while still attached to the parent tree. In some species, e.g., *Bruguiera, Ceriops,* and *Rhizophora,* are viviparous in that the embryo ruptures the pericarp and grows beyond it, sometimes to a considerable extent.[561] In other species, e.g., *Aegialitis, Avicennia,* and *Langvincularia,* the embryo, while developing within the fruit, does not enlarge sufficiently to rupture the pericarp. These genera are termed cryptoviviparous.

Saenger[805] has reviewed the significance of vivipary in mangroves. It has frequently been cited as an adaption to some aspect of the mangrove environment. Its adaptive significance could include rapid rooting,[561] salt regulation,[444] ionic balance,[445] development of buoyancy,[287] and nutritional parasitism.[691] However, the occurrence of apparently successful mangroves without viviparous fruits (e.g., *Osbornia, Sonneratia, Lumnitzera, Xylocarpus,* and *Excoecaria)* makes it doubtful whether the possession of viviparity per se is of any real adaptive advantage.[805] Tidal buffeting and wave-borne objects pose a threat to established seedlings, and it could be expected the smaller the seedling the larger the threat. Thus, viviparity in mangroves may simply be a means of producing a large seedling which is less likely to be damaged by water movements. The propagules of all mangroves are buoyant[287,751] and are adapted to dispersal by water.

Table 13 summarizes the morphological and reproductive features of Australian mangrove genera. On the basis of these adaptations the genera can be classified into colonizing, seral, and "climax" species.[805] Colonizing species could be expected to show the following features: (1) diverse morphological adaptations; (2) long flowering/fruiting periods; (3) unspecialized pollination mechanisms; (4) abundant, long-lived propagules; (5) either widespread dispersal with initial high post-establishment mortalities *(e.g. Rhizophora stylosa),* or less widely dispersed propagules with somewhat reduced initial mortality rates *(e.g., Avicennia marina)* r-strategists; and (6) short juvenile periods with high growth rates. Climax species on the other hand, could be expected to show some, or all, of the following features: (1) fewer morphological adaptations; (2) short flowering/fruiting periods: (3) more specialized pollination mechanisms: (4) few short-lived propagules; (5) propagules positively associated with adults and with low initial mortalities, k-strategists; and (6) long juvenile periods with variable to low growth rates. Seral species would be expected to show features between the two above categories. Using limited reproductive data from Port Curtis, Australia, Saenger[805] considers that *Rhizophora* and *Avicennia* can be classified as colonizers, *Ceriops* and *Aericeras* as seral species, and *Lumnitzera, Aegialitis, Osbornia,* and possibly *Xylocarpus* as climax species.

E. Biomass

In 1974, Lugo and Snedaker[542] reviewed the then available data on standing stock biomass of mangroves from a range of localities. In Table 14 the data they present have been added to from more recent studies. The biomass data show considerable variability, but nevertheless some general trends can be observed, e.g., the tropical forests tend to have the highest aboveground biomass. The variability in the data can in large part be attributed to age, stand history, or structural differences, as can be seen in the Florida forests (riverine, overwash, succession).

Belowground (subsurface roots) biomass data are available from a limited number of localities. Unlike temperate forests where roots seldom make up more than about 20% of the total biomass,[944] mangroves commonly have high root-shoot ratios (Table 12). Unfortunately, good data for root biomass in mangroves are rare because of the difficulty of sampling subsurface roots quantitatively. The very considerable root biomass in mangroves suggests that they could be important in cycling organic and inorganic materials.

F. Primary Production

With a few exceptions[80,114] net primary production in mangroves has been estimated from measurements of photosynthesis and respiration by individual leaves or small branches.[97,302,540] Such estimates generally fall within the range that might be expected for a woody plant with the C_3-pathway of carbon fixation.[129] However, studies which include measurement of respiration of parts other than leaves are the exception rather than the rule. Data given by Lugo et al.[540] for mangroves in southern Florida suggest that 4 to 10% of gross primary production might be lost via respiration by stems or surface roots.

Table 13
SUMMARY OF MORPHOLOGICAL AND REPRODUCTIVE FEATURES IN AUSTRALIAN MANGROVE GENERA

Genus	Aqueous tissue[a]	Root modifications[a]	Salt glands[b]	Coppicing ability or vegetative proliferation[c]	Pollinating agent[d]	Vivipary[e]	High propagule production[f]	High propagule mortality[f]	High propagule dispersal[f]
Acanthus	+	+	+	+	?	++	?	?	?
Aegialitis	-	-	+	-	I, B	++	+++	-	-
Aegiceras	+++	-	++	-	I, B	++	++	+	++
Avicennia	+++	+++	+++	+	I, B	+++	+++	++	+
Bruguiera	+	+++	-	+/-	I, B	+++	++	?	?
Camptostemon	?	++	-	?	?	-	?	?	?
Ceriops	+++	++	-	?	I	+++	++	++	++
Cynometra	+	+	-	?	?	-	?	?	?
Excoecaria	+	-	-	+	W	-	+	?	?
Heritiera	++	-	-	?	?	-	?	?	?
Lumnitzera	+++	++	-	-	I, B	-	+++	?	+
Osbornia	+++	+	-	-	I, B	-	+	?	?
Rhizophora	+++	++	-	-	W	+++	+++	++	+++
Scyphiphora	?	?	?	?	?	-	?	?	?
Sonneratia	+++	+++	-	+	Bt, I	+	?	?	?
Xylocarpus	+	+++	-	-	I, B	+	+	?	?
Nypa	?	+	?	?	I	+++	?	?	?
Hibiscus	++	-	-	?	I, B	-	?	?	?
Amyema	++	+++	-	-	I, B	-	?	?	?
Myoporum	?	-	-	?	I, B	-	++	?	?

a   Aqueous tissue and root modifications: +++ = conspicuous, ++ = well-developed, + = present, and - = absent.

b   Salt glands: +++ = dense, ++ = less dense, + = present, and - = absent.

c   Coppicing ability: + = present and - = absent.

d   Pollinating agent: I = insects, B = birds, W = wind, and Bt = bats.

e   Vivipary: +++ = viviparous, ++ = cryptoviviparous, + = large seeds, and - = small non-viviparous seeds.

f   Propagule production, mortality and dispersal: +++ = high, ++ = medium, + = low, and - = very low.

From Saenger, P., *Mangrove Ecosystems in Australia*, Clough, B. F., Ed., Australian National University Press, Canberra, 1982, 180. With permission.

Table 14
MANGROVE FOREST BIOMASS AND LITTER PRODUCTION ESTIMATES

Locality	Total aboveground biomass		Litter production		Litter as a % of aboveground biomass	Subsurface roots	
	g dry weight m^{-2}	g C m^{-2}	g dry weight m^{-2}	g C m^{-2}		g dry weight m^{-2}	g C m^{-2}
Panama[288]	27,921	11,168	10,210	4084	36%	18,976	7590
Puerto Rico[302]	6285	2514	475	190	7.5%	4997	1999
Florida riverine[859]	9281	3928	4295	1718	43.7%	—	—
Florida riverine[859]	17,390	6956	3393	1357	19.5%	—	—
Florida overwash[859]	11,958	4783	1399	559	11.7%	—	—
Florida succession[859]	812	324	32	13	3.9%	—	—
Thailand[114]	—	—	670	268	—	—	—
Queensland[79]	—	—	960	384	—	—	—
Sydney[321]	—	—	580	232	—	—	—
Westernport Bay, Victoria[128]	8600	3440	260	80	2.3%	14,600	5840
Waitemata Harbour, New Zealand							
Tuff Crater Tall[1040]	—	—	810	324	—	—	—
Tuff Crater Short[1040]	—	—	365	146	—	—	—
Upper Waitemata Harbour[1039]	1556		439	176	11.3%	—	—
Lucas Creek[1039]	2248		564	225	10.0%	—	—

Lugo and Snedaker[542] summarized the then available data for the productivity of mangrove forests. Each of the studies they listed was conducted using the same methodology (carbon dioxide exchange). Values for gross primary productivity ranged from 13.9 g C m^{-2} day^{-1} for red, black, and white mangroves in Fahhahtchee Bay, Florida,[97] to 1.4 g C m^{-2} day^{-1} for a scrub mangrove forest in Dade County, Florida,[542] with a mean of 7.6 g C m^{-2} day^{-1}. Total 24 hr respiration ranged from 9.1 to 0.6 g C m^{-2} day^{-1} with a mean of 4.77 g C m^{-2} day^{-1}. Net primary production ranged from 7.5 to 0 g C m^{-2} day^{-1} with a mean of 4.4 g C m^{-2} day^{-1}.

Clough and Attiwell[129] point out that many estimates of net primary production of mangroves derived from gas exchange measurements neglect the appreciable respiratory losses from stems and roots. There are problems in extrapolating gas exchange measurements carried out on individual leaves and branches to account for variations in boundary conditions for CO_2 exchange within the canopy. Seasonal differences in gas exchange characteristics[605] and marked seasonal variations in leaf growth and litterfall both imply that net primary production may vary appreciably over the course of a full year.

G. Litterfall

Mangrove litter production is of importance in estuarine ecosystems as a primary source of organic detritus. Gill and Tomlinson,[288,289] in their studies of the red mangrove in southern Florida, found that flowering, fruit formation, and leaf fall occurred at measurable rates at all seasons. However, peak rates of leaf fall were observed during the summer months when air temperatures and incident light were at their annual peaks. Similar results have been reported by Heald.[381] In a 2-year study of litter fall at Rookery Bay and Ten Thousand Islands, Florida, Snedaker and Lugo[859] found that leaf fall increased during dry periods.

Mangrove litter comprises leaves, twigs, fruits, and influorescences. There is a distinct seasonal pattern to litterfall.[1022] Woodroffe,[1039,1040] in his study of litterfall in Tuff Crater, Waitemata Harbour, New Zealand (Figure 14), found that 78% of the total litterfall beneath tall mangroves (2 to 4 m) and 69.2% beneath low mangroves (less than 1.5 m tall) were recorded during the months of December to April (Figure 14). Storm events can have a marked effect on mangrove litter production.[321,736] In his Tuff Crater study, Woodroffe[1040] found that a storm in April 1981 with wind speeds reaching up to 113 km hr^{-1} accounted for a particularly large litterfall when over a period of a week 18% of the annual total was recorded.

Litter production estimates for a range of mangrove forests are given in Table 14. They range from 32 g m^{-2} for a succession forest in Florida to 10,210 g m^{-2} for a Panama forest. The contribution of leaf fall to total litterfall has been calculated to be 6.70 t ha^{-2} year^{-1} in Thailand, 8°N[114] 4.75 t ha^{-2} year^{-1} in Puerto Rico,[302] and 5.62 t ha^{-2} year^{-1} in Tuff Crater, New Zealand. Total litterfall has been studied under a range of mangrove species on Hinchinbrook Island, Australia, 18°S.[79] Rates of total annual litterfall on a yearly basis ranged from 365 to 2810 g m^{-2} year^{-1} with a mean of 2280 g m^{-2} year^{-1}. The average litterfall under three species of *Rhizophora* was found to be 9.6 t ha^{-2} year^{-1}, but with variations from year to year, and similar rates were demonstrated for other species.[202] In Tuff Crater, New Zealand, 36°48'S, total annual litterfall was 8.10 t ha^{-2} year^{-1} beneath the tall mangroves of the creek banks and 3.65 t ha^{-2} year^{-1} beneath low studded mangroves on the flats.[1040]

In Florida, 26°N, there is a considerable range of rates of mangrove litterfall; average values of total mangrove litter production are about 8 t ha^{-2} year^{-1}.[542,858] At Roseville, Middle Harbour, Sydney, 34°S, litter production beneath *Avicennia* 8 to 10 m tall has been reported at 5.8 t ha^{-2} year^{-1},[321] and in Westernport Bay, Victoria, 38°S, beneath *Avicennia* 2 to 4 m tall at 2.0 t ha^{-2} year^{-1}.[128] These litterfall data imply that while tropical areas have in general high litterfall rates there is not such a pronounced latitudinal gradient as

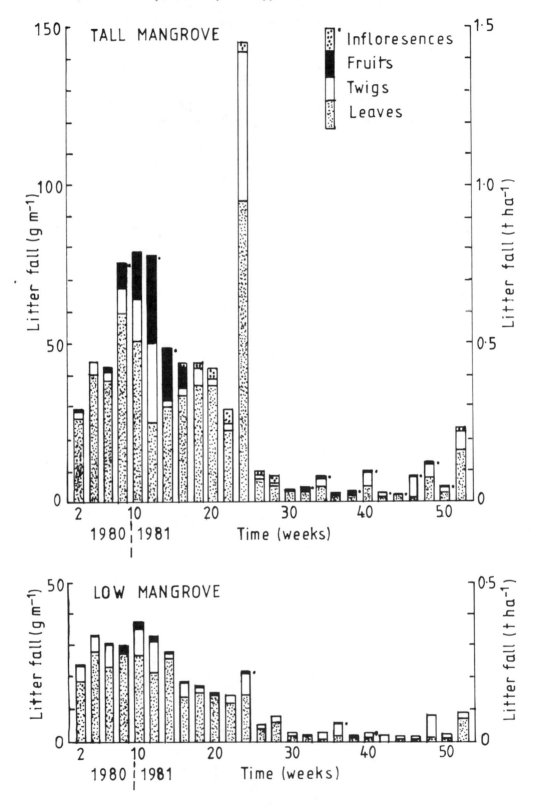

FIGURE 14.   Litterfall beneath tall mangrove (above) and low mangrove (below) in Tuff Crater, Waitemata Harbour (November 4, 1980 to November 3, 1981). (From Woodroffe, C. D., *N.Z. J. Mar. Freshwater Res.*, 16, 180, 1982b. With permission.)

might be expected. In some instances, as in Tuff Crater, mangroves near their latitudinal limit may be at least as productive as mangroves in Queensland and Florida.

H. Factors Affecting Productivity

Assimilation of carbon in photosynthesis is influenced by many factors, some of which have not been explored fully in studies of mangrove productivity. Characteristics of the canopy such as leaf area index, leaf inclination, and the distribution of leaves within the canopy have a dominating influence on primary productivity by modifying light interception, leaf temperature, and boundary layers.[595] In addition, Carter et al.[97] have subdivided the factors regulating mangrove productivity as follows:

1. Tidal factors
   a. Transport of oxygen to the root system
   b. Physical exchange of the soil water solution with the overlying water mass, removing toxic sulfides and reducing the total salt content of the soil water
   c. Tidal flushing interacts with the surface water particulate load to determine the rate of sediment deposition or erosion within a given stand
   d. Vertical motion of the ground water table may transport nutrients regenerated by detrital food chains into the root zone of the mangroves
2. Water chemistry factors
   a. Total salt content governs the osmotic pressure gradient between the soil solution and the plant vascular system, thus affecting the transpiration rate of the leaves
   b. A high macronutrient content of the soil solution has been suggested[496] as enabling the maintenance of high productivity in mangrove ecosystems, despite the low transpiration rates caused by high salt concentrations in sea water
   c. Lugo et al.[543] indicate that allochthonous macronutrients contained in wet season surface runoff may dominate the macronutrient budgets of mangrove ecosystems

Carter et al.[97] also suggested that the gradient of chloride concentration (expressed as the ratio of chloride gradient between sediment and the overlying water mass to the chloride concentration in the sediment water solution), could be considered as an index that integrates the effects of both tidal- and soil-water chemistry factors. They found that with an increase in the chloride ratio the ratio of 24 hr respiration to gross primary production (a measure of the energy used for maintenance) had a slow exponential decrease. Within the range of salinities studied (8 to 30‰), the gross primary productivity of mangroves increased as fresh water became available. Respiration rates along the same gradient, however, also increased. The increase in respiration is a reflection of the amount of physiological work associated with the problems of higher salinity environments.

IV. MACROALGAE

A. Distribution Patterns

Macroscopic algae are in general not well represented in estuaries in comparison with other groups of primary producers. Permanent residents are restricted to a small number of widespread genera which can tolerate the turbidity, silt deposition, and changing salinity patterns. There are, however, some exceptions. Species of brown algae belonging to the genera *Fucus, Pelvetia,* and *Ascophyllum* are abundant on rocky slopes in northern Atlantic estuaries, while smaller plants of *Fucus* and *Pelvetia* mix with the vascular plants of the salt marshes and contribute substantially to salt marsh production. In the tropics there are smaller amounts of brown algae of the genus *Sargassum* growing on hard substrates, while other browns such as *Colpomenia* and *Dictyota* occur as epiphytes on seagrasses such as

Posidonia and *Thalassodendron*. In some Australian estuaries a free-floating form of the normally attached *Hormosira banksii* is common, reproducing by vegetative division.[604]

Red algae are represented mainly by small species such as *Polysiphonia*, *Ceramium*, and *Laurencia* growing as epiphytes on seagrasses, while species of *Bostrychia* and *Calloglossa* grow attached to salt marsh plants and the boles and roots of mangroves. Most common are species of *Gracilaria* which originally grow attached to pebbles, living bivalves, and dead shells, but as they develop they may become free-floating. *G. verrucosa* is abundant in South African estuaries and lagoons where it forms the basis of an agar industry.[122,182,846] Hedgpeth[382] records *Gracilaria* in the Laguna Madre of Texas while the species *Gracilaria secundata* is widespread in New Zealand estuaries.[392,393]

The most common algae are green algae belonging to the genera *Enteromorpha*, *Ulva*, *Ulothrix*, *Cladophora*, *Rhizoclonium*, *Chaetomorpha*, and *Codium*. Filamentous species of *Enteromorpha* and *Cladophora* grow as epiphytes on seagrasses and salt marsh plants. Larger green algae, especially *Ulva lactuca* and *Enteromorpha* spp. can form extensive mats on estuarine mud flats.

The brown alga *Fucus ceranoides* is confined to northern hemisphere estuaries in contrast to the other *Fucus* species which are found only in areas of high salinity.[462] However, on the rocky coasts of the low-salinity Baltic, rocky substrates from 0.5 to 8.5 m are dominated by *Fucus vesiculosus*.[424,425] Figure 15 depicts the biomass and vertical distribution of the algae in the Askö-Landsort area of the Baltic (160 km^2). In the above mean sea level in the splash zone, the rocks are covered by the blue-green alga *Calothrix scropularum*. The mean sea level is dominated by filamentous algae *(Cladophora glomerata, Stictyosiphon*, and *Ceramium tenuicorne)*. A *Fucus* belt dominates to a depth of 5 m and occurs down to 8 m where it is an important substrate for other filamentous algae such as *Ceramium* and *Pilayella littoralis*. At 8 m depth the *Fucus* is replaced by other red and brown algae often forming loose-lying mats entangled with the dense *Mytilus edulis* beds which cover the deeper hard bottoms.

Loveland et al.[539] investigated the species composition, spatial distribution, and seasonal periodicity of the macroalgae in Barnegat Bay, an estuarine system in New Jersey which has been subjected to constant modification from both natural and anthropogenic perturbations. Barnegat Bay is a shallow (1 to 6 m deep) lagoon-type estuary 48 km long and varying in width from 2 to 6.5 km. The salinity range is 12 to 32‰ with an average salinity in the center of the bay of 25‰. Between 1965 and 1973, 116 species of benthic macroalgae were collected from the bay. During the year species diversity increased markedly in the spring as the water temperature and solar radiation increased. Few species persisted throughout the year; Taylor[910] found that only six species, apart from the fucoid genera *Ascophyllum* and *Fucus* which were relatively unimportant, were present year-round. Dominant species were the greens *Ulva lactuca*, *Enteromorpha* spp., and *Codium fragile*, and the reds *Gracilaria tikvahine*, *Ceramium fastigiatum*, and *Agardhiella subulata*. *Ulva lactuca*, *Codium fragile*, *Enteromorpha intestinalis*, and species of *Gracilaria* and *Ceramium* are cosmopolitan estuarine species which have been shown to respond to increased nutrient input by increased growth.[146,485,878,966] Over the period 1969 to 1973 the species listed above comprised 90% of the macroalgal biomass of Barnegat Bay. During the same period the average total dry weight of algae for nine stations in a bay transect showed a significant increase[962] (Figure 16).

Wilkinson[1013,1014] has reviewed the current state of knowledge concerning the distribution and ecology of estuarine benthic macroalgae, although he did not consider species associated with salt marshes. In the Northern Hemisphere the general pattern of distribution is as follows:

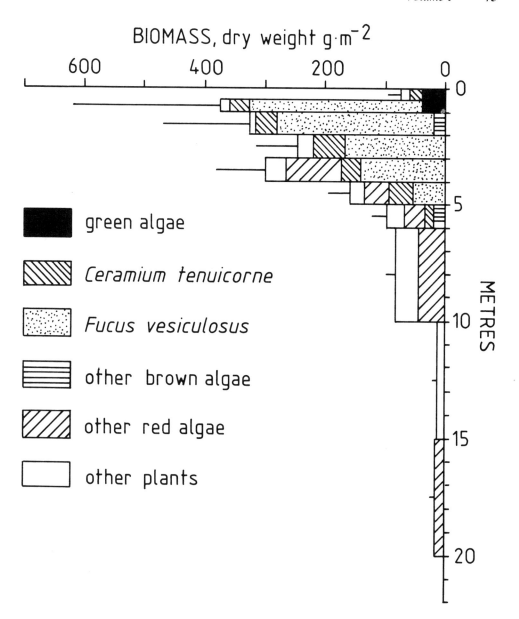

FIGURE 15.   Biomass and vertical distribution of macroalgae in the Askö-Landsort area (160 km²). (From Jansson, A.-M. and Kautsky, N., *Biology of Benthic Organisms,* Keegan, B. F., O'Ceidigh, P., and Boaden, P. J. S., Eds., Pergamon Press, London, 1977, 359. With permission.)

1. Colonization of most of the estuary is by marine species with fresh water species predominating only near the head of the estuary (many algal species have a wide salinity tolerance).
2. There is a progressive reduction in species number going upstream which is brought about by selective attenuation, first of red and then second of brown algae. Green algae, although not necessarily becoming more numerous in terms of species, become relatively more important going upstream.
3. In the midreaches of the estuaries there are a few species confined to brackish water, such as *Fucus ceranoides* and certain *Vaucheria* spp.

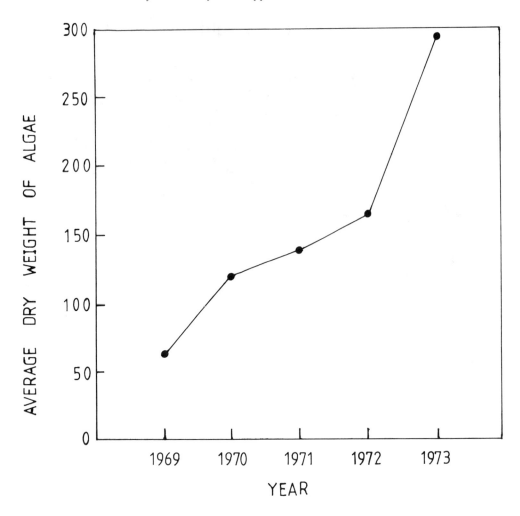

FIGURE 16.   Average total dry weight of algae at nine sampling stations in Barnegat Bay, New Jersey, between 1969 and 1973. (From Lovelund, R. E., Brauner, J. F., Taylor, J. E., and Kennish, M. J., *Ecology of Barnegat Bay, New Jersey,* Kennish, M. J. and Lutz, R. A., Eds., Springer-Verlag, New York, 1984, 88. With permission; after Reference 962.)

B. Primary Production

There have been few studies on photosynthesis and respiration rates of estuarine algae. They include studies on *Polysiphonia* from Gent Bay Estuary,[257] *Hypnea* from a mangrove estuary in Florida,[169] 6 species of algae from an estuary in Oregon,[471] and 6 species of algae from a mangrove and a salt marsh estuary in Florida.[170] In addition, Christie[122] has investigated standing crop and production of *Gracilaria verrucosa* in Langebaan Lagoon, South Africa, and Steffensen[878] and Knox and Kilner[485] have studied annual changes in the standing crop of *Ulva lactuca* and *Enteromorpha ramulosa* in the Avon-Heathcote Estuary, New Zealand. Three of these studies will be discussed below.

In Langebaan Lagoon, which is 14 km in length and about 3.5 km at its greatest width, *Gracilaria verrucosa* is the dominant sublittoral macrophyte although large areas are also exposed at low spring tides. The upper reaches of the lagoon become hypersaline (38.04‰ maximum) in the summer months. Figure 17 depicts the standing crop (ash-free grams per square meter) and energy values (kilojoules per gram per dry mass) in lagoon sections from the entrance (A) to the head (F). Both the area covered and standing crop in each section

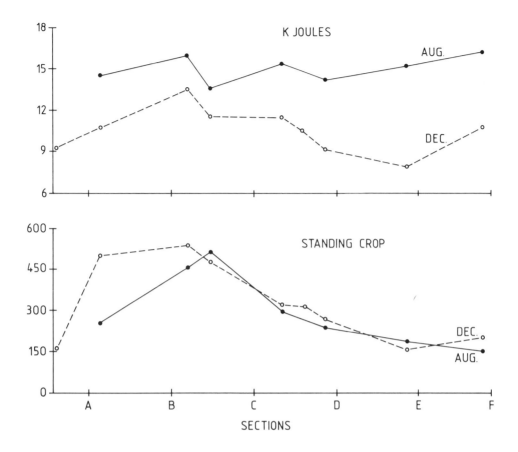

FIGURE 17. Standing crop (ash-free dry weight g m^{-2}) and energy values (kilojoules per gram per dry mass) in each section of Langebaan Lagoon, South Africa. (From Christie, N. D., *Estuarine Ecology with Particular Reference to South Africa*, Day, J. H., Ed., A. A. Balkema, Rotterdam, 1981, 108. With permission.)

declined from the head towards the entrance. The highest standing crop was recorded in section B at 543 g m^{-2} in December and fell to below 200 g m^{-2} in sections E and F. Energy values for August were consistently higher than those for December, and Christie[122] concluded that this probably reflected slower growth rates during winter. Table 15 gives the gross production rates in each section calculated from light and dark bottle experiments. These rates show a similar pattern to standing crop and energy values. Gross production per month as a percentage of the standing crop was highest in the sections near the entrance. The annual total energy production was estimated at 269.88 × 10^9 kJ. The availability of nutrients, especially nitrate, appeared to be one of the important growth-regulating factors. Nitrate values were highest in the two sections at the entrance and decreased up-lagoon. In these entrance sections the *Gracilaria* was black in color and the color steadily changed up-lagoon through brown and brown-green to yellow, lending support to the idea that nitrogen was a limiting factor.

Henriques[392,393] investigated the responses of *Gracilaria secundata* to nutrient enrichment from the Sewage Purification works in Manukau Harbour, New Zealand. *Gracilaria* grows either as isolated plants attached to solid objects such as rocks and shells, or as aggregations of plants in luxurious meadows anchored in the mud. Aggregations of drift plants also occur. Henriques found that the *Gracilaria* meadows covered a total of 152 ha of the Harbour or about 1% of the area. Biomass measurements revealed that the mean dry weight of *Gracilaria* from the meadows was 348 g m^{-2}. He estimated that the amount of above substrate *Gracilaria*

Table 15
GROSS PRODUCTION OF *GRACILARIA VERRUCOSA* IN
LANGEBAAN LAGOON, SOUTH AFRICA, IN AUGUST AND
DECEMBER

Lagoon section	Mouth A	B	C	D	E	Head F
August						
Gross production (% standing crop)	60.2	79.5	80.8	63.2	52.4	30.1
% Organic matter	75.6	80.1	74.3	72.6	68.3	56.6
Energy production (10^3 kJ m^{-2} month^{-1})	2.78	7.82	5.08	2.72	2.21	1.28
December						
Gross production (% standing crop)	67.7	88.4	62.4	40.4	30.8	18.5
% Organic matter	51.6	51.9	55.4	56.8	60.4	61.4
Energy production (10^3 kJ m^{-2} month^{-1})	6.19	11.4	3.32	1.71	0.63	0.65

From Christie, N. D., *Estuarine Ecology with Particular Reference to South Africa*, Day, J. H., Ed., A. A. Balkema, Rotterdam, 1981, 108. With permission.

in the Harbour was of the order of 500 tonnes. Other biomass measurements revealed that about one third of the total dry weight of a *Gracilaria* meadow can be below the surface.

In the Avon-Heathcote Estuary, New Zealand, large populations of the sea lettuce *Ulva lactuca* have become established over the last 40 years. This species was not recorded by Thompson[1094] in 1929, but Bruce,[1065] in 1953, reported that it had been abundant since 1946. Since 1960[485,880] a series of studies have documented the increase in algal density. From these studies (Figure 18) it is evident that increase in algal biomass has been associated with the discharge of increasing amounts of sewage effluent into the estuary.

The link between luxuriant growth of *Ulva* was examined at the beginning of the century[146,505] and has been confirmed by numerous reports since.[86] As early as 1914, Forster[256] demonstrated that growth of *U. lactuca* was stimulated by additions of urea, acetamide, and ammonium nitrate. Anderson[16] showed that nitrate was the preferred nitrogen source for *U. lactuca* except that, when nutrients were at low concentrations, ammonia was preferred. Waite and Mitchell[966] examined the effect of adding $NH_3 - N$ and $PO_4 - P$ in varying combinations, and found that growth in *U. lactuca* was stimulated by the growth of either nutrient. They also noted that ammonia concentrations in excess of 0.9 g N m^{-3} inhibited growth.

In the Avon-Heathcote Estuary where *U. lactuca* was generally the dominant species, a second species of *Enteromorpha ramulosa* was twice as abundant as *Ulva* in the summer of 1969, but since that time it has not occurred in the same abundance. *Enteromorpha ramulosa* is more sensitive to temperature fluctuations (see Figure 18) and a mild preceding winter may be the explanation for its dominance. Laboratory experiments showed that between 15 and 18°C winter plants of *Ulva lactuca* grew 15 to 20 cm in length in 4 to 6 weeks, while below 15°C the rates of growth were very much slower. No growth in *Enteromorpha* was detected below 12°C while some growth in *Ulva* was detected at 10°C.

Figure 18 depicts the seasonal variation in total organic dry weight of the three dominant algal species in the estuary from a total of 37 stations. The seasonal pattern in all three species is evident. However, both *Enteromorpha* and *Gracilaria* die down in the winter, while it will be noted that a substantial *Ulva* biomass persisted over the winter. *Ulva* plants usually develop attached to cockle shells, but as they grow the thalli reach a size and buoyancy that uproots the shells leaving the thalli to drift with the currents. These drifting thalli, and

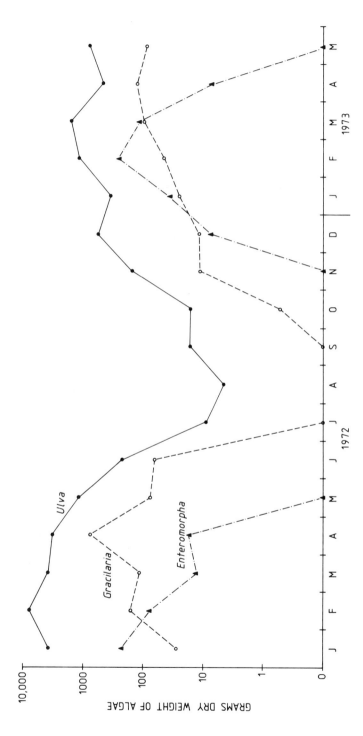

FIGURE 18. Total dry weight of three algal species from 37 sampling sites (each 1 m²) in the Avon-Heathcote Estuary, New Zealand, over the period of January 1972 to May 1973.[878]

fragments broken off from them, continue to grow over the summer and autumn, forming large drifts of unattached plants. These form the bulk of the winter biomass. Summer biomass values of up to 130 g m^{-2} (dry weight) have been recorded.

Steffensen[880] investigated the growth response of *U. lactuca* to different combinations of phosphorus and nitrogen using discs cut from the expanded region of mature plants. Addition of either $NO_3 - N$ or $PO_4 - P$ stimulated growth, optimum levels being 0.6 g m^{-3} for both nutrients. Below the optimum, N and P interacted in their effect on growth; above the optimum, increasing N concentration decreased growth while no change occurred with increasing P. These results are summarized in the response surface in Figure 19 showing the interaction of $NO_3 - N$ and $PO_4 - P$ in stimulating *Ulva* growth.

Over the past 10 years an extensive investigation has been carried out on problem algal growths in the Peel-Harvey Estuary in Western Australia.[193,400] Nutrient cycling in this estuarine system will be discussed in detail in Vol. II, Chapter 1. In the system throughout the 1970s there was a marked increase in the growth of the green alga *Cladophora* aff *albida*. This alga grows as small ball-like clumps of densely branched radiating filaments. These balls (which are 1 to 3 cm in diameter) lie unattached on the estuary floor, where they may form large beds, usually 1 to 10 cm deep. The lower sections of these beds decompose to form an anoxic black cover over the bottom sediments. The Peel-Harvey Estuarine System Study has developed a number of *Cladophora* models to simulate the interaction of *Cladophora* and environmental factors, especially nutrient supply.[400] One of these, GROWMOD (Figure 20) is a suite of programs which model *Cladophora* biomass variation in the field on a weekly basis. The model permits a light-temperature interaction with growth rate and response to inorganic nitrogen and inorganic phosphorus concentrations. Loss processes (decomposition and export) as well as importation of biomass to a growth area are considered. Sections of a bed, or whole beds, are simulated and the important consequences of self-shading and decomposition of buried *Cladophora* balls on total biomass are accounted for. A most important property of the model is that of stochastic (Monte Carlo) simulation. This technique enables all the uncertain model parameters to be randomly varied within predetermined limits, and provides an objective method of model-sensitivity analysis.

Figure 21 depicts simulated and measured *Cladophora* growth rates from an earlier model (PROGRAM B) which computes daily growth rates taking into account changes in the light during the day and the temperature-nutrient-growth interactions discussed above. The agreement between the measured and simulated growth is excellent. Figure 22 compares the observed biomass variation over a 2-year period with the GROWMOD simulated biomass. Again there is good general agreement. The results confirmed conclusions based on studies of the ecology of *Cladophora*: those of chronic phosphorus-limitation of growth, and the importance of adequate levels of light and temperature for growth. They support the importance of winter nutrient input to the *Cladophora* beds of eastern Peel Inlet (see Volume II, Chapter 4, Figure 31). In the absence of such inputs, tissue nitrogen and phosphorus concentrations run down, and growth during the following spring and summer is limited. The GROWMOD simulations indicate that decomposition and high export rates of *Cladophora* to the beaches and sea led to the decline in biomass over the latter part of 1978 and during 1979.

V. SEAGRASS SYSTEMS

A. Introduction

Seagrasses are one of the most characteristic features of shallow coastal seas and estuaries in tropical and temperate zones. According to den Hartog[192] there are 49 species in 12 genera of marine aquatic angiosperms which have the ability to function normally and complete

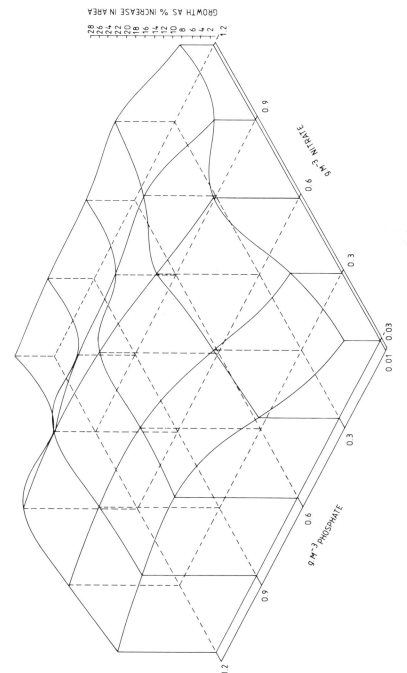

FIGURE 19. Response surface showing the interaction of $PO_4 - P$ and $NO_3 - N$ on the growth of *Ulva lactuca* from the Avon-Heathcote Estuary, New Zealand. (From Steffenson, D. A., *Aquat. Bot.*, 2, 346, 1976. With permission.)

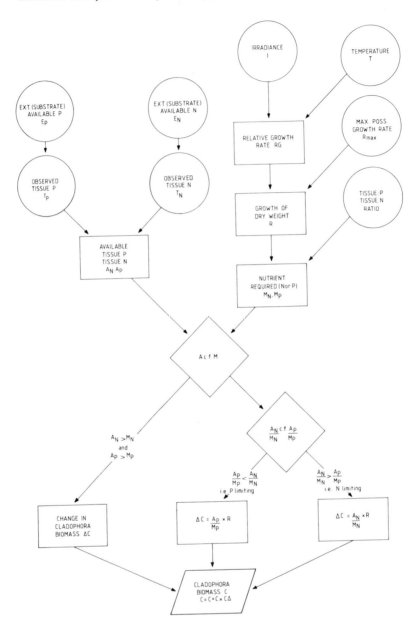

FIGURE 20. Flow chart for the algebraic *Cladophora* growth model (GROWMOD) for the Peel-Harvey estuarine system in Western Australia. (From Hodgkin, E. P., Birch, P. B., Black, R. E., and Humphries, R. B., The Peel-Harvey Estuarine System Study (1976 to 1980), Rep. No. 9, Department of Conservation and Environment, Perth, Western Australia, 1980, 44. With permission.)

their reproductive cycle when fully submerged in a saline medium. Many genera have a world-wide distribution with *Thalassia, Thalassodendron, Cymodocea, Posidonia, Halodule,* and several other genera in warm and tropical waters, and *Zostera, Halophila, Ruppia, Potamogeton,* and *Zannichellia* extending into temperate estuaries. Many of these genera are euryhaline but their extension into lowered salinities is variable. *Posidonia, Thalassia, Cymodocea,* and *Thallasodendron* prefer salinities above 20‰ while *Zostera* and *Halodule* tolerate 10‰; *Ruppia* and *Potamogeton* prefer low salinities while *Zannichellia* extends into fresh water and is seldom found in salinities above 10%.

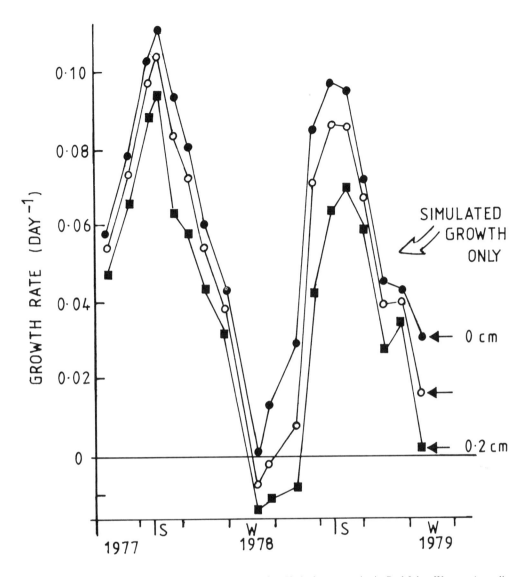

FIGURE 21. Simulated and measured growth rates for *Cladophora* at a site in Peel Inlet, Western Australia. (From Hodgkin, E. P., Birch, P. B., Black, R. E., and Humphries, R. B., The Peel-Harvey Estuarine System Study (1976 to 1980), Rep. No. 9, Department of Conservation and Environment, Perth, Western Australia, 1980, 45. With permission.)

Within any given area the structure of the seagrass system can vary from a few plants or clumps of plants to extensive meadows characterized by a single species. Such meadows are dynamic and highly productive. Two excellent recent reviews of seagrass ecosystems are those edited by McRoy and Helfferich[569] and Phillips and McRoy.[719]

B. Distribution and Zonation

Important factors affecting the colonization and distribution of seagrasses include the morphology and character of the bottom, the sources, routes and rates of sediment transport, and the rates of sediment accumulation.[85] Once established, seagrasses may significantly alter the prevailing sedimentary processes to a degree dependent primarily upon the species composition and the plant density. Alteration of the sediments may lead to an increase or decrease in the colonization pattern with the possibility of partial or complete replacement

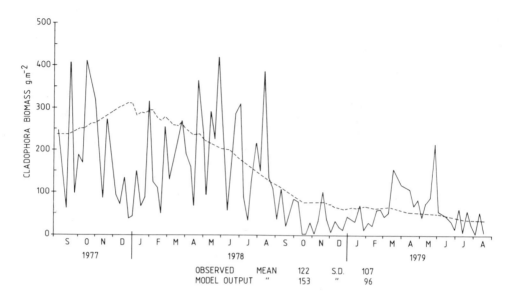

FIGURE 22.   Comparison of observed and GROWMOD simulated biomass variation over 2 years at a site in Peel Inlet, Western Australia. (From Hodgkin, E. P., Birch, P. B., Black, R. E., and Humphries, R. B., The Peel-Harvey Estuarine System Study (1976 to 1980), Rep. No. 9, Department of Conservation and Environment, Perth, Western Australia, 1980, 46. With permission.)

by other species. Individual species tend both to colonize a specific environment and to be mutually exclusive. The *Halolule* to *Thalassia* succession documented by Phillips[716] is a case in point.

Seagrasses alter the prevailing sedimentation processes in a variety of ways, but the major effects are to increase the sedimentation rates, to concentrate preferentially the finer particle sizes (see Figure 9 in Chapter 2), and to stabilize the deposited sediments. They do this by the entrapment of fine waterborne particles by the grass blades, the formation and retention of organic particles produced locally within the grass beds, and binding and stabilization of the substrate by the grass rhizome and root systems.[85] Sediments in seagrass beds are often anoxic below the surface layer and in the case of *Zostera* at least oxygen is conveyed to the roots through a series of intercellular spaces.[567]

The main depth limit to growth is light and abrasion by sand-laden water in more turbulent habitats. Table 16 gives the depth distribution of the principal genera. In many estuaries *Zostera* extends no more than a meter below low water, but in clear estuaries it may extend down to 6 m and according to Ranwell[756] it grows to a depth of 30 m off California. *Halophila* has been recorded from depths as great as 90 m although many species have restricted depth distributions. *Posidonia* descends down to at least 60 m. *Phyllospadix* extends down from the lower eulittoral to 30 m under favorable conditions. In southern Mozambique (Figure 2D) *Halodule uninervis* may extend from MLWN (mean low water neap) to below MLWS (mean low water spring). *Thalassodendron ciliatum* grows among the lower *Halodule* and extends to 5 to 10 m below MLWS. *Cymodocea rotundata, C. serrulata, Zostera capensis,* and *Halophila ovale* grow near MLWS, while *Syringodium isoetifolium* and *Thalassia hemprichii* both grow subtidally.[182] The depth of growth of any particular species is a complex function of water turbidity and sediment regimes, wave action, salinity, etc.

C. Biomass

Seagrasses have a large biomass of leaves, but the majority of the biomass of the plant is in the roots and rhizomes in the sediment and this is very difficult to sample because of the depth of penetration of the root system.[436,1058,1059] There is a bewildering variety of

Table 16

DEPTH DISTRIBUTION OF THE PRINCIPAL SEAGRASS GENERA

	Halo-dule	Zostera subgen Zoster-ella	Zostera subgen Zostera	Cyma-docea	Thal-assia	Syrin-godium	Phyllo-spadix	Enhalus	Posi-donia	Halo-phila	Hetero-zostera	Amphib-olis	Thalasso-dendron
Mideulittoral	+	+	-	-	-	-	-	-	-	+	-	-	-
Belt between MLWN and MLWS	+	+	+	+	+	+	+	+	-	+	+	-	-
Upper sublittoral	+	+	+	+	+	+	+	+	+	+	+	+	+
Lower sublittoral	+	-	+	-	-	-	+	-	+	+	-	+	+

From den Hartog, C., *Helgol. Wiss. Meeresunters.*, 15, 96, Marcel Dekker, New York, 1977. With permission.

Table 17
SOME REPRESENTATIVE SEAGRASS BIOMASS ESTIMATES
(g dry weight m^{-2})

Species	Location	Range	Mean	Ref.
Cymodocea nodosa	Mediterranean	13—340	—	283, 351
Halodule wrightii	North Carolina	103—200	200	197
Halophila engelmannii	Texas		1.6	565
Syringodium filiforme	Florida	15—200	100	1060
Thalassia testudinum	Florida (east coast)	20—1800	125—800	436, 642, 1059
	Florida (west coast)	75—8100	500—3100	37, 716
Zostera marina	Denmark	210—960	490	340, 712, 813
	Nova Scotia	785—1230	1020	575
	New York	250—2060	760	84
	North Carolina	1—550	50—250	197, 928
	Washington	95—540	120	716, 717
	California	6—420	—	455, 963
	Alaska	186—1840	1000	562-564

literature values for plant biomass.[570] Data are available for nine seagrass species, and the bulk of these measurements is concentrated in two species, *Zostera marina* and *Thalassia testudinum*.[1060] For *Zostera marina,* biomass estimates range from 6 to 2060 g dry wt m^{-2} while for *Thalassia testudinum* they range from 71 to 8100 g dry wt m^{-2}. Representative biomass values are given in Table 17. Zieman and Wetzel[1060] list the range of values available at the time. The wide diversity of values is dependent on the habitats sampled, the times of sampling, the number of replicate samples taken, and the methods used.

Few complete measures of all components of the biomass have been reported and the majority of the estimates are values for leaf biomass only. Leaves constitute a variable amount of the weight of the plant depending on depth, substrate, nutrient availability, and season. Burkholder et al.[83] studying *Thalassia* in Puerto Rico found leaf/root and rhizome ratios of 1:3 in fine mud and sand and 1:7 in coarse sand. Various studies have shown that *Thalassia* leaves usually constitute 15 to 22% of the total dry weight of the plant, although this proportion varies between 10 and 45%.[1059] Jones,[436] in a study in Biscayne Bay, Florida, found a relatively constant ratio of 3:2:2 for leaves and short shoots:rhizomes:roots. Burkholder and Doheny,[84] studying *Zostera* on Long Island, found leaf and shoot:rhizome:root ratios of 2:1:2 in sand where leaves were short, and 10:1:2 in muddy substrates where the plants reached maximum development. In a study of seasonal variation *Z. marina* in Denmark, Sand-Jensen[813] found the ratio of leaves:rhizomes varied from 1:2 to 1:1 from winter to summer.

Maximum biomass in any seagrass meadow is related to density as a function of leaf size. There must be some density of leaves that results in increasing self-shading and this then becomes a controlling factor.[570] The leaf area index (LAI) is an estimate of this maximum leaf density.[222] In terrestrial cereal crops, maximum values of LAI areas are as high as nine; for broad-leaved trees in a tropical forest it can reach up to 20.[302] In seagrasses, LAI of over 20 can be reached in a dense *Zostera marina* meadow in Alaska.[564]

D. Primary Production
1. Methods of Estimation

Seagrasses present a number of problems in marine productivity measurements and a variety of techniques have been used. One approach is to measure the changes in the amount of standing stock during the growing season.[729] One of the problems is to account for the loss of plant material during the growing season. In addition many seagrasses, including tropical and often subarctic ones, are perennial and maintain significant winter populations.

Table 18
SOME REPRESENTATIVE SEAGRASS PRODUCTIVITIES

Species	Location	Productivity (g C m^{-2} day^{-1})	Ref.
Cymodoce nodosa	Mediterranean	5.5—18.5	284
Halodule wrightii	North Carolina	0.5—2.0	197
Posidonia oceanica	Malta	2.0—6.0	199
Thalassia testudinum	Texas	0.6—9.0	565, 652
	Florida (east coasts)	0.9—16.0	436, 641, 642, 1057, 1059
	Puerto Rico	2.5—4.5	651
Zostera marina	Denmark	2.0—7.3	711
	Rhode Island	0.4—2.9	138
	North Carolina	0.2—1.7	197, 1017
	Washington	0.7—4.0	717
	Alaska	3.3—8.0	562-564

Consequently this method which was used in early studies has been superseded by other techniques.

Increasingly, production measurements derived from marking techniques are being used and these are a considerable improvement over those obtained from changes in biomass. The marking process fixes the amount of standing crop present at a particular point in time and allows the more exact measurement of plant growth over a defined time interval. The technique involves the marking of grass blades with a small stapler and measuring the growth that has occurred after a specified interval as well as the new leaves produced during the time that has lapsed. Details of the technique will be found in Zieman[1057,1059] and Zieman and Wetzel.[1060]

Changes in dissolved oxygen concentrations in the water have been used by several workers as measurements of the products of photosynthesis and respiration. Either the change in oxygen concentration of water flowing across a seagrass meadow is measured,[630,652] or plants (or parts of plants) are enclosed in bottles[564] or in a type of bell jar placed directly over the bottom[78] and the changes in oxygen concentration are measured. However, there are a number of problems with the use of this technique. Hartman and Brown[373] have shown that in fresh water plants the oxygen resulting from photosynthesis is internally recycled in the lacunar spaces of the leaves. In *Zostera* as much as 50 to 60% of the leaf volume is essentially a gas sac. Thus the relative diffusion rates of oxygen into the water are not correlated directly with the intensity of photosynthesis. Accumulations within the intercellular lacunae can be utilized for photorespiration and mitochondrial dark respiration by both foliage and rooting system during both the photoperiod and in darkness, without effect on the oxygen concentration of the surrounding medium.

Some workers have attempted to estimate the production in seagrass beds by measuring the changes in pH and alkalinity in open, flowing systems.[692] The ^{14}C uptake technique has also been applied to the measurement of productivity in seagrasses.[50,78,565] According to Zieman and Wetzel[1060] the ^{14}C technique appears to yield values close to net productivity.

2. Production Estimates

Table 18 lists some representative estimates of seagrass production. Reviews of the level of productivity are to be found in Phillips,[718] McRoy and Helfferich,[569] and Zieman and Wetzel.[1060] From the data in the table two trends emerge. The values demonstrate that seagrass productivity can rival the most productive agricultural areas.[986] Reliable estimates have shown that *Zostera marina* can attain daily productivities of 8 g C m^2 day^{-1} and estimated yearly production of 500 g C m^{-2}. Due to greater radiation input and much longer growing seasons, tropical seagrasses can have still greater productivities. Communities of *Thalassia testudinum* have been reported to produce up to 16 g C m^{-2} day^{-1}. Using a marking technique, Greenway[331] estimated that the yearly production of *Thalassia* was 825 g C m^{-2} year^{-1}. Even if these values are approximately accurate, tropical seagrass beds rank among the most productive communities existing.

Seagrass beds are often characterized by the presence of highly productive associated species, growing epiphytically and otherwise. For example, Thayer et al.[928] showed that a *Zostera marina* bed in North Carolina produced an average 350 g C m^{-2} year^{-1}, but that associated plants *(Halodule* and *Ectocarpus)* together contributed a further 300 g C m^{-2} year^{-1}. Jones[436] estimated that *Thalassia testudinum* in dense stands in Florida was producing 900 g C m^{-2} day^{-1} while its epiphytes were contributing a further 200 g C m^{-2} day^{-1}. However, these estimates may be in error due to the use of oxygen-exchange techniques. Penhale[705] measured ^{14}C uptake by *Zostera* and its epiphytes separately and estimated that the eelgrass averaged 0.9 g C m^{-2} year^{-1} while the epiphytes averaged 0.2 g C m^{-2} year^{-1}.

Productivity estimates have been concerned primarily with aboveground production, and the belowground production needs to be accounted for. Zieman[1059] found that *Thalassia* leaves usually constituted 15 to 22% of the total dry weight of the plant, while Patriquin[698] estimated that the short shoots and rhizomes accounted for only 10 to 13% of the net production. Sand-Jensen[813] found that for *Zostera* in Denmark, the living rhizomes increased from 100 to 200 g dry wt m^{-2} throughout the growing season, while leaves and flowering short shoots increased from 50 to 230 g m^{-2}.

VI. EPIPHYTIC ALGAE

A. Distribution Patterns

Epiphytic macroalgae have been discussed in the previous section and here we will be concerned with the epiphytic microalgae. Many species of benthic diatoms occur in estuaries as epiphytes on rock, shells, and other hard substrates, or as epiphytes on the leaves of *Zostera, Thalassia, Posidonia, Ruppia,* and other seagrasses, and on the submerged portions of emergent vascular plants. This component of the primary production mix has been little studied apart from a number of investigations on seagrass epiphytes.[367]

The epiphytic flora can be extremely diverse and include as many as 100 species of microalgae and small macroalgae.[192,470,952] Most of this flora is dominated by a few species. The leaves of *Zostera* in Alaska are at times covered by a dense felt that primarily consists of the diatom *Isthmia nervosa.* The biomass of this epiphyte can be considerable and at maximum development can amount to as much as 50% of the total leaf plus epiphyte dry weight.[569]

In a brackish lagoon in Australia, Wood[1032] compared the communities of microalgae from macrophytes *(Zostera capricorni, Z. muelleri, Posidonia* sp., *Ruppia maritima, Halophila ovale,* and macroalgae) with mud surfaces and the surrounding water column. The number of species collected in each group was as follows: epiphytes: 39, phytoplankton: 45, and mud algae: 63. The species found in the guts of fish were more frequently epiphytic species than the mud (benthic) microalgae, leading to the conclusion that these algae were more important than the other groups in the diets of these animals.

Kita and Harada[470] compared the species composition of phytoplankton in a *Zostera* bed near Seto, Japan, with the microalgae on the blades of the plants. They found that the two populations were distinct with little overlap. The overwhelming majority of the epiphytes were diatoms, generally *Cocconeis scutellum* and *Nitzchia longissima.* The standing crop increased towards the top of the blade, averaging 0.1 mg dry wt cm^{-2}. In a study in the Yaquina Estuary, Oregon, Main and McIntire[571] identified 221 diatom taxa on the blades of *Zostera marina.* The initial development of the epiphytic community on *Zostera marina* has been examined in detail by Sieburth and Thomas.[840] They found few bacteria or algae on the surface of young *Zostera* and suggested that it was necessary for one species of pennate diatom, *Coscinasterias scutellum*, to form a crust on the leaf surface before other species could develop.

B. Primary Production

Where epiphytic algal production has been measured the values show that it can be significant when compared with that of the host plants and the total ecosystem. Marshall[580] estimated this productivity to be 20 g C m^{-2} year^{-1} for an area in Massachusetts. In a more detailed study of the epiphytes of *Thalassia* in Florida, Jones[436] found considerable seasonal variation in epiphyte productivity; peak rates occurred in Feburary and March, and July and October, with very low and sometimes undetectable rates of net production in the intervening months. The early peak was attributable to a spring bloom associated with the seasonal warming trend; in the July and October peak the blooms followed hurricanes. He estimated the net epiphyte production in summer to be 0.9 g C m^{-2} day^{-1} and in winter 0.2 g C m^{-2} day^{-1}. The total annual production of epiphytes was estimated at 200 g C m^{-2} year^{-1}; this value was 20% of the estimated net production of *Thalassia* in the region. Thayer et al.[928] showed that a *Zostera marina* bed in North Carolina produced an average 350 g C m^{-2} year^{-1}, while the associated epiphytic algae (both microalgae and fine macroalgae) contributed a further 300 g C m^{-2} year^{-1}.

In Beaufort, North Carolina, Penhale and Smith[706] measured epiphyte production at 73 g C m^{-2} year^{-1} which, averaged over the total estuary, contributed 13 g C m^{-2} year^{-1} (8.5% of the total annual primary production), while in Flax Pond, New York, Woodwell et al.[1041] recorded an epiphyte production averaged over the total area of 20 g C m^{-2} year^{-1} (3.7% of the total annual primary production).

McRoy and Goering[568] have speculated that the epiphyte load on seagrasses is inversely related to the nutrients available in the water column. It has been demonstrated that carbon and nitrogen could be transferred from the solution surrounding the roots of *Zostera* through the plant to the diatoms on the leaves.[568] Harlin[366] also observed the transfer of carbon and phosphorus from the leaf to an epiphyte. This suggests that the nutrient pool of the sediments is available for the growth of the epiphytic flora and is a possible explanation for the heavy epiphyte loads observed on *Thalassia* in waters that contain very low or undetectable concentrations of nitrogen and phosphorus. In addition there can be appreciable nitrogen fixation by epiphytic blue-green algae.[299]

It is thus clear that the contribution of one fifth to one third of total community production[436,705] is significant to the ecosystem. The turnover of the epiphytic algae is greater than that of the host and there is a need for further year-round studies such as those conducted by Penhale[705] to determine the extent of the contribution they make to community metabolism.

VII. BENTHIC MICROALGAE

A. Introduction

Along with the epiphytic microalgae, the benthic microalgae (also termed epibenthic algae or epilyptic algae) are the least studied of the estuarine primary producers. Estuarine sand and mud habitats harbor diverse assemblages of pennate diatoms, blue-green algae, and flagellates.[228] In the salt marshes of Georgia, the benthic microalgal flora includes a diverse assemblage of several hundred species of pennate diatoms that compose 75 to 93% of the total microalgal biomass.[1015] Williams found that an average of 90% of the cells belonged to one of four genera: *Cylindrotheca*, *Gyrosigma*, *Navicula*, or *Nitzchia*. Filamentous blue-green algae *(Anabaena oscillarioides, Microcoleus lyngbyaceous, Schizothrix calicola)* and a single species of *Euglena* constitute most of the remainder of the microalgal community.

In an intertidal mud flat in the Avon-Heathcote Estuary, New Zealand, McClatchie et al.[549] identified 64 diatom species. The dominant genera were *Nitzchia* (11 species), *Navicula* (10 species), *Achnanthes* (6 species), and *Amphira* (4 species). The number of diatom taxa were comparable with the range found in North American estuaries. Sullivan[894] found between 57 and 62 species in the edaphic communities associated with vegetated areas in a Delaware salt marsh, whereas a bare bank lacking macroscopic vegetation supported only 43 species, and a salt pan only 30 species. Diversity calculated for the Avon-Heathcote Estuary (H' = 3.4577 ± 0.172 SD bits per individual) agreed well with Sullivan's[894] bare bank community (H' = 3.604 ± 0.239). Numerous studies[555,894-896] have demonstrated marked differences in the association of dominants, the number of endemic species, number of taxa, diversity, and evenness between different habitats in an estuary.

The large nonflagellated euglenoid, *Euglena obtusa* Schmidt, is a cosmopolitan species in estuaries occurring in fine muds and especially in areas with high organic and nutrient inputs.[485,877,879] Palmer and Round[685] found in a study in the Avon River estuary, England, that *E. obtusa* had a pattern of vertical migration within the mud, moving down to 2 mm prior to being covered by the tide or in response to reduced light. Diatoms also undergo similar vertical migrations which in some cases appears to have characteristics of an endogenous circadian rhythm.[77,686] A common characteristic of these algae is that of leaving a trail of mucus when moving through the sediment.[130] The mechanism of the motility of these algae is controversial.[405]

B. Primary Production
1. Methods of Estimation

Standing crop estimates are based either on cell numbers or cell volume on a surface area basis (e.g., mm^{-2})[1015] or by measuring the chlorophyll *a* content of a measured volume or weight of sediment. There are difficulties with chlorophyll determinations as the sediments contain high concentrations of chlorophyll degradation products which can cause serious errors in data obtained through standard methods for measuring chlorophyll. Whitney and Darley[997] have recently developed a technique that overcomes some of the problems.

While the microalgal standing stock is small, the turnover rate is high. Much of the microalgal production in salt marshes occurs when the macrophyte plants are dormant, thus increasing the relative contribution of the algae to the total energy flow during the winter months.[955,1071]

There are many technical problems that are encountered in estimating production in benthic microalgae. This stems from the characteristics of the environment in which they live. Pomeroy et al.[732] describe it in the following terms:

> The microalgae live in and on the top few millimetres of sediment, a habitat whose microenvironment is very difficult to describe. The interface represents a boundary between a dark, nutrient-rich, anaerobic sediment and either an illuminated, aerobic, comparatively nutrient-poor water column or, at ebb tide, the atmosphere. This microenvironment is extremely patchy and is subject to rapid and extreme variation, being directly affected by many factors. It is influenced by variations in tidal exposure, sedimentation, higher plant cover, and surface and subsurface herbivores and detritovores. These factors in turn affect light intensity, temperature, pH, salinity, levels of organic and inorganic nutrients, intensity of grazing, and the stability of the sediment surface. The habitat of the epibenthic algae is virtually impossible to define or reproduce adequately; thus, when attempting to measure the performance of the algae in their native habitat, we must maintain the integrity of the natural relationships of the surface layer of the sediment.

Production estimates are also complicated by the vertical migrations mentioned above. Because of strong light attenuation in muddy sediments (99% in 0.2 to 1 mm), the migration rhythm has a pronounced effect on photosynthesis.[709,1015] Production is usually estimated by the ^{14}C technique with intact sediment cores being incubated in special chambers *in situ*.[157] Changes in oxygen concentration have also been used.

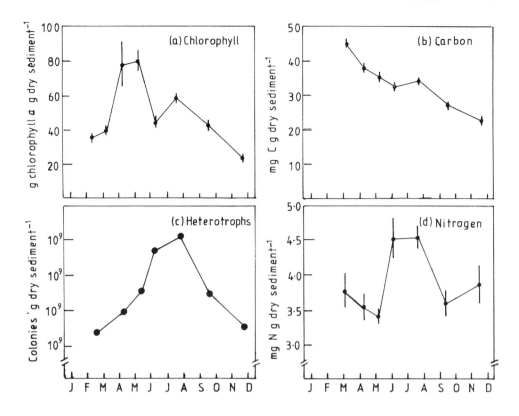

FIGURE 23. Seasonal cycle of (a) chlorophyll *a*, (b) carbon content, (c) numbers of aerobic heterotrophs, and (d) nitrogen content in the top 0.5 cm of the sediments in the River Lynher estuary, England. The vertical bars are two standard errors. (From Joint, I. R., *Estuarine Coastal Mar. Sci.*, 7, 190, 1978. With permission.)

2. Biomass

One of the most intensive studies of the standing stocks of benthic microalgae is that of Williams[1015] on the Sapelo Island salt marshes. He found that diatom cell numbers and total cell volume tended to be highest, 3100 to 6400 cells mm^{-2} and 7.4 to 22.7 mℓ m^{-2}, respectively, in the tall *Spartina* near the creeks, decreasing towards the creek bottom and also towards the high marsh. In the vegetated portion of the marsh, diatom cell numbers fluctuated annually with winter values that averaged about ten times higher than the summer values. Total diatom volume also exhibited an annual cycle, but the variation in the volume was less pronounced than variation in numbers because the small diatoms (< 3000 μm^3) which made up the majority of the assemblage were those most affected by the summer decline in numbers. These conclusions contrast with those of Riaux[785] who studied the benthic microalgal community in an estuary on the North Brittany. Diatoms longer than 20 to 30 μm were considered as microphytobenthos, and those smaller as nannobenthos. Riaux found that the spring and autumn months were characterized by an increase in microphytobenthos biomass (in terms of chlorophyll *a* content), as well as by an increase in the nannobenthos. In contrast, in winter there was a sharp decrease in nannophytobenthos. The differences between the two regions may be due to the shorter winter days and the lower winter sea and air temperatures in Brittany.

Joint[432] investigated the production of benthic microalgae in a mud flat in the River Lynher estuary, Cornwall, England. The seasonal cycle of chlorophyll *a* content of the surface sediments is shown in Figure 23. The increase in chlorophyll *a* during April coincided with an increased rate of photosynthesis as shown in Figure 24. The increase in the standing stock of the chlorophyll *a* between March and April was 39 μg g^{-1} dry sediment, equivalent

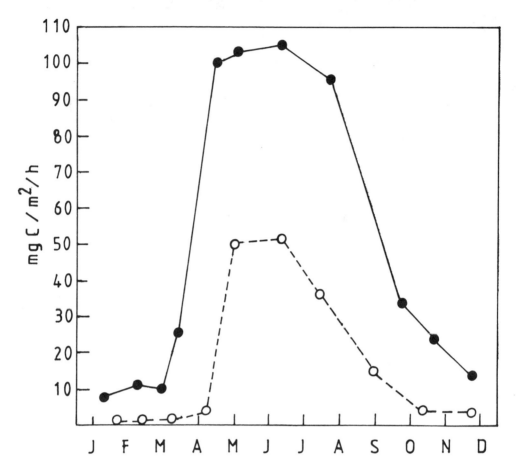

FIGURE 24.   The rates of primary production in the surface sediments (●) and the water column (○) during 1974 in the River Lynher estuary, England. (From Joint, I. R., *Estuarine Coastal Mar. Sci.,* 7, 187, 1978. With permission.)

to an increase in biomass of 12.5 g C m^{-2}, if a carbon to chlorophyll *a* ratio of 50 is assumed; the calculated photosynthetic production for the same period was 20 g C m^{-2}. There was a decrease in the chlorophyll *a* content of the surface sediment during May but this increased again in July. The decrease in chlorophyll *a* content was assumed to be due to the turnover of the sediment by animal activity, as the maximum number of heterotrophs occurred at this time. This was supported by studies of depth profiles of chlorophyll *a* and phaeopigments, while the maximum chlorophyll *a* values were in the top 2 cm, appreciable levels were recorded down to 14 + cm and larger quantities of phaeopigments were found at the same depths.

A number of estimates of benthic microalgal chlorophyll *a* levels are listed in Table 19. A wide range of values have been recorded dependent upon tidal levels, season, geographic locality, and type of estuarine system. For the three New Zealand estuaries the ranges differ and are in accord with the trophic status of the three systems; Delaware Inlet (range 12.5 to 30.5) is oligotrophic, the Upper Waitemata Harbour (range 0.6 to 67) is mesotrophic, whereas the Avon-Heathcote Estuary (range 9.3 to 109.8) is eutrophic. In Netarts Bay, Oregon, microalgal biomass expressed as the concentration of chlorophyll *a* in the top cm of sediment varied seasonally (Figure 25). An analysis of variance indicated that there were significant differences among mean concentrations associated with the effects of sediment type, tidal height and time, and a two-way interaction between sediment type and tidal

Table 19
BENTHIC MICROALGAL CHLOROPHYLL *a* LEVELS
MEASURED IN A RANGE OF ESTUARIES

Locality	Chlorophyll *a* (mg Ch a m^{-2})	Biomass (g C m^{-2})	Ref.
Upper Waitemata Harbour	0.6—6.7	0.072—8.04	72, 477
Avon-Heathcote estuary, Christchurch	9.3—109.8	1.12—13.18	447
Delaware Inlet, Nelson	12.5—30.5	1.5—3.66	447
Nanaimo River estuary, British Columbia	2.6—10.6	0.32—1.27	617
Netarts Bay, Oregon	Means	—	168
Sand	46.2	—	
Fine sand	74.7	—	
Silt	93.7	—	

Note: Most of the available data for other estuaries is expressed in terms of μg chlorophyll a/g dry sediment and thus are not easily translated into mg Ch a m^{-2}. Chlorophyll *a* values have been converted to biomass estimates assuming a carbon to chlorophyll *a* ratio of 120:1.

height. The highest concentrations occurred generally in the finer sediments and the lowest in the coarser sediments. Mean concentrations of chlorophyll *a* in the top cm of sediment for the entire study period were 46.2 mg m^{-2} (sand site), 74.7 mg m^{-2} (fine sand site), and 93.7 mg m^{-2} (silt site). The highest concentrations were recorded in silt at MLLW (mean lower low water) + 2 m. Other studies have reported increasing chlorophyll *a* concentrations with increasing silt content.[91,130,132,501] The most pronounced peaks in chlorophyll *a* levels in the MLLW ± 2 m and + 1 m occurred in the fall (autumn) period whereas at MLLW + 1.5 m there was a spring peak. This pattern of temporal variation in chlorophyll *a* concentration was similar to that observed by Colijn and Dijkema[132] in the Wadden Sea.

3. Production Estimates

Comparing the results of benthic microalgal productivity in different estuaries is difficult because of the different study techniques that have been used and the differing environments of the estuaries studied. Nevertheless from Table 20 it can be seen that benthic microalgae do make significant contributions to total estuarine primary production.

Pomeroy's[728] study in the Sapelo Island salt marshes was the first attempt to describe the seasonal primary productivity of benthic microalgae and of the factors influencing it. He measured productivity at low tide, using a flowing-air system with CO_2 absorption columns. High tide was simulated by placing bell jars filled with sea water over the exposed sediments and monitoring dissolved oxygen. In the bare creek bank zone hourly rates of photosynthesis during low-tide conditions were highest in the winter. Based on his investigations, Pomeroy estimated annual gross algal production at 200 g C m^{-2} and net algal production at not less than 90% of this value. Gallagher and Daiber,[1071] using dissolved oxygen changes over flooded cores in the laboratory, estimated gross production for a Delaware salt marsh at 80 g C m^{-2}, which was about one third of net angiosperm aboveground production in that particular marsh. Van Raalte et al.[955] measured benthic microalgal productivity in the vegetated areas of a Massachusetts salt marsh using ^{14}C incubation in the field. Brief spring and autumn (fall) peaks in productivity coincided with blooms of filamentous green algae. Algal production was estimated at 105 g C m^{-2} or about 25% of the aboveground macrophyte production for the marsh.

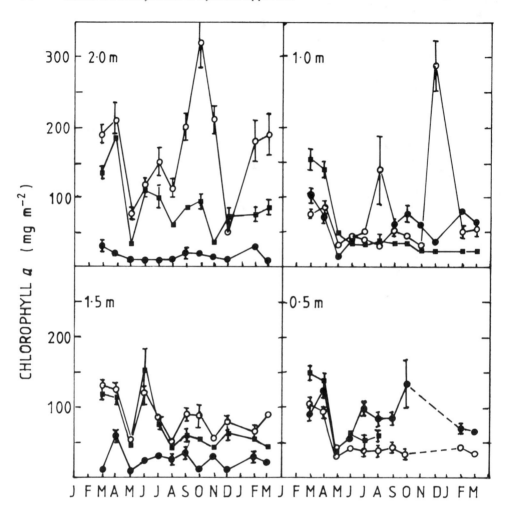

FIGURE 25.   Concentrations of chlorophyll *a* at Sand (●), Fine sand (■), and Silt (○) sites at 2.0, 1.5, 1.0, and 0.5 m above MLLW in Netarts Bay, Oregon. Means ± 1 standard error. (From Davis, M. W. and McIntire, C. D., *Mar. Ecol. Prog. Ser.*, 11, 108, 1983. With permission.)

In their investigation of the effects of physical gradients on the production dynamics of benthic microalgae in Netarts Bay, Oregon, Davis and McIntire[168] found that the maximum hourly rate of microalgal production occurred in the summer when *Enteromorpha prolifera* sporlings were abundant in the sediment (Figure 26). When calculated rates from chambers which contained *E. prolifera* were included, the *Fine Sand* had the highest mean hourly rates of gross primary production (47 mg C m^{-2} hr^{-1}), followed by the *Sand* site (37 mg C m^{-2} hour^{-1}), and the *Silt* site (25 mg C m^{-2} hr^{-1}). If calculated rates from chambers which contained *E. prolifera* were excluded, the three study sites had similar mean hourly rates of gross primary production; 28, 28, and 25 mg C m^{-2} hr^{-1} for sand, fine sand, and silt, respectively. The maximum hourly community oxygen uptake by the sediments occurred during the summer when temperatures were relatively high (Figure 26). Annual rates of community primary production and oxygen uptake are shown in Table 21. From these data it can be seen that the fine sand site generally had the highest primary production and community oxygen uptake. Although these were obvious effects of season, sediment type, and tidal height on the production dynamics of the benthic microalgae, these relationships were not linear and suggested that other factors were involved in the control of the observed

Table 20
**ESTIMATES OF BENTHIC MICROALGAL PRODUCTION FROM
VARIOUS LOCALITIES**

Locality	Gross production (g C m^{-2} year^{-1})	Net production (g C m^{-2} year^{-1})	Ref.
Barataria Bay, Louisiana	362	244	75
Sapelo Island, Georgia			732
Bare Creek bank	—	43—1156	
Levee	—	96—201	
High marsh	—	52—61	1071
Delaware salt marsh	—	40	
Massachusetts salt marsh	—	105	955
Intertidal sand flat, Washington	80	53—74	688
Yaquina Bay, Oregon	—	0—163	794
Netarts Bay, Oregon			
Sand	74.9—204.5	15.7—64.6	168
Fine sand	145.7—162.6	48.5—59.8	
Silt	63.7—87.9	14.5—31.7	
Nanaimo River estuary, British Columbia	—	4.22—55.46	617
River Lynher estuary, England	—	143	432
Ythan estuary, Scotland	—	31	501
Western Wadden Sea, Netherlands	—	101 ± 58.5	91
Northern Wadden Sea, F.R.G.	115	99	1063
Upper Waitemata Harbour, New Zealand	87—385	60—262	477
Avon-Heathcote estuary, New Zealand	70—201	49—140	447
Delaware Inlet, Nelson, New Zealand	210—367	147—257	447

patterns. Possible mechanisms include the effects of infauna[167] and epifauna,[447,537] possible lack of light limitation associated with tidal inundation, and heterotrophic nutrition of benthic algae which would modify the effects of physical factors on photosynthesis.[2,558]

Benthic microalgal production in the Sapelo Island marshes has recently been reinvestigated using ^{14}C incubation techniques.[732] Average hourly production from these studies is given in Table 22. The overall annual estimate of net productivity of approximately 190 g C m^{-2} was close to Pomeroy's[728] estimate. More than 75% of this algal production occurs when the marsh is exposed at ebb tide and at this time the bare creek bank was the most productive (Table 22). Conversely, it is the least productive when submerged at high tide. Working in the River Lynher estuary. Joint[432] found that the rate of photosynthesis decreased rapidly as the mud flat was submerged and was not detectable, i.e., was less than 0.25 mg C m^{-2} hr^{-1}, only 30 min after flooding. Migration of the diatoms into the sediment and the turbidity of the estuarine water probably account for the low rates when the tidal flats are submerged. When the water depth was less than 0.5 m the light intensity at the sediment surface was 76% of that at the sea surface; 10% illumination was measured at a depth of 2 m. In the high marsh at Sapelo Island algal production was nearly equal under high- and low-tide conditions.

From the data in Table 20 it can be seen that there is a general trend of increasing production in warmer waters, e.g., mean net production was 31 g C m^{-2} in the Ythan Estuary, Scotland, 99 in the North Wadden Sea, and 143 in the River Lynher estuary. Values recorded in the New Zealand estuaries are comparable to those recorded in the southern U.S. estuaries.

C. Environmental Regulation of Distribution, Abundance, and Production

There are many potential factors limiting the standing crop biomass and productivity level of estuarine benthic microalgae. Some of these such as sediment type, intertidal height, and

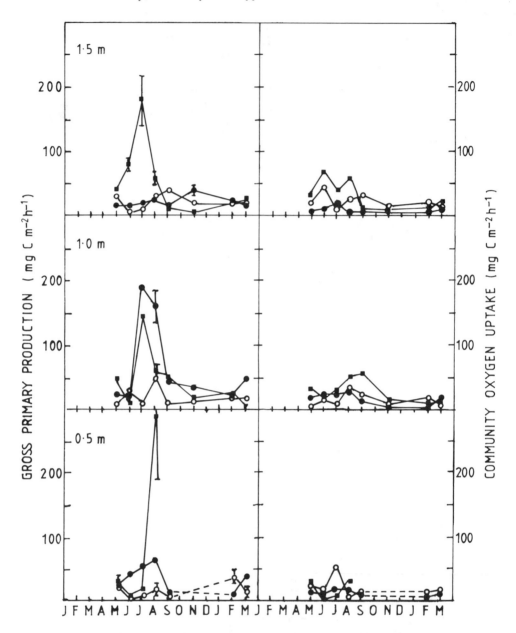

FIGURE 26.   Rates of gross primary production and community oxygen uptake in carbon equivalents at Sand (●), Fine sand (■), and Silt (○) sites at 1.5, 1.0, and 0.5 m above MLLW in Netarts Bay, Oregon. Values for primary production: means ± 1 standard error, and means for community oxygen uptake. (From Davis, M. W. and McIntire, C. D., *Mar. Ecol. Prog. Ser.*, 11, 108, 1983. With permission.)

seasonal changes in light intensity have been touched upon above. In all studies production occurred predominantly when the sediments were exposed to the air with low values when the sediments were submerged. The tidal regime may also affect productivity indirectly through its influence on other parameters, including salinity, pH, temperature, light intensity, and nutrients.

The salinity of the surface layer of estuarine sediments varies from that of the overlying water at flood tide, to increasing salinity following evaporation at low tide, to dilution during rains at ebb tide. In the River Lynher estuary, Joint[433] recorded salinity changes greater than

Table 21
CALCULATED ESTIMATES OF ANNUAL SEDIMENT-ASSOCIATED MEAN MICROALGAL BIOMASS AND TOTAL PRIMARY PRODUCTION IN NETARTS BAY AT THE SAND, FINE SAND, AND SILT SITES

Site (m)	Biomass	GPP	RESP	NPP	OUPTK	Nonalgal OUPTK
Sand						
0.5	6.22	107.5	73.1	34.4	109.6	36.5
1.0	6.34	204.5	139.9	64.6	140.0	23.1
1.5	1.94	74.9	59.2	15.7	50.0	13.3
Fine sand						
0.5	5.21	151.3	99.6	51.8	116.2	44.6
1.0	4.31	145.7	97.3	48.5	248.1	150.9
1.5	5.09	162.6	102.8	59.8	225.1	122.2
Silt						
0.5	3.94	64.0	49.5	14.5	130.5	81.0
1.0	6.09	63.7	45.7	18.1	162.8	117.2
1.5	5.00	87.9	56.3	31.7	174.6	118.4

Note: Tidal levels are 0.5 m, 1.0 m, and 1.5 m above MLLW. Variables, expressed as g C m^{-2}, include: mean microalgal biomass (Biomass), gross primary production (GPP), microalgal respiration (RESP), net primary production (NPP), community oxygen uptake (OUPTK), and nonalgal community oxygen uptake (Nonalgal OUPTK). Nonalgal OUPTK equals OUPTK — RESP.

From Davis, M. W. and McIntire, C. D., *Mar. Ecol. Prog. Ser.*, 13, 103, 1983. With permission.

Table 22
AVERAGE HOURLY BENTHIC MICROALGAL PRODUCTIVITY IN VARIOUS AREAS OF THE SALT MARSHES OF SAPELO ISLAND, GEORGIA
(mg C m^{-2} hr^{-1})

Area	Exposed	Submerged
Bare creek bank	132	2
Levee (tall *Spartina*)	11	23
High marsh (short *Spartina*)	7	6

From Pomeroy, L. R., Darley, W. M., Dunn, E. L., Gallagher, J. L., Haines, E. B., and Whitney, D. M., *The Ecology of a Salt Marsh*, Pomeroy, L. R. and Wiegert, R. G., Eds., Springer-Verlag, New York, 1981, 57. With permission.

20% over a tidal cycle. However, estuarine benthic diatoms appear to be particularly tolerant of such salinity changes.[556,1048] Fourteen species of diatoms isolated from Sapelo Island marshes grew well in salinities of 10 to 30‰ and several grew well over a range of 1 to 68‰.[1016]

In the Sapelo Island marshes the pH of the marsh surface generally is between 7 and 8, but during low tide algal photosynthesis can increase it to 9.[728,894] It is possible that inadequate supplies of CO_2 and HCO_3^- under these conditions could limit photosynthesis.[732]

Seasonal variation of temperature does not appear to have a marked effect on benthic microalgal production. Pomeroy[728] and Van Raalte et al.[955] note that photosynthetic rates are independent of temperature at suboptimal temperatures. Tidal ebb and flow have an ameliorating effect on sediment temperatures. Temperature may also indirectly affect the benthic microalgae by influencing the activity of grazers. Williams[1015] attributed the warm water decrease in diatom standing crop to increased activity of grazers rather than to any direct effect of temperature. The impact of grazers will be considered in Volume II, Chapter 2, Section VII. C.

Benthic microalgae growing on intertidal areas are exposed to considerable variations in light intensity; in addition to diurnal changes in solar radiation the light regime varies from day to day because the period of tidal exposure changes with tidal periodicity. Benthic microalgae are much less sensitive to high light intensities than are phytoplankton. Taylor[911] found very little photoinhibition at "full sunlight" in experiments with diatoms from a Massachusetts intertidal sand flat, and that photosynthesis was saturated at about 16% full sunlight. Cells receiving only 1% incident solar radiation were able to fix carbon at 35% their maximum rate. These results have been confirmed by other studies.[91,131,1015] Other data are somewhat contradictory.[1,609] Joint[433] points out that care needs to be taken in extrapolating from laboratory culture experiments to field conditions.

Field experiments at Sapelo Island[158] showed that benthic microalgae were not limited by light at the average intensity (about 25%) that penetrates the canopy of the marshes, although limiting conditions are approached during the summer months. It was also demonstrated that the winter algal assemblage was better shade-adapted than the summer assemblage.

Limitation by nutrients has also been investigated in a number of studies. Van Raalte et al.[955] found that nutrient enrichment in the vegetated portion of a Massachusetts salt mash stimulated productivity of benthic microalgae. In a study in a Delaware marsh, Sullivan and Daiber[898] found that removing vascular plant cover increased benthic microalgal chlorophyll concentration. Nutrient enrichment of clipped plants did not result in any further significant increases in chlorophyll, except for those plants enriched with nitrogen in the summer months. One problem with such experiments is that fertilization increases the growth of the vascular plants and thus reduces the light intensity reaching the sediment surface. Darley et al.[158] carried out some short-term experiments at Sapelo Island incubating sediment cores in the field and fertilizing them only with nutrient solution. In the short *Spartina* marsh (Figure 27), the algal standing crop, as chlorophyll *a*, and productivity of algae both increased significantly when the cores were fertilized daily with nitrogen, or with a complete nutrient solution containing nitrogen. Similar results were obtained in winter experiments, suggesting that the algae were nitrogen-limited both in summer and winter. In the bare creek bank marsh zone, similar experiments indicated that the algae there were limited by the grazing activity of snails and fiddler crabs. Nutrient enrichment only stimulated algal production following an algal bloom that was the result of grazer removal, and which had depleted the relatively high standing stock of nutrients in the surface creek bank sediments.

D. Biotic Regulation of Distribution, Abundance, and Production

From a number of studies it is clear that the species composition of the benthic microalgal community is important in determining the overall productivity. The impact of sporlings of the green alga *Entermorpha prolifera* on increasing production rates in an Oregon marsh has been discussed above.[168] Working in Delaware Inlet, New Zealand, Gillespie and MacKenzie[293] found that the highest rates of benthic microalgal $^{14}CO_2$ fixation occurred at sandy sites colonized primarily by the flagellate *Euglena obtusa,* with sometimes occasional blooms of the blue-green alga *Oscillatoria ornata*. Rates of fixation at these sites were generally 10 to 20 times greater than among the microalgal populations of other habitats. The highest rate of fixation observed (216 mg C m^{-2} hr^{-1}) occurred under bloom conditions of *Euglena*. Rates observed in the other habitats ranged from 1 to 5 mg C m^{-2} hr^{-1}.

The importance of biotic interactions in regulating benthic microalgal populations is not

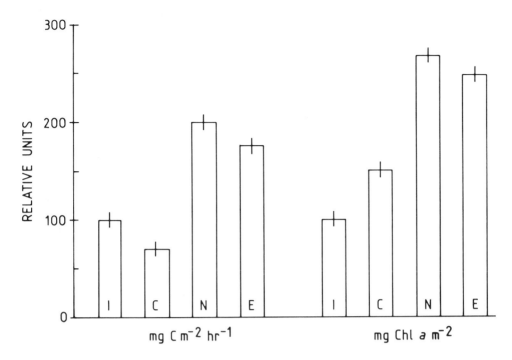

FIGURE 27.   Summer enrichment experiment in short *Spartina* marsh at Sapelo Island. Initial (I) values at 14.9 mg C m^{-2} hr^{-1} and 35.5 mg chlorophyll *a* m^{-2}. Sediment cores were incubated for 8 days in fiddler crab enclosures at 50% ambient light intensity with nutrient solution added at low tide daily. Controls (C) received sea water. Nitrogen-enriched cores (N) received 3 μm N as NH$_4$Cl in sea water. Completely enriched cores (E) received a complete nutrient solution. Bars = 95% confidence limits. (From Pomeroy, L. R., Darley, W. M., Dunn, E. L., Gallagher, J. L., Haines, E. B., and Whitney, D. M., *The Ecology of a Salt Marsh*, Pomeroy, L. R. and Wiegert, R. G., Eds., Springer-Verlag, New York, 1981, 61. With permission.)

well understood. Many epifaunal and infaunal deposit feeders ingest and assimilate microalgae.[244,512,688,812] Experimental manipulations of estuarine gastropods *Hydrobia* spp.,[244,514,535] *Nassarius obsoletus*,[680,988] *Bembicium auratum*,[69] *Illyanassa obsoleta*,[137,215,514] and *Amphibola crenata*[71,446,447,549] have demonstrated changes in microalgal populations. The role of these mud flat snails will be further discussed in Volume II, Chapter 2, Section VII. C. Experimental studies of infaunal regulation of benthic microalgae include the investigation of White et al.[992] who found that the sand dollar *Mellita quinquiesperforata* had no significant effect on chlorophyll *a* concentrations, and that of Coles,[130] who investigated the seasonal abundance of benthic microalgae in The Wash on the east coast of England. Coles recorded low numbers of microalgae on the inner sand flats compared with the persistently high numbers on the upper mud flats and pioneer salt marshes. The differences appeared to be primarily the result of grazing by large numbers of deposit-feeding macroinvertebrates, which occurred on the inner sand flats, and not the result of differences in the nature of the sediment. In one experiment where populations of invertebrates (especially those of the amphipod *Corophium*) were killed there was a dramatic explosion of diatom numbers, which reached an average of 95 × 10^4 diatoms cm^{-2} within a week compared with only 5 × 10^4 diatoms cm^{-2} on the surrounding sand flats.

Davis and Lee[167] carried out a series of experiments in Yaquina Bay, Oregon, to determine the rate of recolonization of benthic microalgae and the effects of infauna on microalgal biomass and production. Estuarine sediment was defaunated and transplanted to the field and laboratory. Microalgal colonization in the field was rapid, with chlorophyll *a* levels returning to control levels by day 10, while infaunal densities returned to control levels within 40 days. Removal of infauna in the laboratory, primarily tanaids, increased benthic microalgal growth. After 40 days chlorophyll *a* was four times greater and gross primary production two times greater in the defaunated sediment than in the controls (Figure 28). These results indicated that natural densities of infauna can control both microalgal biomass and production.

VIII. PHYTOPLANKTON

A. Introduction

The contribution of phytoplankton to the overall primary production of estuaries is dependent upon a number of factors among which salinity, temperature, availability of light (as influenced by turbidity) and nutrients, and the configuration of the water basin are important. In estuaries that drain to a system of low-tide channels on the ebb tide, the phytoplankton make a much smaller contribution than in deep-water estuaries in which exposed mud flats form a small percentage of the total area. The views of workers on the importance of the role of phytoplankton have been colored by the type of estuary that is being investigated.

Grindley,[338] working in South African estuaries states that: "The significance of phytoplankton in estuaries is somewhat controversial; while it may play a major role and be the basis of heterotrophic life in the open sea, this appears to be unusual in estuaries. In most estuaries, the primary production of phytoplankton is insignificant in comparison to that of attached plants and organic detritus derived from them." On the other hand Boynton et al.[65] state that: "Phytoplankton production is of central importance in estuarine ecosystems because of its role in supporting many food webs."[457,927,1074]

Early workers in turbid, shallow estuaries in the southeastern U.S. regarded photosynthesis as a minor source of fixed carbon when compared to photosynthesis of the marsh grass. This stemmed from the work of Ragotzkie[753] who found that the critical depth (that depth at which the integrated mean light intensity reaches the compensation point) was usually above the bottom of the estuary and so he concluded that phytoplankton production was negative. However, several recent studies in the same area, including those in the Duplin River, Sapelo Island, have found that phytoplankton is a significant source of organic matter for the estuarine water food web, if not for the salt marsh food web as well.[732,825,933]

There is considerable literature on marine neritic and oceanic phytoplankton, and the factors underlying production are well understood.[696] In contrast to this the processes governing phytoplankton production in estuaries are not as well comprehended. Williams[1018] concluded that phytoplankton production in some shallow estuaries along the southeastern coast of the U.S. tended to follow the seasonal cycle of water temperature and that available nitrogen (rather than phosphorous) commonly limited production. Riley[791] compared the magnitude of productivity in selected estuarine, coastal, and marine systems and concluded that the supply of nutrients to surface waters was of overwhelming importance. Recently Nixon[627] reviewed productivity patterns in a group of estuarine systems and concluded that high rates of production common to estuarine systems were the result of relatively complete and rapid heterotrophic nutrient recycling.

B. Distribution and Geographic Variation in Species Composition

Due to the fluctuating temperatures and salinities that are found in estuaries the phytoplankton tends to be both euryhaline and eurythermal. The phytoplankton of the lower reaches of estuaries is dominated by diatoms, and dinoflagellates are less abundant although they may be important at certain seasons. Small nannoflagellates are usually abundant in the upper reaches. Neritic species from adjacent coastal waters penetrate to varying degrees dependent on their euryhalinity and the number of neritic forms becomes further reduced up an estuary. In the upper reaches, the plankton community may include characteristic

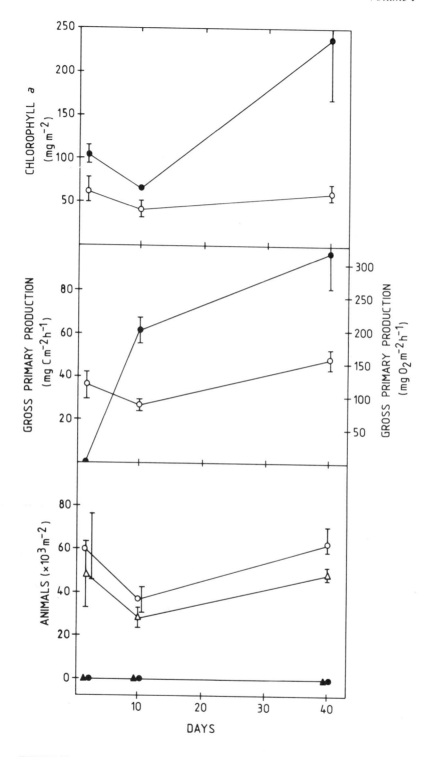

FIGURE 28.   Results of the laboratory defaunation experiment sampled at days 1, 10, and 40, showing concentration of chlorophyll *a* and gross primary production in control (○) and defaunated sediments (●); total infaunal density in control (△) and defaunated (▲) sediment. Values are means ± 1 standard error. (From Davis, M. W. and Lee, H., II, *Mar. Ecol. Prog. Ser.*, 11, 230, 1983. With permission.)

estuarine species not normally found in the open sea or the fresh waters of the rivers. Few fresh water species will tolerate even very low salinities and they consequently have only a very minor role in estuaries.

Of the marine estuarine diatoms *Skeletonema costatum* is an abundant and widespread species, as are *Nitzchia closterium* and *Thalassiosira decipiens*. In South African estuaries species such as *Coscinodiscus grandii, Rhizosolenia setigera, Chaetoceros lorenzianum, Biddulphia mobiliensis,* and *Actinoptychus splendens* are common.[338] Some dinoflagellates such as *Prorocentrum micans* and *Peridinium* spp. are temporarily autochthonous in South African estuaries and elsewhere. The phytoplankton communities in Georgia[732] and North and South Carolina estuaries[1061] are dominated by diatoms such as *Skeletonema costatum, Rhizosolenia* sp., *Asterionella* sp., and *Coscinodiscus* sp. Several species of dinoflagellates in the 15—25 μm range may sometimes dominate the community. Among the autochthonous brackish water diatom species *Chaetoceros danicum* is important in European estuaries, while *Chaetoceros subtilus, Nitzchia longissima,* and *Melosira dubia* occur in South African estuaries.

Nannoplankton, principally small flagellates and dinoflagellates may play an important role particularly in the upper part of estuaries. In South African estuaries common brackish water species include *Cryptomonas* spp., *Heteromastrix longifilis, Emiliana huxleyi,* and *Pyramimonas orientalis*.[338] The relative importance of the larger forms of phytoplankton and nannoplankton varies greatly. The mix appears to depend on the environmental conditions while competition for nutrients appears to determine the succession. Dinoflagellates and nannoflagellates grow best at high temperatures and may bloom (particularly in the early autumn) in the upper reaches of calm, slow-flowing estuaries.

In the Cananeia estuary, Brazil,[497] nannoplankton composed of microflagellates and small diatoms made up an average of 87.2% of the phytoplankton. Hulbert[417] has shown that in some very shallow estuaries such as Moriches Bay near New York and Salt Pond at Woods Hole, Massachusetts phytoplankton cells exceed $10^9 \ \ell^{-1}$. The dominants are nannoplankton. These sink more slowly than neritic marine species and their monopolization of the nutrient supply appears to favor their dominance. Nutrient levels are kept down to almost undetectable levels continually so that diatoms introduced from the sea are reduced and they also tend to sink to the bottom in the shallow water. McCarthy et al.[548] found that phytoplankton which passed through 35 μm mesh (nannoplankton) were responsible for 89.6% of the phytoplankton productivity in Chesapeake Bay. On one summer cruise the nannoplankton was responsible for 100% of the primary productivity. At that time the < 10 μm fraction was responsible for 94% of the productivity.

C. Primary Production

Early studies of estuarine phytoplankton productivity were carried out by measuring the changes of oxygen in light and dark bottles. More recent work generally employed the ^{14}C method; this method is believed to approximate net production. However, it has recently been questioned by critics[286,435,826] who point out that the amount of community respiration apparently exceeds the estimated levels of photosynthesis. As Pomeroy et al.[732] point out, recent work using the ^{14}C method has produced higher estimates than the earlier studies did, however, if the criticisms have merit, future estimates will be higher. Thus, it is fair to say that the present estimates of phytoplankton photosynthesis are still conservative figures.

1. Biomass

Estimates of phytoplankton biomass are difficult since the seston, or suspended organic particles, included particulate organic matter (detritus) and benthic microalgae resuspended from the surface sediments as well as true phytoplankton. Odum and de la Cruz,[639] when measuring the seston in a creek on a Georgia salt marsh, estimated that only 10% was

Table 23
PHYTOPLANKTON CHLOROPHYLL *a* LEVELS MEASURED IN VARIOUS ESTUARIES

Locality	Chlorophyll *a* (mg m^{-3})	Biomass[a] (g C m^{-3})	Ref.
Upper Waitemata Harbour	2.2—8.7	0.264—1.05	72
Langebaan Lagoon, South Africa	0.52—5.07	0.063—0.608	122
Sydney Harbour	1.8—17	0.215—2.04	761
Nanaimo River estuary, British Columbia	2.6—10.6	0.312—1.272	617
River Lynher estuary, England	0.5—11.4	0.06—1.368	432
Sapelo Island, Georgia	10—20	1.2—2.4	732

[a]   Estimated from the chlorophyll *a* values assuming a carbon to chlorophyll ratio of 120:1.

Table 24
ESTIMATES OF GROSS AND NET PHYTOPLANKTON PRODUCTION FROM VARIOUS ESTUARIES

Location	Production (g C m^{-2} year^{-1}) Gross	Net	Ref.
Narragansett Bay	—	300	264
Duplin River, Georgia	248	—	753
Beaufort Channel, North Carolina	255	—	1019
Bogue Sound, Newport, North Carolina	100	—	1017
Cove Sound, North Carolina	—	67	926
North Inlet, South Carolina	—	346	825
Doboy Sound-Duplin River, Georgia	—	375	732
Barataria Bay, Louisiana	598	412	184
Nanaimo River estuary, British Columbia	—	7.5	617
Langebaan Lagoon, South Africa	—	56—314	122
Fafa Lagoon, Natal	—	2.8—65.7	665
Cochin Backwater, India	14—575.4	—	747
River Lynher estuary, England	—	81.7	432
Sydney Harbour	—	11—127	761
Upper Waitemata Harbour	200	140	72

plankton, of which two thirds to three quarters was phytoplankton, i.e., about 7% of the total mass of seston. Thus, measures of phytoplankton pigment concentration are generally used as estimates of phytoplankton biomass. Table 23 gives chlorophyll *a* levels as measured in a number of estuaries from a range of geographic localities. Values range from 0.5 to 20 mg Ch *a* m^{-3}. There are wide variations in individual estuaries, depending on season and site within an estuary. Levels tend to be lower in estuaries at higher latitudes.

2. Primary Productivity

Estimates of gross and net production in estuaries from a range of geographic areas as shown in Table 24 vary greatly. In high latitudes light intensity may be critical while in tropical areas other factors such as seasonal nutrient or salinity fluctuations may be more important. Furnas et al.[264] found an annual carbon production in temperate Narragansett Bay of 308 g C m^{-2} of which 42% occurred in July and August. In the Sapelo Island marshes[732] investigators found that the highest photosynthetic rates for phytoplankton occurred in the water over the marsh on spring high tide. However, the number of daytime hours per year when the marshes were inundated were so small that the increase did not greatly influence the total phytoplankton production within the system. The annual production in the Duplin River and the adjacent Doboy Sound, Sapelo Island, was estimated to be 375 g C m^{-2}.

In Cochin Backwater, a tropical estuary in India, phytoplankton production far exceeded the rate of production by zooplankton herbivores.[747] Daily production ranged from 0.2 to 1.5 g C m^{-2}, and the annual net production was 124 g C m^{-2} year^{-1}. During the monsoon season when salinity was in the 10 to 20‰ range, peak gross production was 280 g C m^{-2} year^{-1}. Consumption by zooplankton was only 30 g C m^{-2} year^{-1}. Thus, most of the production adds to the detritus in the system.[748]

In the Paramatta Estuary and Sydney Harbour in Australia[761] primary production, chlorophyll *a*, and cell counts increased dramatically towards the upper reaches (Figure 29). In the inner estuary (salinity range 29 to 33‰), values as high as 175 mg C m^{-3} hr^{-1} and 19 mg chlorophyll *a* m^{-3} have been recorded with means of 127 g C m^{-3} hr^{-1} and 16 mg chlorophyll *a* m^{-3}. In the middle estuary (salinity range 33 to 34‰) the mean primary production was 54 g C m^{-2} hr^{-1} and chlorophyll *a* was 8 mg m^{-3} while in the lower estuary (salinity range 34 to 35‰) the comparative figures were 11 g C m^{-2} hr^{-1} and 1.8 mg m^{-3}. Nannoflagellates increasingly dominated the phytoplankton communities towards the inner estuary. Cell densities in excess of 110×10^6 cell ℓ^{-1} were recorded in the upper estuary. Nannoplankton/microplankton ratios increased from 1 in the outer estuary to ratios in excess of 4.2 in the inner estuary. Shannon-Weaver diversity indices decreased systematically in the estuary from 4.0 near the mouth to 2.4 near the head and were inversely related to community biomass.

Measurements of primary productivity in the hypersaline Fafa Lagoon on the Natal coast in South Africa were made using both the oxygen method and the ^{14}C method.[665] Values obtained with the ^{14}C method ranged from 0.32 to 7.50 mg C m^{-3} hr^{-1} while those obtained with the oxygen method ranged from 1.65 to 29.93 mg C m^{-3} hr^{-1}.

3. Factors Regulating Estuarine Phytoplankton Production

Boynton et al.[65] have developed a conceptual model (Figure 30) of the factors influencing phytoplankton production. Inputs of energy and materials or physical characteristics (morphology) common to all estuarine sources are shown as circles. Rectangles represent the mechanisms through which the inputs affect primary production, for example, turbidity is shown as influencing primary production and it in turn is caused by algal biomass and sediment from both external (riverine and others) and internal (resuspension due to winds and tides) sources. Data were collected for 63 estuarine systems covering latitude, insolation, temperature, extinction-coefficient, mean depth, stratification depth, mixed depth, critical depth, surface area, drainage area, fresh water input, tidal range, salinity, nutrient concentrations, nutrient loading rate, chlorophyll *a* concentration, and phytoplankton production rate.[454] These data were analyzed statistically to test hypotheses and factors regulating temporal patterns.

Prior to the statistical analysis[65] the estuarine systems were classified into four groups (fjord, lagoon, embayment, or river dominated). *Fjords* were defined as having a shallow sill and deep basin waters with slow exchange with adjacent sea waters.[742] *Lagoons* were taken as those systems which are shallow, well-mixed, slowly flushed, and only slightly influenced by riverine inputs. *Embayments* were considered to be deeper than lagoons, often stratified, only slightly influenced by fresh water inputs, and having good exchange with the ocean. The category *river dominated* contains a more diverse group of systems, but all members of the group are characterized by seasonally depressed salinities due to riverine inputs and variable degrees of stratification. This subjective analysis was tested with the aid of a step-wise discriminate analysis. The distribution of the groupings in discriminate space agreed reasonably well with the classification criteria (Figure 31).

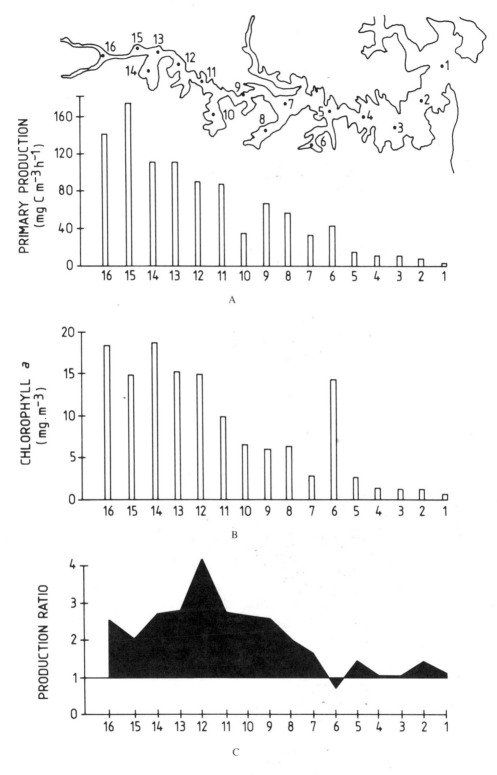

FIGURE 29.   (A) Primary production rates; (B) chlorophyll *a* standing crop; and (C) nannoplankton/microplankton production ratios in the Parramatta Estuary and Sydney Harbour. (After Relevante, N. and Gilmartin, M., *Aust. J. Mar. Freshwater Res.*, 29, 9, 1978. With permission.)

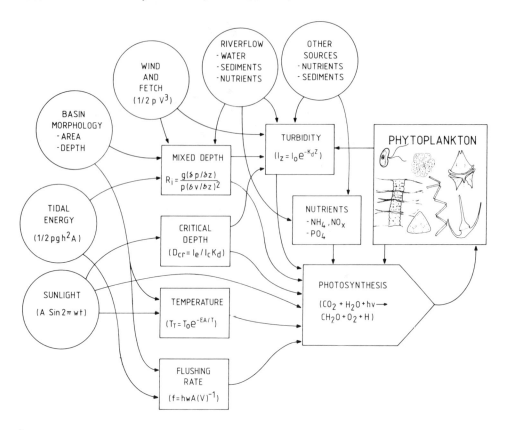

FIGURE 30.   A conceptual model indicating sources of materials and mechanisms influencing phytoplankton production rate in estuarine systems. (After Boynton, W. R., Kemp, W. M., and Keefe, C. W., *Estuarine Comparisons,* Kennedy, V. S., Ed., Academic Press, New York, 1982, 71. With permission.)

a. Production, Biomass, and N:P Ratios

Average daily phytoplankton production for 45 estuarine systems are shown in Figure 32. Average seasonal rates ranged from near 0 to 2.5 g C m^{-2} day^{-1}. The mean annual average rate was 0.52 g C m^{-2} day (190 g C m^{-2} $year^{-1}$), a value intermediate between the 100 g C m^{-2} $year^{-1}$ reported by Ryther[803] for coastal areas and 300 g C m^{-2} $year^{-1}$ estimated for upwelling areas. Despite the large range in fresh water input, physical morphology, insolation, and other factors, maximum production rates always occurred during the warm periods of the year while minimum rates generally occurred during the winter. According to Boynton et al.[65] this pattern has been interpreted to indicate that temperature-regulated metabolism strongly influences nutrient recycling processes and planktonic growth rates which are in turn important factors maintaining high photosynthetic rates.[218,252,627,1017]

Annual means for river-dominated estuaries, embayments, lagoons, and fjords were 0.58 ± 0.37, 0.36 ± 0.23, 0.49 ± 0.23, and 0.62 ± 0.53 g C m^{-2} day^{-1}, respectively. This indicates that estuaries with different physical characteristics commonly have comparable rates of primary production, suggesting that there are system-specific biotic and physical mechanisms operating in different estuaries.[65]

Chlorophyll *a* concentrations were highest during the warm season of the year and in general they paralleled the phytoplankton production values. Ratios of dissolved inorganic nitrogen (DIN = NO_3^- + NO_2^- + NH_4^+) to dissolved inorganic phosphorus (DIP) were low during periods of high production, except in highly eutrophic systems. The significance of this ratio lies in the fact that algal production is constrained by (among other things) the

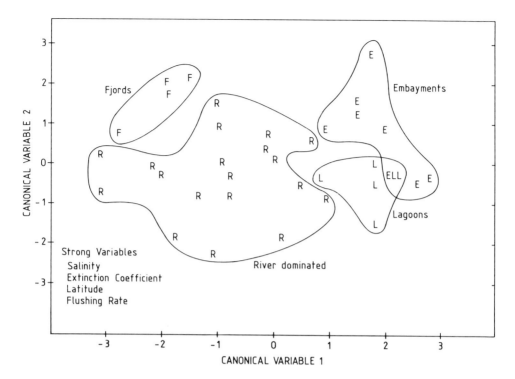

FIGURE 31. Results of a typical discriminate analysis where four types of estuarine systems are depicted in discriminate space. Solid lines enclosing each group are provided as a visual aid and do not necessarily represent statistically significant differences between groups. (From Boynton, W. R., Kemp, W. M., and Keefe, C. W., *Estuarine Comparisons,* Kennedy, V. S., Ed., Academic Press, New York, 1982, 73. With permission.)

requirement for nitrogen and phosphorus in proportions (atomic) of 16:1, respectively.[760] Water column nutrient concentrations of DIN:DIP less than the ''Redfield Ratio'' would indicate that nitrogen is less abundant than phosphorus in terms of phytoplankton demand, while values in excess of the ratio indicates that phosphorus is less abundant.

Data on seasonal mean N:P ratios from 28 estuarine ecosystems show that the ratios are less than the algal concentration ratios of 10 to 20:1. At peak production nitrogen is consistently less abundant than phosphorus in nearly all systems. Exceptions are those systems that are heavily enriched by diffuse or point sources of nutrient enrichment such as sewage inflows. Actual concentrations vary considerably between various estuaries. In addition there may be substantial annual excursions in the N:P ratios, particularly in the river-dominated group. In Chesapeake Bay during periods of high river flow ratios > 60:1 have been recorded and these were due to very high concentrations of nitrogen rather than low concentrations of phosphorus.

Boynton et al.[65] found that in general phytoplankton production and biomass exhibited weak correlations with a variety of physical and state variables in the estuarine systems they examined. They concluded that this perhaps indicated the significance of rate processes as opposed to standing stocks in regulating phytoplankton production. They also examined the temporal patterns of primary production in one large estuary, mid-Chesapeake Bay. Between January 1972 and December 1977, Mihursky et al.[593] measured phytoplankton production at six stations in this bay (Figure 33). Over the 6-year period there was considerable variation in estimates of integrated annual production, the peak year (1973) being some 2.3 times that of the least productive year (1977). Furthermore the annual peak productivity varied by a factor of 3.6 over the 6-year period. While the amplitude of peak production was

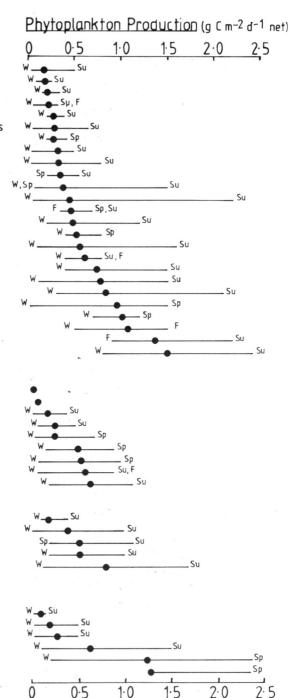

Phytoplankton Production (g C m⁻² d⁻¹ net)

$$\text{Phytoplankton Production (g C m}^{-2}\text{ d}^{-1}\text{ net)}$$

RIVER DOMINATED

- St. Lawrence River, Canada
- Upper San Francisco Bay, Calif.
- Fraser River, British Columbia
- Upper Patuxent River, Maryland
- St. of Georgia, British Columbia
- Western Wadden Sea, Netherlands
- Swartvlet, South Africa
- Waccasassa River, Florida
- Eastern Wadden Sea, Netherlands
- Meyers Creek, New Jersey
- Upper Chesapeke Bay, Maryland
- Hudson River, New York
- Long Island Sound, New York
- Duwamish River, Washington
- Cochin Backwater, India
- Barataria Bay, Louisiana
- Lower San Francisco Bay, Calif.
- Mid-Patuxent River, Maryland
- Raritan Bay, New Jersey
- Narragansett Bay, Rhode Island
- Buzzard Inlet, British Columbia
- Apalachicola Bay, Florida
- Mid-Chesapeake Bay, Maryland
- Pamlico River, North Carolina
- Altamaha River Mouth, Georgia

EMBAYMENTS

- Central Kaneohe Bay, Hawaii
- S. East Kaneohe Bay, Hawaii
- Roskeeda Bay, Ireland
- Funka Bay, Japan
- Loch Ewe, Scotland
- St. Margarets Bay, Nova Scotia
- Sheepscott River, Maine
- Bedford Basin, Nova Scotia
- Port Hacking Basin, Australia

LAGOONS

- Beaufort Sound, North Carolina
- High Venice Lagoon, Italy
- Chincoteague Bay, Maryland
- Peconic Bay, New York
- North Inlet, South Carolina

FJORDS

- Baltic Sea
- Loch Etive, Scotland
- Kungsbacka Fjord, Sweden
- Byfjord, Sweden
- Indian River, British Columbia
- Puget Sound, Washington

FIGURE 32.   Summary of average daily phytoplankton rates (solid circle) in 45 estuarine systems. Horizontal bars indicate annual ranges. Season in which maximum and minimum rates occurred is also indicated (W, winter; Sp, spring; Su, summer; F, fall). (From Boynton, W. R., Kemp, W. M., and Keefe, C. W., *Estuarine Comparisons*, Kennedy, V. S., Ed., Academic Press, New York, 1982, 75. With permission.)

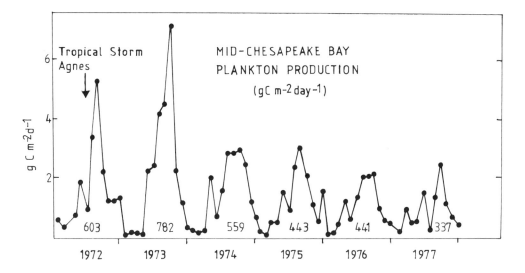

FIGURE 33. Mean monthly phytoplankton production rates at six stations in central Chesapeake Bay, 1972 to 1977. Values below peaks represent estimates of annual phytoplankton production (g C m^{-2} year^{-1}). (From Boynton, W. R., Kemp, W. M., and Keefe, C. W., *Estuarine Comparisons*, Kennedy, V. S., Ed., Academic Press, New York, 1982, 81. With permission; based on data from Reference 593.)

different in each year, the timing, shape, and frequency of events was nearly identical among the six annual sequences.

b. Responses to Nutrient Inputs

Patterns evident in Figure 33 suggest the possibility that changes in annual winter-borne nutrient load may explain some of the yearly variation in phytoplankton production. Boynton et al.[65] note that nutrient input as a result of a major storm (Agnes) in June 1972 resulted in inputs of nitrogen 2 to 3 times higher than in preceding or subsequent years. While production in that year was high, peak production occurred in the following year when nutrient inputs were more typical.

While most of the annual nutrient load to Chesapeake Bay occurs during the late winter and early spring,[342] maximum productivity rates are not apparent until much later in the year after high nutrient concentrations associated with riverine sources have decreased to low levels. Boynton et al.[65] suggest that high summer rates of production appear to be supported primarily on recycled nutrients (most importantly nitrogen); some fraction of which is introduced to the estuary during the spring runoff period. In 14 estuarine systems where N and P loading-rate data were available, there was a reasonably good relationship between N loading ($r^2 = 0.60$) and annual production but not for P loading ($r^2 = 0.08$). Based on these findings Boynton et al.[65] concluded ''We are led to believe that nitrogen rather than phosphorus dynamics are of more central importance in regulating phytoplankton production in estuarine systems and that rate processes, such as nutrient loading rates and recycling rates, seem to be more useful than standing stock values for predicting productivity.''

They point out that some of the residual variability in the nutrient loading-phytoplankton production relationship may be accounted for by including variables such as insolation, euphotic depth, flushing rate, and nutrient-recycling rates. They also stress the need for the development of carefully designed decade-long monitoring programs as a tool for addressing questions such as year-to-year variability responses to catastrophic events (e.g., hurricanes), and eutrophication tendencies of the phytoplankton component of estuarine ecosystems.

IX. RELATIVE CONTRIBUTIONS OF THE VARIOUS PRODUCERS

In Table 25 information is listed from five estuarine studies where data are available for the major primary producers. The different systems vary widely in the relative contributions of the primary producers. Phytoplankton contribution varies from 43.3% in Beaufort to 2.2% in Flax Pond. Epiphytes have only been considered in two studies where they are estimated to contribute 8.5% (Beaufort) and 3.7% (Flax Pond), respectively. In Flax Pond fucoid macroalgae contribute 20.5%, but macroalgae are only a minor component in two other estuaries. Epibenthic algae are significant contributors in the Nanaimo River estuary (39.5%) and in the Upper Waitemata Harbor (35.9%); in two other studies they contribute only 5.6% (Flax Pond) and 10% (Sapelo Island). While *Spartina* contributes only 9.8% in Beaufort, the emergent vascular plants are the major contributors with 70.7% *(Spartina)* in Flax Pond, 84% *(Spartina)* in Sapelo Island and 31.4% *(Avicennia*—mangrove) in the Upper Waitemata Harbour. In the Nanaimo River estuary, *Zostera* contributes 42.1%.

The major sources of carbon available to consumer species in the Nanaimo Estuary intertidal area are shown in Figure 34. River input is the major contributor. Phytoplankton and macroalgae are relatively minor contributors. Epibenthic algae (benthic microalgae) and the eelgrass *Zostera* are about equal in productivity. While the *Carex* production is high the marsh is poorly flooded, and Naimen and Sibert[617] assumed that little, if any, of the production was exported to the intertidal flats.

It can be seen from the data in Table 25 that each system has its own characteristics dependent on the mix of primary producers. In some of the systems not all of the potential contributors have been taken into account, for example, in Beaufort, North Carolina, the benthic microalgae have not been estimated. If they had been included the percentage contributions would be rather different.

Table 25
NET PRIMARY PRODUCTION OF PARTICULATE MATERIAL AND DISSOLVED ORGANIC MATTER IN VARIOUS ESTUARIES

Estuary	Source of production	Production (g C m^{-2} year^{-1}) Particulate	Dissolved	Production (g C m^{-2} of surface water year^{-1}) Particulate	Dissolved	Ref.
Beaufort, North	Phytoplankton	66.0	3.3	66.0 (43.3%)	3.3	706
Carolina	*Zostera marina*	330.0	5.0	58.0 (38.0%)	0.9	
	Epiphytes	73.0	1.5	13.0 (8.5%)	0.3	
	Spartina alterniflora	249.0	62.3	15.0 (9.8%)	3.6	
	Total			152.6	8.1	
Flax Pond, New	Phytoplankton			11.7 (2.2%)	n.e	1041
York	Epiphytes			20.0 (3.7%)	n.e	
	Epibenthic algae			30.0 (5.6%)	n.e	
	Fucoid algae			75.0 (20.5%)	n.e	
	Spartina					
	Aboveground			292.0 (54.5%)	n.e.	
	Belowground			108.0 (20.2%)	n.e.	
	Total			535.0		
Sapelo Island,	Phytoplankton	375.0		79 (6%)		732
Georgia[a]	Epibenthic algae	190.0		150 (10%)		
	Aboveground					
	Spartina alterniflora	55.2		608 (42%)		
	Juncus roemenianus	880.0				
	Other species	15404.0				
	Belowground					
	Spartina alterniflora	770.0		608 (42%)		
	Juncus roemenianus	1340.0				
	Other species	1320.8				
	Total			1445		
Nanaimo River es-	Phytoplankton	7.5	n.e	7.5 (11.8%)	n.e	617
tuary, British	Epibenthic algae	25.1	n.e	25.1 (39.5%)	n.e	
Columbia[b]	Macroalgae	4.2	n.e	4.2 (6.6%)	n.e	
	Zostera	147.7	n.e	26.8 (42.1%)	n.e	
	Carex	564.0	n.e	0	n.e	
	Total			63.6		
Barataria Bay,	Phytoplankton			167.2	n.e	184
Louisiana	Epibenthic algae			195.2	n.e	
	Epibenthic algae			10.3	n.e	
	Spartina alterniflora			607.2	n.e	
	Total			879.9		
Upper Waitemata	Phytoplankton	120	20	120.0 (25.3%)	20.0	478
Harbour, New	Epibenthic algae	145.6	24.4	145.6 (30.8%)	24.4	
Zealand	Macroalgae	1.4	0.14	1.4 (0.3%)	0.14	
	Mangrove *(Avicennia)*	638.75	—	148.49 (31.4%)		
	Total			473.49	44.54	

Note: Data expressed both in terms of g Cm^{-2} year^{-1} for the areas covered by the producers and in terms of the total high tide area of the estuary.

[a] Salt marsh plants other than *S. alterniflora* over a small area only and this is not accounted for in the production budget. Production of DOM not directly measured.

[b] Naimen and Sibert[617] assumed that there was no export of *Carex* primary production to the estuarine mud flats.

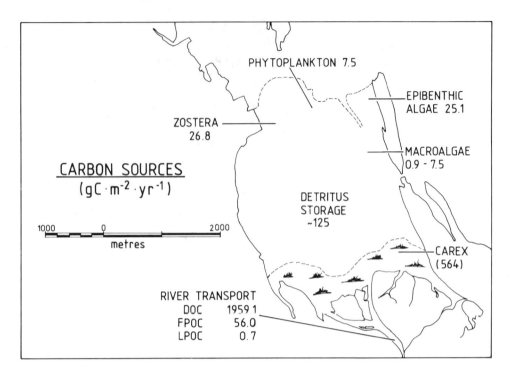

FIGURE 34.   Sources of carbon (g C m^{-2} year^{-1}) available to the Nanaimo Estuary intertidal mud flats. (From Naimen, R. J. and Sibert, J. R., *J. Fish. Res. Board Can.*, 36, 517, 1979. With permission.)

Chapter 4

DETRITUS AND THE ROLE OF MICROORGANISMS

I. INTRODUCTION

Estuaries are broadly open systems exchanging nutrients, producers, and consumers with adjacent fresh water and salt water areas, as well as with adjacent marshes and swamps.[161] The inherent feature of the community metabolism is the role played by organic detritus in its various forms. Current views on the ecology of estuaries, especially those fringed with salt marshes or mangroves, emphasize the importance of the decomposer-based food web. As Newell[624] states: ''It has become increasingly apparent during the past decade that although phytoplankton is the basis for production in oceanic waters, food chains in the intertidal zone and shallow coastal waters are based primarily on the production of attached plants.'' These may be utilized directly in the form of the diatoms of the benthic microalgae, small epiphytes attached to sand grains, or epiphytic microalgae growing on macrophytes. Grazing herbivores, however, are thought to ingest only a small fraction of the annual macrophyte crop, while the remainder, as we have seen in the previous section, decays to form dissolved and particulate organic matter.

The dissolved organic matter together with that released from the phytoplankton, epiphytic microalgae, and macrophytes may then be absorbed either directly by animals, or serve together with the particulate material as a substrate for heterotrophic metabolism. Bacteria and fungi rapidly colonize the particulate detritus as it becomes available and through their activities convert the high-fiber, low-nitrogen, dead plant material into variously sized particles of high nutritional value. These microbe-laden particles are either incorporated into the marsh surface sediments, or transported by tidal action to the adjacent estuarine waters and near-shore environments where they become available to filter-feeding organisms. Some settle to the bottom, where together with those in the marsh sediments, they are consumed by surface and sediment deposit feeders.

It is generally accepted that consumers receive most of their nutrition from the attached microorganisms and not directly from the decomposing plant particles. Egestion of the stripped detrital particles returns them to the aquatic environment where they are recolonized and the process repeated over and over again. Abrasion, maceration by consumers, and the continued action of decomposers results in smaller and smaller fragments. At some point in the sequence the most minute of the fractions are considered to become unidentifiable and are often referred to simply as amorphous aggregates or organic material of undetermined origin.[786]

Heterotrophic microorganisms are the major link in the mineralization and transformation of organic matter thus recycling inorganic nutrients which then become available to the primary producers. Two groups of microorganisms are involved, aerobic and anaerobic (Figure 1). The left-hand side of the figure is divided into aerobic and anaerobic environments on the basis of the 'oxidizing potential' or Eh. Aerobic environments are the water column and the surface sediments (see Chapter 1, Section II.B), while the anaerobic environment comprises the sediments below the RPD (redox potential discontinuity) layer. The important aspect of the anaerobic zone is that it represents a storehouse of chemical energy which may have some ability to feed back into the food chain.[864,868] In such zones different groups of bacteria exist which can decompose organic material by using sulfate and nitrate as a source of oxygen, and in the process form reduced substances such as CH_4, H_2S, and NH_4. The latter can be utilized by other bacteria, some of which are strictly chemoautotrophs since they can use CO_2 as a source of carbon, and inorganic compounds as a source of energy.

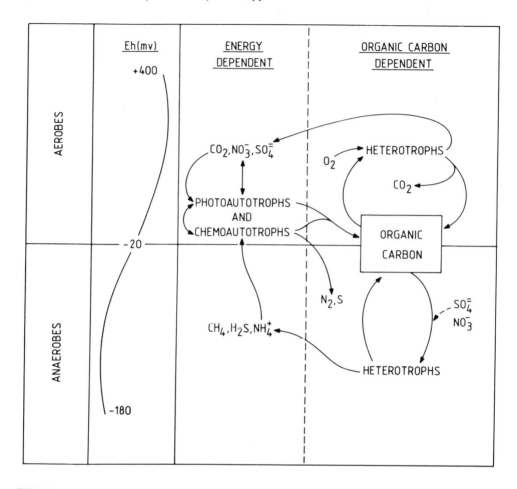

FIGURE 1.   Organic carbon and energy-dependent cycles in marine aerobic and anaerobic environments. (Redrawn with permission from Parsons, T. R. et al., 1977, *Biological Oceanographic Processes,* 2nd ed., copyright 1977, Pergamon Press.)

The top of Figure 1 is divided into energy-dependent reactions or organic-carbon dependent reactions. Processes can either be described in energy units, or quantities of organic carbon.

II. PARTICULATE ORGANIC MATTER (DETRITUS)

Darnell[160] defines organic detritus (POM) broadly as "all types of biogenic material in various stages of microbial decomposition which represent potential energy sources for consumer species." Thus defined, organic detritus includes all dead organisms as well as secretions, regurgitations, and egestions of living organisms, together with all the subsequent products of decomposition which still represent potential sources of energy such as proteins or amino acids. Organic detritus consists of a range of particles from very large to very small and it is often convenient to distinguish coarse POM (that material retained by filters with apertures of 1 μm in diameter) and fine or subparticulate (sometimes called nanno-detritus) POM (material which passes through such filters).

As we have seen, the great bulk of POM is derived from primary production and comes from two sources:[161]

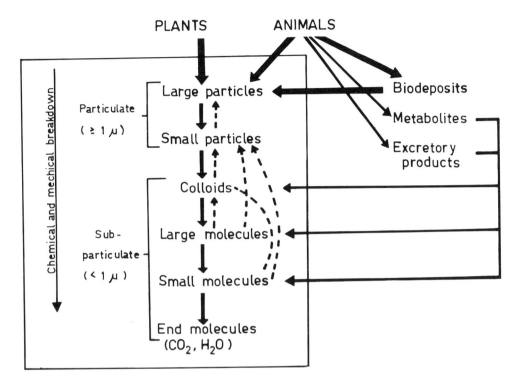

FIGURE 2.   Schematic view of the detritus formation and decomposition. (After Darnell, R. M., *Estuaries*, Publ. No. 83, Lauff, G. H., Ed., American Association for the Advancement of Science, Washington, D.C., 1967b, 377. With permission; copyright 1967, AAAS.)

1. Autochthonous sources

    a. Plankton (including phytoplankton and bacteria)
    b. Marginal emergent vegetation (salt-marsh plants and mangroves)
    c. Submerged macrophytes (seagrasses)
    d. Macroalgae
    e. Periphyton-epiphytic algae growing on the stems and leaves of emergent and submerged macrophytes and other surfaces
    f. Benthic microalgae (diatoms, flagellates, and blue-green algae)
    g. Sediment bacteria

2. Allochthonous sources

    a. Fresh water swamp vegetation
    b. River-borne phytoplankton and organic debris, including sewage
    c. Beach and shore material washed in by the tides and particularly during storms
    d. Windblown terrigenous material, especially leaves and pollen grains

As previously mentioned, POM can be thought of as particulate and subparticulate (Figure 2), the former being composed of large (the origin being recognizable) and small particles (the origin being generally unrecognizable). Subparticulate POM includes colloidal micelles as well as chemically reduced organic molecules. Colloids may include molecular aggregates or large molecules, such as proteins, carbohydrates, lipids, etc. Smaller molecules may exist

Table 1
SEDIMENT PARTICULATE ORGANIC CARBON MEASUREMENTS
FROM SELECTED ESTUARIES

Locality	mg C g^{-1}	Depth of sediment	Ref.
Salt marshes			
Providence River, Rhode Island (3 marshes)	52	Top 5 cm	630
Narragansett Bay, Rhode Island (7 marshes)	41	Top 5 cm	630
North Carolina (6 marshes)	16	Top 15 cm	76
Barataria Bay, Louisiana	80—160[a]	—	1069
Lewes, Delaware	40—240	—	538
Estuarine sediments			
Narragansett Bay, Rhode Island (4 stations)	11	—	828
Central Long Island, New York	40—70	Top 10 cm	8

[a] Annual range, referring to the range measured over a year.

as dissolved liquids (bichromes, vitamins, amino acids, sugars, urea, nitrites, nitrates, etc.), or as dissolved gases (methane, ammonia, hydrogen, sulfide, etc.). Thus there is a continuous gradient from small molecules to large organic particles, most of which represent a potential energy source.

Figure 2 depicts a generalized scheme of biological decomposition and organic detritus formation. Biological decomposition involves both mechanical and chemical breakdown. The latter is brought about primarily through the processes of hydrolysis and oxidation. According to Darnell[161] three agents are likely to be involved: autolysis, where there is breakdown of tissues by their own enzymes; chemical effects of the passage through the guts of consumers; and the chemical activities of heterotrophic microorganisms (bacteria and fungi). A number of factors also tends to increase particle size. Colloidal micelles tend to adhere by agglomeration and coacervation, forming larger dispersed particles.

A. Quantities of Particulate Organic Matter (POM)

There are relatively few available measurements of POM in estuarine waters. Measured concentrations of POM in suspension in marine areas ranges from 0.01 to 2.0 g C m^{-2} with the highest concentrations occurring inshore.[576] Higher values have been recorded in estuaries. In the Sapelo Island marsh area in Georgia, the annual recorded average is 9 g C m^{-2}.[1008] In most estuaries much of the input of dead plant material sediments fairly rapidly to the bottom and becomes incorporated in the sediments, where much higher levels have been recorded than those found in the water. Estimates of sediment particulate organic carbon vary widely ranging from 100 g C m^{-2} for a New England salt marsh to 18,200 g C m^{-2} for the Sapelo Island marshes. Other data available are expressed in terms of mg C g^{-1} (dry weight) of sediment and are not readily comparable. Nixon[626] (Table 1) lists values ranging from 16 mg C g^{-1} for six North Carolina marshes[76] to 240 mg C g^{-1} for a marsh at Lewes, Delaware.[538] Nixon[626] concluded that marsh sediments were richer in organic carbon than those found in the sediments of open estuarine areas and offshore zones. The distribution of organic carbon appears to vary considerably with depth in marshes (Figure 3)[949] and in subtidal sediments.[828,884] On the basis of available data it appears that much of the organic carbon buried in estuarine sediments is decomposed. However, a small amount of the more refractory carbon is effectively trapped within the sediment and removed from the estuarine system. The amount of carbon so removed is a function of the carbon content of the deep sediment, the accretion rate, and the sediment density. Nixon[626] concluded that the quantity buried appears to lie between 50 and 400 g C m^{-2} year^{-1}. He further calculated that the

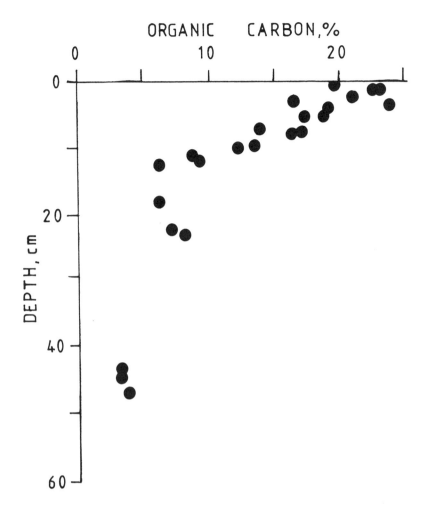

FIGURE 3. Distribution of organic carbon with depth in the sediment of a stunted *Spartina alterniflora* marsh in Delaware. (From Nixon, S. W., *Estuarine and Wetland Processes*, Hamilton, P. and MacDonald, K., Eds., Plenum Press, New York, 1980, 455. With permission.)

annual marsh sediment input is likely to be somewhere between 750 to 1500 g m^{-2}, and that the organic carbon content of this material is at least 50 mg C g^{-1}, giving an organic carbon input associated with sediment trapping of 37 to 75 g C m^{-2} year^{-1}. He also calculated that of 500 g C m^{-2} year^{-1} fixed on a marsh in plant biomass, 150 g (30%) would be buried.

B. River Input of Organic Carbon

River transport can be an important source of carbon to estuaries.[379,616,694,883,901,902] Naimen and Sibert[616] examined the timing and magnitude of carbon movements from the Nanaimo River in British Columbia to the mud flats of the estuary. Dissolved organic carbon concentrations in the river ranged from 6 to 14 g m^{-3} dependent on the season with a mean of 6.44 g m^{-3}. When compared to that in many streams and rivers the concentrations of organic carbon in the Nanaimo River is not especially large, but due to the large annual discharge, the river contributed 76.5% of the total carbon input to the estuary. Most of the annual carbon input occurs during autumn freshlets. From October to December the intertidal mud flat received 70% of the annual dissolved organic carbon (DOC: < 0.5 μm) river export, 73% of the annual fine particulate organic carbon (FPOC: > 0.5 μm, < 1 mm) export, and

93% of the annual large particulate organic carbon (LPOC: > 1 mm) export. DOC contributes by far the greatest amount of carbon each year, nearly 2000 g C m^{-2}. FPOC contributes 56 g C m^{-2} year^{-1}, while LPOC contributes only 0.7 g C m^{-2} year^{-1}. The total POM concentrations were only 2.8% of the DOM concentrations. On the other hand Quasim and Sankavanayanan[749] found that POM derived from terrestrial and aquatic macrophytes constituted the greater part of the POC in suspension in a tropical estuary, with the contribution from the phytoplankton being less than 1%.

There are several primary sources of DOC in rivers: leaves which leach once they fall into the river, periphyton which release significant amounts of DOC, forest soils from which carbon is leached, and atmospheric input via rain. POM in river inflow is derived from leaves and other plant litter washed or blown into the rivers, organic matter derived from marginal macrophyte vegetation or submerged macrophytes, and river phytoplankton. The input of organic matter from these sources is dependent upon the amount of fresh water inflow in relation to the tidal volume of the estuary, the nature of the catchment, and the type of river. It is clear that in some estuaries it is the major input of organic carbon, e.g., in the Nanaimo River estuary it represents 75% of the carbon input.

As colloidal and particulate matter carried by fresh waters enters the saline environments the surface charge on the particles approaches zero[619] and the particles begin to coagulate.[216,489] As a result all of the particulate matter sediments within the estuary.[349,808] This process is effective even for such major rivers as the Mississippi and the Amazon.[596]

Recent studies[434,786] have demonstrated that the bulk of organic matter in the open waters of estuaries consists of amorphous aggregates not identifiable as decomposition products of macrophytes such as marsh grasses and mangroves. Odum and de la Cruz[639] separated suspended particles in a Georgia marsh into three size fractions: "coarse detritus", retained by a netting of 0.239-μm apertures; "fine detritus", passing through the 0.239-μm net but retained by a net of 0.064 μm; and "nannodetritus" passing through the 0.064-μm aperture but retained by a millipore filter with 0.045-μm pore size. The coarse and fine fractions, comprising 1 and 4% of the total particulates, respectively, were identified as fragments of vascular plants, while the nanofraction (95% of the total) comprised the amorphous aggregates mentioned above.

Ribelin and Collier[786] have drawn attention to the common phenomenon of floating surface films in estuaries and hypothesized that they originated from material produced by benthic microalgae, both diatoms and filamentous blue-green algae. These films are lifted from the surface of the intertidal sediments by the flood tides. The organic material in the surface films could also arise from the DOM released by phytoplankton or macrophytes or during the breakdown of POM. These sources coupled with river inputs of DOM, contribute to the pool of fine particulate matter (nannodetritus) in the estuarine water via processes converting DOM to particulate form. As noted above, the colloidal material in river input flocculates and adsorbs dissolved material to form larger particles.[972] There are also a variety of other mechanisms that assist in the conversion process, all of them involving collection of the surface-active material at a gas-liquid interface and compression of the interface.[576] The surfaces of bubbles and the surface film of the water itself are sites of aggregation of surface-active molecules and breaking waves are excellent sites for producing the necessary force of compression.[312,582,791] There is thus a complex DOC-POC equilibrium which leads to the formation of amorphous organic aggregates. Paerl[682,683] has drawn attention to the fact that bacteria form large quantities of extracellular particulate materials that are used to bring about adhesion of bacteria into clumps and adhesion to substrates.[585] These materials can also contribute to the pool of nannodetritus. These processes, plus the fact that much of the labile excreted photosynthate, are utilized by microheterotrophs almost as rapidly as they are being produced,[1003] are the reason for the low standing stock of DOM in the water.

Thus, primary production in estuarine ecosystems produces large quantities of organic matter that is not utilized directly by herbivores but is converted into detritus which is decomposed into finer POM particles and DOM. While some of these breakdown processes occur within the water column, most occur in the surface sediments. One of the central questions concerning the functioning of estuarine-marsh ecosystems is whether the plant organic matter is largely processed *in situ*, or whether a major proportion is exported to the adjacent waters.[732] Another question which is basic to our understanding of the functioning of the ecosystem is the role of the detritus and its associated microbial community in the estuarine food web. Both of these questions will be discussed below.

C. Heterotrophic Utilization of Organic Matter in the Initial Biodegradation Process

As we have seen, most of the primary production within estuarine ecosystems (especially that of the macroalgae, marsh plants, seagrasses, and mangroves) is not utilized directly by herbivores but enters the detrital food chain. The most detailed studies on the initial phases of the biodegradation of dead plant material have been made on marine grasses, particularly *Spartina, Zostera,* and *Thalassia.* There is also a fair amount of data available for other plants and for debris at various stages of decomposition. Odum and de la Cruz[639] studied the rate of decomposition of a variety of salt marsh plants including *Juncus, Distichlis, Spartina, Salicornia,* and the fiddler crab *Uca.* They found that *Uca* decomposed completely after 180 days in the field, while the plants showed variable rates of breakdown; after 300 days, the residues expressed as a percentage of the initial dry weight of the original material were: *Salicornia* 6%, *Spartina* 42%, *Distichlis* 47%, and *Juncus* 65%. These residues were then carried off the marsh into tidal creek water where the organic particulate fraction ranged from 2 mg ℓ^{-1} at the midflood tide to 20 mg ℓ^{-1} at midebb and was found to comprise 93.9% of fragments of *Spartina alterniflora*, 4.8% of macrophytic algae, and only 1.3% of animal debris.

They also found that as the detritus derived from *Spartina* decayed progressively into smaller particles it became richer in protein (Figure 4). The small suspended particles were 70 to 80% ash but the organic portion comprised as much as 24% on an ash-free dry weight basis, compared with 10% in living *Spartina* and 6% in the dead grass as it entered the water. Oxygen consumption per gram was found to be five times as great in this 'nanno-detritus' as compared with coarse detritus. Odum and de la Cruz[639] suggested that the increase in protein represented a microbial population which was utilizing the carbohydrate component, while the crude fiber remained as a substrate for attachment, and that the increased protein of the decomposing particles rendered them a potentially richer food resource than the original *Spartina* grasses from which they were derived.

The results that Odum and de la Cruz[639] obtained have been amply confirmed by subsequent studies. Investigations of detritus formation in the turtlegrass *Thalassia testudinum*[83,229,231,473,1057] and the eelgrass *Zostera marina*[84,371,372,706] have included observations on the components of the microheterotrophic population. Fenchel[229] studied the microbial communities living on detrital particles derived from *Thalassia testudinum.* He found that the microbial community was a complex one whose population density and rate of oxygen consumption were similar in all fragments when expressed on a unit area basis. Field samples of detritus on average harbored 3×10^6 bacteria, 5×10^7 flagellates, 5×10^4 ciliates, and 2×10^7 diatoms, and consumed 0.7 to 1.4 mg O_2 g^{-1} hr^{-1}. The total surface area per gram of particulate debris depends on the diameter of the particles and is greater in finer particles than in large ones. Particles with a median diameter of 0.1 mm will have numbers of microorganisms about one order of magnitude higher than particles with a median diameter of 1 mm, because the microorganisms are found principally on the outer surface of the particles. Figure 5 shows the relationship between the number of microheterotrophs and particle diameter and also the oxygen consumption of the detritus fragments at 24°C in the dark.

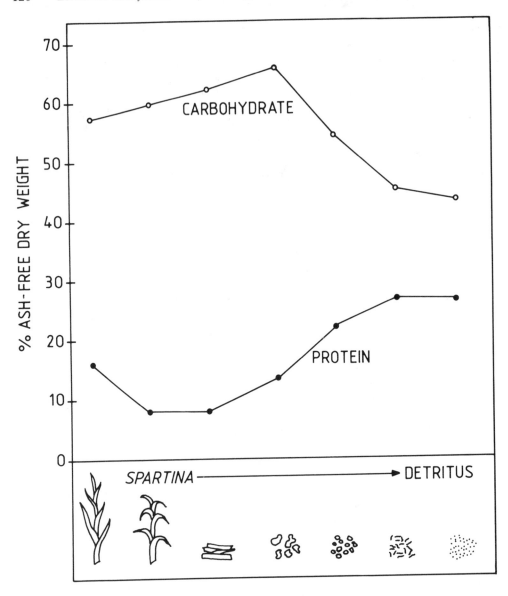

FIGURE 4. Diagram illustrating the successive stages in the disintegration and subsequent degradation of organic debris from *Spartina*. The increase in protein from 10% in the fresh plant to 22% in the detritus is associated with the establishment of a community of microorganisms on the surface of the finely divided organic debris. (From Odum, E. P. and de la Cruz, A. A., *Estuaries*, Publ. No. 83, Lauff, G. H., Ed., American Association for the Advancement of Science, Washington, D.C., 1967, 383. With permission; copyright 1967, AAAS.)

Harrison and Mann[371,372] have made detailed studies on the changes associated with detritus formation for the eelgrass *Zostera marina* which proved much more resistant to decay than *Spartina*. They found that detritus formation from dead leaves began with fragmentation, leaching of the water soluble components, and autolysis of the cell contents by actions of the enzymes of the plants. This was followed by the establishment of a microbial community. Starting with dried ground plant material they found that in the laboratory no more than 20% of organic matter was lost after incubation for 100 days at 20°C, with ample micro-organisms and nutrients provided. Leaching was of major importance in the initial stages of decomposition, accounting for 73 to 92% of the loss in predried leaves and for 55 to 77% of leaves which were not dried prior to culture. As in the case of the biodegradation

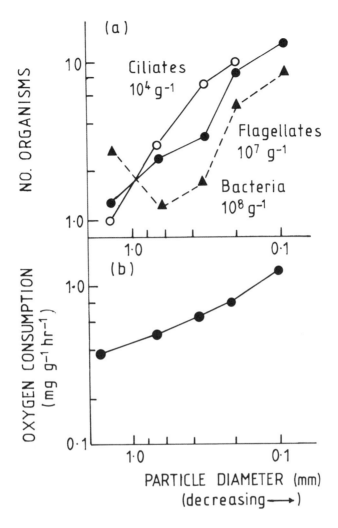

FIGURE 5. The relationship between particle size. (a) The numbers of organisms on their surfaces and (b) the oxygen uptake of the detritus. (From Mann, K. H., *The Ecology of Coastal Waters*, Blackwell Scientific, Oxford, 1982, 141. With permission.)

of other grasses, Harrison and Mann[372] found that there was an increase of from 12.5 to 50% in the particulate nitrogen component of the incubated detritus. They also found that the rate of loss of organic matter from the detritus was enhanced by the addition of micro-flagellates and ciliates in spite of the fact that their presence resulted in reduced bacterial populations.

In contrast to the laboratory experiments listed above other experiments have involved the placing of weighed samples of plant material in litter bags in the natural environment. The bags have a mesh of specified size which permits small particles to wash away after partial decomposition. Zieman[1057] found that *Thalassia* leaves in litter bags lost weight at a rate three times as fast as those incubated in experimental tanks. Figure 6 depicts the rates of loss for a range of emergent macrophytes. Losses were lowest in *Juncus* and highest in *Thalassia*. One of the factors influencing the rate of loss is the activity of any invertebrate 'shredders' which may gain entry to the bags and accelerate the reduction of particle size. Fenchel[229] showed that the amphipod *Parhyalella whelpleyi* greatly accelerated particle size reduction in *Thalassia*, and in his experiments almost doubled the rate of oxygen uptake of

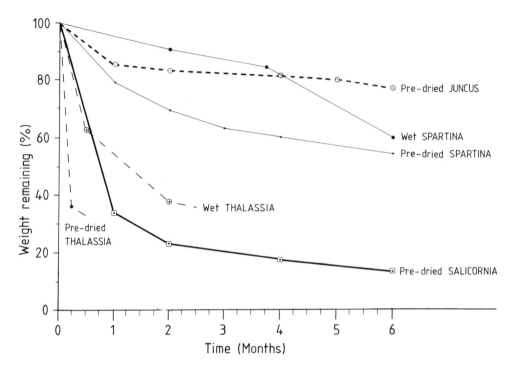

FIGURE 6.   Rates of disappearance of named marine macrophytes from litter bags of 2.5 mm mesh. (From Wood, J. E. F., Odum, W. E., and Zieman, J. C., in Lagunas Costeras, un Symposia, Symp. Int. Lagunas Costeras, UNAM-UNESCO, Mexico, 1967, 495. With permission.)

the detritus in 4 days due to the increase in microbial populations accompanying reduced particle size.

Woodroffe[1040] has studied *in situ* decomposition of mangrove *(Avicennia marina)* litter in Tuft Crater, Waitemata Harbour, New Zealand, using a litter bag technique. The leaves degraded rapidly at first losing approximately half their weight in 42 to 56 days and more slowly from then onwards. Within the initial period amphipods were observed to be abundant in the bags and they were an important agent in the breakdown of the leaves[62,380] along with the leaching of soluble components. Similar studies in other parts of the world have looked at the initial period of rapid weight loss. In Thailand, leaves of *Avicennia* decomposed much more rapidly, losing half their weight in 20 days.[62] At Roseville, Sydney, Australia, in winter, leaves of *Avicennia* were found to lose weight more slowly.[321]

One of the most interesting aspects of the breakdown of detritus concerns the demonstrated increase in nitrogen content. This increase must arise from a variety of sources.[472] Senescent leaves of *Zostera* and *Thalassia* contain 1.0[371] and 0.8%[232,472] nitrogen, respectively. Also the difference in nitrogen content between mature *Zostera* leaves prior to senescence and senescent leaves was found to be small,[371] suggesting tightly bound nitrogen compounds. Therefore, the bacteria, in order to decompose plant detritus, must assimilate inorganic nutrients from the water and in this way enrich the detritus with nitrogen.

Garber[278] used ^{15}N-labeled particulate organic matter obtained from cultured marine diatoms *(Skeletonema costatum)* to determine the fate of particulate organic nitrogen at the surface of 10 to 12 cm deep sediment cores. The time-course distribution of the tracer was determined in inorganic-N and organic-N compartments in the sediment and the free water. The data provided evidence that 5 to 10% of the nitrogen in freshly deposited POM was conveyed to the deeper layers of the sediment and that both particulate and dissolved forms of nitrogen were carried downward from the sediment surface. The experiments further

Table 2
ANNUAL INPUTS AND OUTPUTS (in mmols m⁻²
year ⁻¹) OF CARBON, NITROGEN, AND
PHOSPHORUS TO SEDIMENTS IN
NARRAGANSETT BAY, R.I.

	Carbon	Nitrogen	Phosphorus
Particulate inputs to benthos	13,071	1,728	189
Dissolved fluxes	11,250	1,532	170
Permanent burial	988	113	15
Macrofaunal production available to predators	833	83	4

From Kelly, J. R. and Nixon, S. W., *Mar. Ecol. Prog. Ser.*, 17, 157, 1984. With permission.

provided evidence that freshly remineralized NH_4^+ was the dominant form of nitrogen leaving the sediment. Observed rates of NH_4^+ release suggested that 10 to 50% of the NH_4^+ flux from the sediment was due to rapid nitrogen remineralization at the sediment-water interface. Rate constants determined for the decomposition of the labeled organic-N suggested that the 'half-life' of organic nitrogen at the surface of shallow coastal marine sediments is in the order of 1 to 2 months in the spring and 2 to 3 weeks in the fall.

The amounts of ^{15}N that were taken up by the benthic macrofauna (the bivalve *Yoldia limatula,* the polychaetes *Nepthys incisa* and *Maldanopsis* sp., and the gastropod *Nassarius trivittatus*) were determined on the isotopic composition of animals captured when the cores were sliced. Garber[278] concluded that a small fraction, perhaps a few percent, of the organic nitrogen conveyed to the sediment surface in the fallout of detritus is incorporated into benthic biomass, with the rest being rapidly remineralized at the sediment surface.

Kelly and Nixon,[456] in experimental laboratory microcosm experiments using intact sediment cores of a benthic community from 7 m in mid-Narragansett Bay, Rhode Island, have determined the influence of the rate of supply of organic matter on benthic metabolism and nutrient regeneration. Enrichment experiments of 3 to 4 months duration were carried out at 15°C using seston filtered from bay water. Replicate benthic microcosms were either 'starved', given regular (every 3 days) organic inputs, or exposed to large 'pulses' of organic matter equivalent to 3 to 5 months of metabolic loss. With particulate starvation, sediment oxygen uptake and inorganic nitrogen release rates decreased only slowly, approximately halving in 80 to 120 days. O_2 uptake, CO_2 release, and NH_4^+ release all increased immediately in response to pulse input, subsequently declining exponentially towards the original levels within 1 to 2 months. Rates increased gradually under regular organic additions and declined quickly when inputs were discontinued. Kelly and Nixon found that within 2 months about 24 to 30% of the organic nitrogen and 11 to 20% of the organic carbon experimentally deposited at the sediment surface had returned to the overlying water as dissolved inorganic decomposition products. They calculated that, for the *in situ* mid-Narragansett Bay sediment community, over 80% of the annual C and N deposition (Table 2) would be mineralized and returned to the overlying water.

Anderson and Hargrave[14] have recently studied the effect of *Spartina alterniflora* detritus on the metabolism of a muddy intertidal flat in the upper end of the Bay of Fundy, Nova Scotia. Aboveground, harvested, air dried, living biomass of *Spartina* was buried at a depth of 1.5 cm at the rate of 400 g m⁻². Total carbon dioxide release to the overlying water during the 3 months following detritus enrichment increased 5.9 times over the control area while oxygen uptake only increased 1.4 times. The ratio between carbon dioxide release

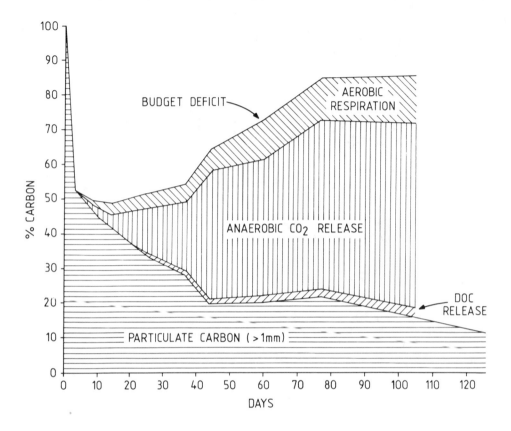

FIGURE 7.   Carbon budget for decomposition of *Spartina* detritus added to sediment. Particulate carbon (as recognizable pieces of *Spartina* >1 mm) remaining in the sediment, DOC released, anaerobic carbon dioxide released, aerobic respiration, and carbon deficit calculated by difference from the initial amount of carbon added to the sediment on basis of mean values derived from individual sampling dates. Cumulative values for DOC, anaerobic carbon dioxide, and oxygen flux were too small to be plotted on day 3. (From Anderson, F. Ø. and Hargarve, B. T., *Mar. Ecol. Prog. Ser.*, 16, 167, 1984. With permission.)

and oxygen uptake varied between 1 and 5 and was generally 1.6 to 3 times higher in the enriched area than in the control. Decomposition of the *Spartina* added to sediment was rapid with about 50% of the mass being lost during the initial leaching phase. Loss of ash-free dry weight from the particulate matter after the leaching period was exponential with a half-life of 44.6 days and a decay coefficient of 0.016. This was similar to values reported for *Spartina* enclosed in litter bags.[1082] However, other studies have shown considerably lower decomposition rates.[13,411,584,639]

A budget for organic carbon loss from the buried *Spartina* detritus was calculated from measurements of particulate carbon remaining over time, loss of DOC, loss of carbon dioxide, and aerobic respiration estimated from oxygen uptake (Figure 7). After 105 days of decomposition 84.4% of the original amount of particulate carbon (> 1 mm) had been lost. Of this, 67.3% had been metabolized to carbon dioxide and 2.8% was lost as DOC and as fine particulate matter (< 1 mm). Anderson and Hargrave[14] consider that a greater fraction would have been lost if carbon dioxide release had been underestimated due to chemoautotrophic fixation. The budget deficit was particularly high in the beginning and if chemoautotrophic fixation of carbon dioxide had been most rapid during the early stages of decomposition then the low rates of carbon dioxide observed early in the experiment might actually have been higher. Aerobic respiration estimated from oxygen uptake corrected for chemical oxidation accounted for 21% of the total carbon dioxide production. Thus, 79% of the carbon dioxide originating from *Spartina* during detritus decomposition over 3 months was produced by anaerobic metabolism. The importance of anaerobic decomposition after enrichment of the sediment with detritus was apparent in the development of a low Eh in the detrital layer.

The three experiments detailed above all demonstrate relatively rapid rates of the decomposition of organic matter and remineralization of nutrients in shallow coastal and estuarine sites. As we have seen, rates of input of organic matter in these situations are high as demonstrated by the high organic content of the sediments and the high rates of metabolic activity (as determined by rates of oxygen consumption) reported by many investigators.[631,689,690,855,856,913,1067] The remineralization of deposited organic detritus has been shown to make a significant contribution to the nutrients required to support primary production in the water column of these coastal systems.[458,627,1085]

III. DISSOLVED ORGANIC MATTER

There are two pools of DOM within estuarine ecosystems, the water column pool and the sediment pool. Most studies of carbon flow within ecosystems have largely neglected the contribution made by DOM. Two exceptions are the studies of the salt marshes in the Duplin River estuary at Sapelo Island, Georgia,[735] and of the Nanaimo River estuary, British Columbia.[617] Sources of DOM in estuarine systems include exudates from phytoplankton, benthic microalgae, epiphytes, macroalgae, submerged macrophytes, release of DOM from the microbial breakdown of particulate organic matter, and river input.

A. Sources of Dissolved Organic Matter (DOM)

Excretion of DOM — Excretion of a portion of the photoassimilated carbon from algae and estuarine macrophytes is now generally accepted although the quantities and rates of excretion are still debated.[105,253,390,463,987] Fogg et al.[254] found that 0 to 50% of photoassimilated carbon in microalgae was released during *in situ* studies conducted in marine and fresh waters. Since that time Thomas,[934] Choi,[113] and Berman and Holm-Hansen[49] have recorded amounts within this range in marine phytoplankton. Hellebust[389] estimated that 10% of the photosynthate of phytoplankton is excreted as DOC and this is a figure that is often used in calculating carbon budgets.

Epibenthic algae are another potential source of DOM. Their excretion has proved difficult to measure since the sediment bacteria would be expected to assimilate any labile organic material nearly as fast as it is released.[732] Darley et al.[157] suggest that the amount lost is very small, approximately 1% of total fixation, but this represents only the nonassimilated excreted matter that remains in the sediment after 15 min.

Khailov and Burlakova,[463] Siebruth,[834] and Moebus and Johnson[600] reported 23 to 40% of photoassimilated carbon excreted from various macroalgae. These results contrast with those of Majock et al.,[572] which indicated that only 0 to 4.4% of photoassimilated carbon was lost from a group of red and green algae.

It has also been demonstrated that estuarine macrophytes lose DOM to the water. *In situ* rates of excretion from *Najas flexis* and *Scirpus subterminalis* were found to be 5.4% of the nonphotoassimilated carbon on an annual basis.[408,990] Brylinsky[78] demonstrated a loss of 1 to 3% of photoassimilated carbon in several species of tropical seagrasses. As mentioned previously, it has been demonstrated that *Spartina alterniflora* can release 0.4 to 6.1 g C m^{-2} $year^{-1}$.

A dense complex epiphytic microbial community is associated with the submerged parts of macrophytes. A variety of nutrient interactions involving carbon has been demonstrated between the plants and the epiphytic community.[7,366,568] The microalgae in this community are another potential source of DOM.

As we have seen, excretion of photoassimilated organics increases as macrophytes senesce,

Table 3
DOC CARBON EXPORTED FROM MARSHES EXPRESSED AS A PERCENTAGE OF AERIAL NET *SPARTINA* PRODUCTION

Marsh location	DOC export (g m^{-2} year^{-1})	Aerial net *Spartina* production (g C m^{-2} year^{-1})[a]	Precent of production exported as DOC
New York	8.4[b]	372.2	2.3
Delaware	38[c]	252	15.1
Virginia	80[c]	599.4	13.3
	25	599.4	4.2
South Carolina	416[d]	455	91.4

[a]   See Reference 630; assumes a 45% C content.
[b]   See Reference 1097.
[c]   See Reference 626.
[d]   See Reference 123.

From Chrzanowski, T. H., Stevenson, L. H., and Spurner, J. D., *Mar. Ecol. Prog. Ser.*, 13, 173, 1983. With permission.

due to autolysis and leaching of cellular constituents. Similar release occurs during the breakdown of macroalgae.[463,1077]

As discussed above, river input is an important source of carbon input. Fredericks and Sackett[261] estimated that the dissolved organic carbon (DOC) added annually to the Gulf of Mexico by runoff was 0.6×10^{13} g C, about two thirds of the total phytoplankton production in the area.

DOC concentrations in estuarine systems can be two to ten times that of particulate organic carbon (POC).[361] Despite its importance as a major component of the total organic carbon (TOC) pool, annual transport fluxes between marshes and adjacent estuarine waters were available for only five areas when Nixon[626] carried out a review in 1980. Four marsh systems exported DOC at rates ranging from 8 to 80 g C m^{-2} year^{-1}, while in Barataria Bay the export rate was considerably higher, approximately 140 g C m^{-2} year^{-1}.[361]

Recently Chrzanowski et al.[123] investigated the transport of organic carbon through a major creek draining 1800 ha of the North Inlet marsh ecosystem in South Carolina. DOC concentrations were variable, ranging from 0.9 to 13 g m^{-3} of the water with as much as 2.5 g m^{-3} of variation during a 1.5-hr period. Net transports ranged from approximately 5 to 480 g DOC s^{-1}. Annual budgets revealed a DOC export rate as high as $7.5 \pm 1.8 \times 10^9$ g C year^{-1}, corresponding to 416 g DOC m^{-2} year^{-1}. These export rates are high when compared with data from the northern U.S. *Spartina* marshes (see Table 2). In Table 3 the DOC export as a percentage of aerial net *Spartina* production can be seen to vary from 2.3 to 15.1%. Chrzanowski et al.[123] calculated the net aerial *Spartina* production of the North Inlet marsh at about 455 g C m^{-2} year^{-1}. The DOC export rates recorded appear to be unreasonable in comparison with the *Spartina* production and alternative explanations must be sought for the extreme export rates recorded. Either the high density quarterly sampling used by Chrzanowski et al. were inadequate for extrapolating to annual budgets, or there are other sources of DOC production, such as intermittent inputs from runoff or ground water flow; belowground *Spartina* production, phytoplankton, epiphytic, and benthic microalgae may be of greater importance. It is clear that further more-detailed investigations are needed to elucidate the flux of DOC in estuarine ecosystems.

B. Heterotrophic Utilization of DOC
A variety of organic compounds are excreted by algae and macrophytes. Carbohydrates

and organic acids dominate those reported for algae.[390,838] Simple carbohydrates and amino acids were noted in axenic cultures of *Najas flexis*,[989] and certain tropical seagrasses.[78] Heterotrophic utilization of amino acids, lipids, and other dissolved organic matter such as glucose and especially phosphate and bicarbonate[279] plays an important role in making such energy available in a particulate form to other organisms.

Pütter[746] first suggested that DOM in water might serve as a nutrient source when directly absorbed by metazoan organisms. Stephens[881,881a] demonstrated the uptake of various labeled dissolved organic compounds by numerous soft-bodied marine and estuarine invertebrates. Other studies such as those of Reisch,[769] Taylor,[909] Southward and Southward,[870-872] and Jørgensen[442] demonstrated an ability of nereids and Pogonophora to absorb amino acids directly from solution. However, there are methodological problems in interpreting the data from these experiments and, apart from the Pogonophora, it is uncertain as to whether uptake of DOM is of nutritive value to the organisms concerned. Stephens[882] has recently reviewed the trophic role of DOM. He concluded that there is a significant nutritional input to many marine organisms of free amino acids (FAA), but that the case for a trophic role of FAA is fundamentally an indirect one.

Crawford et al.[151] have shown that the heterotrophic production of particulate material from amino acids can represent a significant proportion of the total production of estuarine waters. They studied the amino acid flux in estuarine microorganisms in the Pamlico River estuarine river system, North Carolina, and found that the particulate production averaged 0.79 μg C ℓ^{-1} hr^{-1} over the year (ranging from 0.06 to 2.37 μg C ℓ^{-1} hr^{-1}). This represents approximately 10% of the rate of production by the algae during the summer months.

Other studies[220,360] also confirmed that bacteria in the water column make a small but significant contribution to the production of inshore waters by the utilization of DOC, either liberated directly from photosynthetic organisms, from zooplankton or benthos, or biodegradation of detritus from macroalgae, salt marsh plants, mangroves, and seagrasses.

IV. ROLE OF MICROORGANISMS

Heterotrophic microorganisms in detritus-dominated estuarine ecosystems are a major link in the mineralization and transformation of organic matter. These microorganisms can be subdivided on a metabolic basis into two groups, aerobes and anaerobes. Aerobic environments comprise the water column and the superficial sediments. Because of the predominantly fine sediments in salt marshes and many estuarine intertidal flats and bottom substrates, only the top few millimeters are oxidized. Aerobic zones are also found along the surface of animal burrows and the rhizosphere of living plant roots. Although most of the sediment is anaerobic, all materials entering or leaving the anaerobic sediments must pass through the aerobic zone, the zone in which much of the nutrient transformation and most of the rapid decomposition of organic matter take place. Aerobic microorganisms include bacteria, fungi, protozoans, and microscopic metazoans. Anaerobic microorganisms are predominantly bacteria.

Mann[576] listed the processes in which microorganisms are involved as follows: (1) release of much of the primary production as DOM or POM and colonization of the latter by microbes; (2) uptake of DOM by bacteria with the production of living and dead POM; (3) decomposition of the less refractory portions of POM of plant origin by microorganisms; (4) consumption of bacteria in large quantities by protozoan, planktonic filter feeders, and benthic detritivores; (5) conversion of some DOM to particulate form by physiochemical processes, and subsequent colonization by microorganisms; sedimentation of the POM onto the intertidal flats, or the bottom in deeper water; and (6) subsequent incorporation into aerobic food chains on the sediment surface, or decomposition by anaerobic organisms in the deeper layers of the sediment.

A. Microbial Processes in the Water Column

1. Microbial Standing Stocks

It has proved difficult to obtain accurate measurements of either the biomass or the activity of microorganisms in the water column. Several recent volumes have addressed methods for the determination of the characteristics of microbial communities.[144,437,527] These involve either direct observations of the organisms in question or an indirect approach involving measures of the chemical components of the microorganisms.

Direct methods include culturing or microscopic observation of the sample. Culture methods rely on questionable assumptions and are generally considered inadequate for community assessment, though they are important for the enumeration of specific populations.[144] For bacteria, the direct method most used is the acridine orange direct count method (AODC) which uses a fluorescent dye and epi-illumination.[398] The use of AODC in sediments is difficult since many cells are attached to the sediments. Scanning electron microscopy is also used as an aid.

Perhaps the most common indirect method for analysis of total living biomass is the use of firefly lantern extracts in the determination of adenosine triphosphate (ATP) and other cellular nucleotide triphosphates.[403] This method is rapid, sensitive, relatively inexpensive, but fraught with problems of both analysis and interpretation, particularly when used in sediments.[799] One particular problem is the ratio of carbon to ATP. Karl[450] has pointed out that there is wide variation in this ratio but suggests that a ratio of 250:1 is reasonable, at least in planktonic systems. Christian et al.[119] who have used the technique in the Sapelo Island studies consider that the technique remains useful, if interpreted with care.

There have been a limited number of studies of the planktonic bacteria of estuaries. In their study of Newport River estuary, North Carolina, Palumbo and Ferguson[687] concluded that the upper reaches of the estuary were a source of bacteria which then mixed with coastal sea water as a conservatively distributed 'substance' in the lower reaches of the estuary. Bent and Goulder[46] studied numbers and activity of bacteria for a 2-year period in the Humber Estuary in northeast England. In this highly polluted estuary, the seasonal pattern indicated a low in the summer and high numbers for autumn to spring. This pattern is a consequence of the fact that most of the bacteria were attached to suspended particles. The Fraser River estuary in British Columbia is also characterized by high turbidity;[42] here total counts showed an inverse trend with salinity, although heterotrophic activity of the bacteria was generally highest in the plume water of intermediate salinity. Numerous other studies have indicated that heterotrophic activity, on a per cell basis, is higher in estuarine waters than in connecting fresh or marine waters.[320,885,1045]

Hanson and Snyder[359] estimated microbial biomass in the surface waters from three locations in the Sapelo Island marshes (see Chapter 3, Figure 1): Doboy Sound, 2 km from its mouth; the Duplin River, approximately 0.5 km from its mouth; and Study Creek, a first-order marsh creek 2 km from the mouth of the Duplin River. ATP concentrations varied seasonally, and concentrations peaked during the warmer months (Figure 8). Although POC concentrations and viable bacterial densities decreased with distance from the marsh, ATP concentrations were similar at all locations. In Study Creek and Duplin River water, approximately 85% of the total ATP was associated with particles <64 μm and 67% with particles <10 μm. In Doboy Sound 66% of the total ATP was associated with particles <64 μm and 33% with particles <10 μm.

One of the most comprehensive studies of the distribution of planktonic bacteria is that of Wright and Coffin[1046] in the Essex, Parker, and Ipswich River estuaries, and the Ipswich Bay in northern Massachusetts. Figure 9 depicts the pattern of seasonal development of estuarine and coastal water bacteria in the Essex River. In winter, bacterial counts and activity were uniformly low throughout the Essex River. During spring there gradually developed a peak of bacteria in midestuarine waters which was sustained throughout the

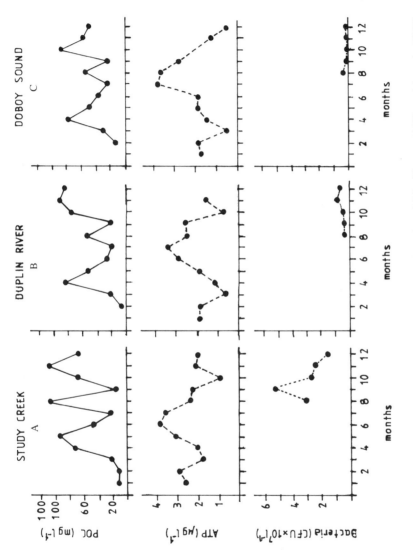

FIGURE 8. Seasonal and spatial relations of particulate organic carbon (POC), biomass (ATP), and bacteria (CFU) concentrations in unfractionated water samples in Study Creek (A), Duplin River (B), and Doboy Sound (C), during 1976. (From Christian, R. R., Hanson, R. B., Hall, J. R., and Wiebe, W. J., *The Ecology of a Salt Marsh*, Springer-Verlag, New York, 1981, 116. With permission.)

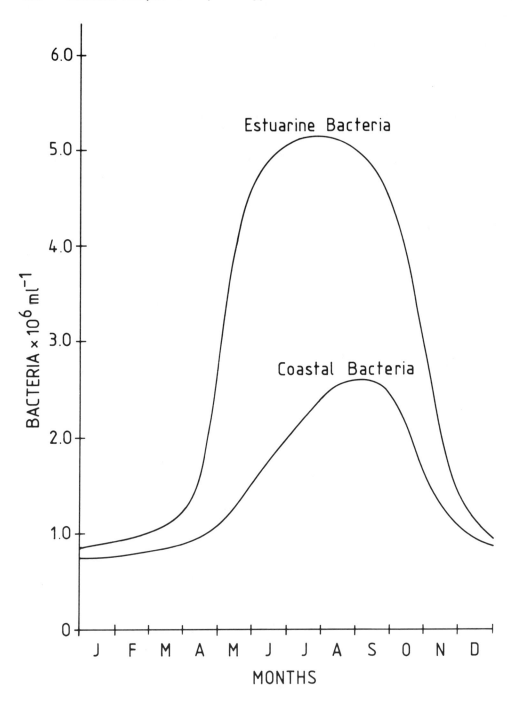

FIGURE 9.   Seasonal development pattern of estuarine and coastal water bacterial flora (total counts) in the Essex River — Ipswich Bay system. (From Wright, R. T. and Coffin, R. B., *Mar. Ecol. Prog. Ser.*, 11, 209, 1983. With permission.)

summer and subsided in the autumn (fall). The seasonal range in these waters was tenfold (0.7 to 70 $\times$ 10^6 mℓ^{-1}) while the range in coastal waters was only fourfold.[1045] Table 4 presents mean data for upper, mid, and lower estuary and coastal water masses. Activity and counts appear to be related and follow the same pattern of highest values in midestuary

Table 4

SALINITY, HETEROTROPHIC ACTIVITY, AND DIRECT COUNTS OF BACTERIA FOR FOUR WATER MASSES IN THE ESSEX RIVER ESTUARY, IPSWICH BAY, MASSACHUSETTS

Water mass	Salinity (ppt.)	Total activity ($\times 10^{-3}$ hr^{-1})	Total counts ($\times 10^{6}$ mℓ^{-1})
Upper estuary	10.0 (10.2)	46.0 (50.3)	2.19 (1.50)
Mid estuary	24.6 (6.7)	70.5 (67.6)	3.12 (1.67)
Lower estuary	29.1 (2.4)	40.9 (36.9)	2.47 (1.19)
Coastal	30.9 (1.35)	29.2 (35.8)	1.51 (0.66)

Note: Means with standard deviations in parentheses.

From Wright, R. T. and Coffin, R. B., *Mar. Ecol. Prog. Ser.*, 11, 209, 1983. With permission.

with lower values in the upper and lower estuary and still lower values in the coastal waters. The seasonal development of the estuarine bacterial flora is strongly linked to temperature, a relationship also reported for the bacterio-plankton of a salt marsh coastal system in South Carolina.[1023]

Although the relationship between temperature and bacterio-plankton is highly significant, it is not necessarily direct. Bacteria have a capability for rapid growth.[262,590] Among the factors responsible for this is the availability of a dissolved organic substrate. As we have seen, the sources of such substrates are exudates or leachates from phytoplankton, epiphytic algae, benthic microalgae, macroalgae, and estuarine macrophytes. These substrates are more readily available during the summer period. Turner's[941] studies have indicated that in estuaries with a high marsh-to-water area, heterotrophic activity is most strongly linked to leachate derived from *Spartina* photosynthesis, which in turn is seasonally controlled by temperature. This is a possible explanation of the midestuarine peak in bacterial numbers and the difference between these numbers and those in the adjacent coastal waters.

One further factor influencing the distribution of water bacteria in estuarine ecosystems is the availability of suspended particulate substrates. Most investigators have found that roughly 80% of the bacteria in the water columns of the open ocean are to be found floating freely, but Wangersky[972] argues that those attached to particulate matter are much more likely to be metabolically active. Hanson and Wiebe[360] have suggested that there may be differences between inshore waters and offshore waters in respect of the proportion of bacteria attached to detritus particles. As an index of bacterial activity they used the rate of uptake of ^{14}C-glucose. On the edge of the continental shelf in the U.S., they found 80% of the uptake was by particles less than 3 μm in diameter, and 93% was by particles less than 8 μm. Hence they showed that bacteria not attached to particles comprised 80% of the activity, which is in conflict with the findings of Wangersky[972] quoted above. At 8 km from the shore Hanson and Wiebe[360] found that particles under 8 μm accounted for only 35 to 45% of the uptake. At 4 km from the shore, and in tidal creeks, over 80% of the activity was associated with particles greater than 3 μm in diameter. Almost all of the uptake by this fraction was found to be carried out by bacteria attached to detritus. There were marked tidal cycles, with the greatest activity occurring near low ebb-tide (Figure 10). The authors were aware of one problem with these experiments which they stated as follows: "It is possible that many of the bacteria in the sea do not utilise glucose, or utilise it at very low rates, and that bacteria which actively take up glucose are preferentially attached to particles. This would alter the picture we have described and it is well to keep this possibility in mind."

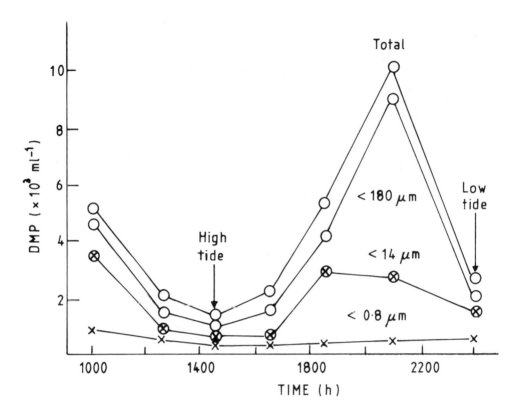

FIGURE 10.   Heterotrophic activity in creek water in the Sapelo Island salt marsh, Georgia, in various size fractions over one tidal cycle. (From Christian, R. R., Hanson, R. B., Hall, J. R., and Wiebe, W. J., *The Ecology of a Salt Marsh*, Springer-Verlag, New York, 1981, 121. With permission.)

From the studies quoted above it is clear that there is some controversy as to whether the bacteria in the water column are found attached to organic particles,[161,639] or whether they are predominantly free-living,[120,434,1002] and much of the evidence is contradictory. Goulder[319] found that the majority of bacteria in a turbid estuary are attached. Jannasch and Pritchard,[423] in an experimental study with bacterial cultures, found that the density of added inorganic particles as well as the concentration of dissolved organics influenced the degree of attachment. However, the attached and unattached estuarine microbial plankton populations showed different degrees of cellular activity with relatively more heterotrophic activity being found with attached bacteria.[119] In order to be active, bacteria in an estuary may require a particle for attachment, and particle origin (plant or fecal matter), as well as particle size, may influence both biomass and metabolic activities of the associated microorganisms. Siebruth[835] has proposed that small bacteria observed by epifluorescence microscopy, are never attached to particles and utilize DOC, and that the larger bacteria, capable of culture, are those which do attach to particles and which utilize POC.

Linley and Field[519] have studied the nature and ecological significance of bacterial aggregation in a nearshore upwelling system. Five types of aggregates, gelatinous aggregates, flakes, fragments, and bacterial clumps, and films were found. Figure 11 summarizes some of the hypotheses concerning the formation of the various types of aggregates. In estuaries the main sources of organic matter are macrophytes (marsh plants, seagrasses, and macroalgae), phytoplankton, benthic microalgae, and animals. The animals release particulate and dissolved organic matter through secretion, mucus production, break-up and dissolution of feces, and through spillage during ingestion. High molecular weight components such as

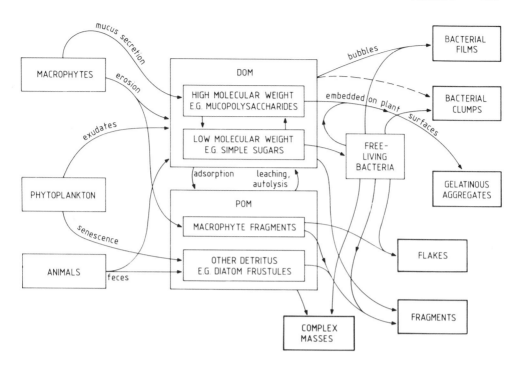

FIGURE 11. Suggested formation of various types of aggregate from three main sources of dissolved and particulate organic matter. (From Linley, E. A. S. and Field, J. G., *Estuarine Coastal Shelf Sci.*, 14, 9, 1982. With permission.)

mucopolysaccharides are likely to be important since gelatinous aggregates that appear to be formed from mucus were found by Linley and Field[519] to form a large proportion of the aggregates. Bacterial films, as previously noted, may be formed by the adsorption of free bacteria onto the DOM coating bubble surfaces.[1093] Flakes, fragments, and complex masses appear to be derived primarily from macrophyte fragments and other algal particles. A feature of the aggregates is their adsorption of DOM and in many cases the majority of the bacteria appeared to be associated with the DOM. Thus, in addition to decomposing and enhancing the nutritional value of particulate debris, bacteria (which would otherwise be inaccessible to many consumers) may become associated with DOM, transforming it into nitrogen-rich particulate form for suspension feeders.

Heterotrophic activity can be estimated by measuring the concentration, distribution, and rate of change of dissolved oxygen in the water. In the waters of the Duplin River (Sapelo Island) oxygen consumption, most of which is probably bacterial, increases in a linear manner from mouth to headwaters.[258] This suggests that there is a significant input of labile organic matter from the marsh, in addition to organic matter produced by the phytoplankton. A substantial part of the bacterial respiration was found to be associated with particles.[119] Thus, there is evidence of a net export of organic matter and energy from the marsh. However, most of it is consumed within the Duplin River and its tributary creeks, except for refractory lignocellulosic POC and humic and fulvic DOC.[733,1006]

Wright and Coffin[1046] conclude that in estuarine waters, a particular population density of bacteria must be viewed as a steady-state system, fluctuating within bounds set by factors that are contributing to bacterial production (e.g., substrate supply) and other factors operating to reduce bacterial numbers. Among the latter, grazing by microzooplankton,[344] macrozooplankton,[387] and benthic fauna.[1047] The best evidence for the importance of grazing comes from experiments which demonstrate rapid increases of bacteria following filtration through a 1- or 3-μm filter, compared with very little change in control samples.[263,265]

2. Role of Microorganisms in Estuarine Water Food Webs

Traditionally bacteria have been regarded as remineralizers, responsible for converting organic matter to inorganic, and recycling nutrients to primary producers. While this is an important role in the food web, evidence is accumulating that bacterial production is much greater than had been previously assumed.[434,435,836] The role of DOM is also being increasingly recognized. Watson[981] has argued that the efficiency of utilization of ingested food at any trophic level was only 50 to 60% and that 10 to 20% was required for growth with another 30 to 40% for maintenance. Therefore, at each trophic level, 40 to 50% of the ingested food is excreted as DOC or released as POC in feces, and hence a very large proportion of the primary production would seem to be available for bacterial utilization. There is also evidence that the DOC is being utilized effectively by microorganisms rather than being simply respired. Wiebe and Smith[1003] showed that in kinetic tracer experiments of about 10-hr duration, DOM was converted to microbial biomass with an efficiency of about 97 to 99%.

Pomeroy[730] has drawn attention to the possibility that Protista are major consumers of bacteria in the sea. Support for Pomeroy's views comes from several sources. Eriksson et al.[219] found that ciliates, microrotifers, and nauplii constituted 15% of the biomass of zooplankton in a shallow coastal area off Sweden, but estimated their energy requirements were 65% of the total. Beers and Stewart[40] reached similar conclusions regarding two sites in the Pacific Ocean and Margalef[578] has suggested that ciliates contribute as much, or more, to organic production as do net zooplankton. Fenchel[238-242] has shown that heterotrophic microflagellates in the size range 3 to 10 μm are effective bacteriovores in the sea, capable of filtering 12 to 67% of the water column per day.[837,868] Sorokin[865] has suggested that bacteria in the sea normally occur in aggregates large enough for net zooplankton to feed on directly.

In 1979 in a development of his views, Pomeroy[731] presented a compartmental model of energy flow through a continental shelf ecosystem, demonstrating the various pathways by which energy may pass from primary producers to terminal consumers, and showing that pathways involving dead organic matter, bacteria, and their predators (especially Protozoa) could play a major role. In Figure 12 his model has been further developed by Pace et al.[679] This model involves abandoning classical ideas of trophic levels and regarding food webs as anastamozing structures that defy classification under trophic levels. Pomeroy[731] demonstrated that it was possible for energy to flow through either the grazers or the alternate pathways to support all the major trophic groups at a reasonable level, and to maintain fish production at about the levels that commonly occur.

Pomeroy's ideas have been expanded and modified in two recent papers, those of Newell and Linley,[1084] on the significance of microheterotrophs in the decomposition of phytoplankton, and Pace et al.[679] on a simulation analysis of continental shelf food webs. Pace et al.[679] have expanded and developed Pomeroy's[731] 1979 mass balance model into a simulation model to evaluate pathways of energy flux (Figure 12). The simulation results indicated that only slightly more than 50% of the annual net phytoplankton was grazed and that bacterial production was equivalent to zooplankton production. Newell and Linley,[1084] from a study of standing stocks off the western approaches to the English Channel during late August concluded that the microheterotrophic community of protozoans and bacteria comprised as much as 46% of the total consumer carbon at a mixed water station, 60% at a frontal station, and 21% at a deeper stratified water station. Thus, in the late stage of a phytoplankton bloom 50 to 60% of the carbon flow was estimated to enter the bacteria, 34 to 41% being subsequently respired by the bacteria, and a further 9 to 12% by the heterotrophic microflagellates. In their model simulations Pace et al.[679] found that benthic production appeared to be related both to the quantity of primary production and the sinking rates of the phytoplankton. From these and other studies[1081,1083] it is clear that heterotrophs become increas-

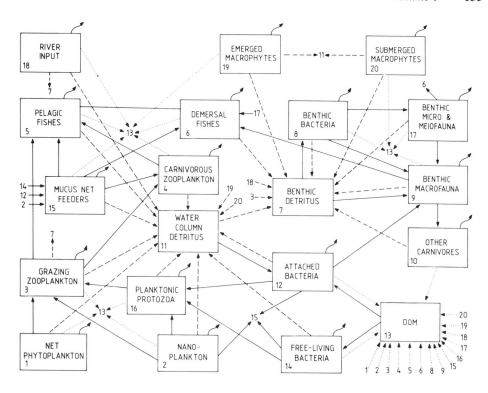

FIGURE 12.   Diagram of the energy fluxes in an estuarine ecosystem. Solid lines = trophic flows, dotted lines = dissolved flows, broken lines = detrital flows, and wavy lines = respiratory flows.[679]

ingly important as one progresses from the open ocean across the continental shelf to coastal and estuarine waters.

Experiments by Thingstad[25] showed that as a phytoplankton bloom declined, bacterial populations built up and these in turn declined when flagellates became abundant. Newell[624] summarizes the results of microcosm experiments on the degradation of DOM, kelp debris, animal feces, phytoplankton, and *Spartina* debris which all showed the same successional pattern in natural sea water. In all these studies there is a remarkably similar pattern, with heterotrophic microflagellates controlling bacterial numbers with a lag of some 3 to 4 days between bacterial and flagellate peaks.

Figure 13 from Azam et al.[25] presents a modification of the Sheldon et al.[1089] particle-size model to illustrate the role of bacteria and other microbes in the water column. The main feature of this model is that organisms tend to utilize particles one order of magnitude smaller than themselves. DOM released by phytoplankton (in estuaries this would be supplemented by DOM from other sources), and to a smaller extent by animals, is utilized by bacteria. When sufficient DOM is supplied for their growth, bacteria (0.3 to 1 μm) are kept below a density of 5 to 10 $\times$ 10^6 cells mℓ^{-1}, primarily by heterotrophic flagellates which reach densities of 3 $\times$ 3^3 cells mℓ^{-1}. Flagellates (3 to 10 μm) probably also feed on Cyanobacteria in the same size range (0.3 to 1 μm). Flagellates both autotrophic and heterotrophic are in turn preyed upon by microzooplankton in the same size range as the larger phytoplankton (10 to 80 μm). Thus, energy released as DOM is rather inefficiently returned to the main food chain via a microbial loop of bacteria-flagellates-microzooplankton.

An important consequence of the "microbial loop" stems from the ability of bacteria to absorb mineral nutrients from the water.[25] Their small size and large surface-to-volume ratio allow absorption of nutrients at very low concentrations, giving them a competitive advantage over phytoplankton as documented in CEPEX bag and laboratory experiments.[3,946] Bacteria

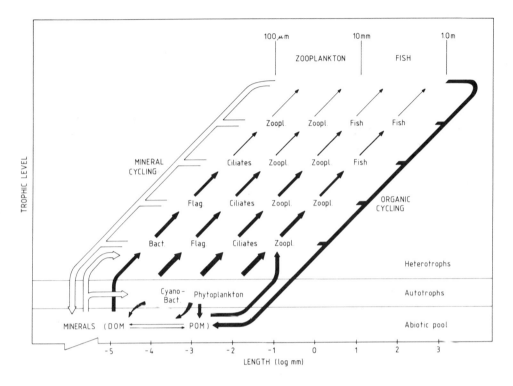

FIGURE 13.   Semiquantitative model of planktonic food chains. Solid arrows represent flow of energy and materials; open arrows, flow of materials alone. It is assumed that 25% of the net primary production is channeled through DOM and the "microbial loop", bacteria (Bact.), flagellates (Flag.), and other microzooplankton (e.g., ciliates). It is further assumed that the most efficient predator size ratio is 10:1, hence the slope of the lines relating trophic status to log body length is 1:1. The food chain base represents a size range of three orders of magnitude (smallest bacteria 0.2 μm, largest diatoms 200 μm), therefore, any trophic level will have a size-range factor of 10^3 and conversely, each size class of organisms (100 μm) will represent at least three trophic levels. The thickness of open arrows (left) represents the approximate relative magnitude of minerals released in excretion at each trophic level; corresponding organic losses (feces, mucus, etc.) are shown on the right hand side. (From Azam, F., Fenchel, T., Field, J. G., Gray, J. S., Meyer-Reil, L. A., and Thingstad, F., *Mar. Ecol. Prog. Ser.*, 10, 260, 1983. With permission.)

sequester mineral nutrients efficiently and are consumed by flagellates with a carbon assimilation efficiency of some 60%,[239] the remaining 40% of carbon being egested as feces. Furthermore, heterotrophic flagellates and other microzooplankton excrete nitrogen and, since their C:N ratios are similar to that of their food (3.5:4),[243] an amount of nitrogen corresponding to some 25% respired carbon must also be excreted. Thus, while bacteria excrete minerals and respire carbon, they compete efficiently to regain mineral nutrients; as a consequence heterotrophic flagellates and microzooplankton play a major role in remineralization.[576]

B. Microbial Processes in the Sediments

Estuarine sediment systems are complex receiving energy from two sources: light and organic matter either DOM produced *in situ* (dead roots and rhizomes of macrophyte plants), or imported in the form of detritus or DOM leached by benthic microalgae or from plant roots. The processes of synthesis through oxidative and reductive pathways which these energy sources drive are interrelated and their relative importance depends on the availability of light, oxygen, and various organic and inorganic hydrogen acceptors (Figure 14). The sediment stratification that results has been discussed previously in Chapter 2, Section III.

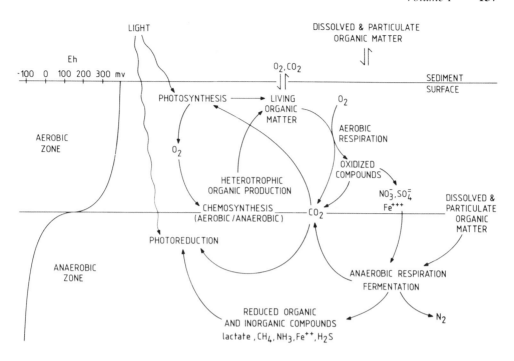

FIGURE 14. Interrelations between photosynthetic, heterotrophic, and chemosynthetic processes which occur in sediment. Photosynthesis and photoreduction only occur in the presence of light. *Aerobic metabolic processes:* heterotrophic production (oxidation of reduced simple organic compounds with possible reduction of external CO_2); photosynthesis (reduction of CO_2 to carbohydrates using H_2O and light); aerobic respiration (reduction of oxygen to water with organic compounds as electron donors); aerobic chemosynthesis (oxidation of CH_4, H_2S, NH_3, Fe^{++}, and H_2 to form organic carbon compounds by fixation of external CO_2). *Anaerobic metabolic processes:* anaerobic respiration (oxidized inorganic end products of aerobic decomposition used as hydrogen acceptors for the oxidation of organic matter); fermentation (organic compounds used as hydrogen acceptors to produce CO_2, H_2O and reduced organic compounds lactate, glycollic acid, H_2S, and NH_3); photoreduction (reduced compounds in the presence of light used to reduce CO_2 to carbohydrates with H_2S, SO_3, S, H_2 or reduced organics serving as a hydrogen donor); anaerobic chemosynthesis (oxidize inorganic compounds H_2, H_2S, Fe^{++}, $NO_2^{=}$, and use energy to reduce CO_2 to carbohydrates).[228,696] (Redrawn with permission from Parsons et al., 1977, *Biological Oceanographic Processes*, 2nd ed.; copyright 1977, Pergamon Press.)

The processes and the organisms involved have been reviewed by Fenchel[228] and Fenchel and Riedl.[245]

In the upper part of the aerobic zone as detailed in Section I, photosynthetic unicellular algae (diatoms, coccolithophorids, euglenoids, dinoflagellates, phytomonids, and other small flagellates) as primary producers, contribute significantly to the overall primary productivity of the estuarine ecosystem. In this layer organic detritus derived from the multiplicity of sources outlined previously is decomposed aerobically by bacteria and probably also fungi, resulting in the formation of bacterial and fungal biomass on the one hand, and oxidized inorganic compounds with oxygen as the terminal hydrogen acceptor (CO_2, N_2O, $NO_3^{=}$, etc.) on the other.

Below the redox potential discontinuity layer, however, conditions are anaerobic and here the limiting factor for the utilization of the potential energy of organic material is the availability of hydrogen acceptors. Heterotrophic bacteria, by various types of fermentation processes, utilize organic compounds (fermentation) as hydrogen acceptors while at the same time oxidizing other organic substances and producing, besides CO_2 and H_2O, reduced compounds such as lactate, alcohols, H_2S, NH_3, etc. Other anaerobic bacteria utilize certain inorganic compounds as hydrogen acceptors for the oxidation of organic material. In this process reduced inorganic compounds (CH_4, NH_3, H_2S) which diffuse upwards are produced.

The reduced compounds thus formed, both inorganic or organic, still contain potential energy which may be utilized by chemoautotrophic bacteria in the presence of oxygen and by photoautotrophic bacteria in the presence of light. The chemotrophs obtain their energy by the oxidation of inorganic substances (H_2S, Fe^{++}, H_2, NH_3, NO_2, etc.) and the energy thus obtained is used to reduce CO_2 to carbohydrates. The photoautotrophs (purple and green sulfur bacteria and some protophytes) utilize reduced compounds for the reduction of CO_2 to carbohydrates in a photosynthetic process in the presence of light, instead of using H_2O as in normal photosynthesis (photoreduction). The purple and green sulfur bacteria use H_2S, SO_3^2 and S, and some forms may also use H_2 or reduced organic compounds as hydrogen donors in this process. Sulfate reduction, carried out by the bacterium *Desulfovibrio* sp., is a dominating process in the chemistry of the anaerobic sediment layers.[228,230] On the other hand, the chemoautotrophic white sulfur bacterial species of *Thiobacillus, Thiovolum, Macromonas, Boggiatoa*, etc. oxidize H_2S to elemental S or sulfate. When the anaerobic zone reaches the surface of the sediment they become visible to the naked eye as a white scum on the surface.

The organic matter which fuels the above processes includes living and recently dead plant fragments, plant detritus in various stages of colonization by microbes, DOM and amorphous organic matter derived from it, dead animals and parts thereof, and fecal pellets of planktonic and benthic animals. Various invertebrates may use this material as a food source (see Vol. II, Chapter 2, Section VII for a further discussion). Some like amphipods and surface feeding snails process the material at the surface, but others cause it to be buried in the sediments. Filter-feeding bivalves, for example, may filter material at or just above the surface, then egest feces just below the surface. Detritus also becomes buried through the activities of burrowing invertebrates.

1. Measurement of Microbial Activity

Techniques for the estimation of microbial biomass were discussed above and brief mention has been made of some measures of bacterial activity. Here they will be discussed in more detail. Four basic approaches have been used to study decomposition and microbial activity in estuarine sediments.[411] Two of these litter bag studies and seasonal studies of the abundance of living or dead roots and rhizomes measure the disappearance of organic matter. Two other approaches are to measure consumption of electron acceptors (O_2, NO_3^-, SO_4^-, CO_2), and turnover and mineralization of specific dissolved organic compounds such as amino acids, glucose, or acetate. As Howarth and Teal[413] point out it is important to remember that these approaches give qualitatively different types of information.

The assessment of microbial growth or metabolic transformation of energy sources in natural populations is a central problem in aquatic microbiology.[422] Techniques employing radioisotopes are the only methods for studying certain processes and the controls of those processes in the sediments of aquatic environments. For example, processes such as the breakdown of organic matter into fermentation products can be studied by measuring the turnover of added ^{14}C-acetate or ^{14}C-glucose combined with measures of concentrations of these compounds even though the concentrations in the sediments do not change. However, many problems are encountered in the application of these techniques. These include disturbances due to mixing of sediments, problems in label injection, sorption of the isotope onto sediment particles making it unavailable to microbes, exchange of isotopes with roots, transformation of added labeled compounds, and isotope dilution with time.[411] Many of these problems are easily overcome in the laboratory. But as Christian and Wiebe[121] point out, *in situ* experiments are necessary and must be used to check extrapolations from laboratory and microcosm studies.

Processes of mineralization of organic compounds are redox reactions involving transfers of electrons from the organic matter being oxidized to an electron acceptor. Examination of the dynamics of the electron acceptors is a potentially powerful tool for studying decomposition processes.[411] Electron acceptors usually considered are oxygen, nitrate, sulfate, and carbon dioxide.[243,863] Each electron acceptor is studied by a different technique, and the problems of interpreting microbial activities from the electron acceptor dynamics vary with the electron acceptor.

Measurement of oxygen uptake rates is the most widely used technique. Sediment oxygen uptake reflects the sum of respiration by animals, plants, and heterotrophic microorganisms, consumption by chemosynthetic bacteria, and consumption by purely chemical reactions such as the nonbiotically autolyzed oxidation of sulfides. For understanding controls on decomposition in marshes we are interested only in oxygen consumption by heterotrophic organisms, yet consumption by other processes is large and cannot be ignored.[411] Antibiotics have been used by Hopkinson et al.[407] to separate out consumption by heterotrophic microbes, but they are not completely successful. In addition to oxygen exchange across the sediment surface in marsh areas, oxygen also moves down the internal spaces of marsh grasses to roots and rhizomes.[914] This adds a further complication in these situations.

2. Microbial Standing Stocks

Rublee[799] has summarized data on direct counts of total bacteria in estuarine intertidal and subtidal sediments of a range of estuaries in North America. Numbers ranged from about 4 to 17×10^9 cells cm^{-3} in surface sediments and decreased with increasing depth of sediment. Seasonal distribution of bacteria numbers has been reported in a study of a North Carolina marsh.[800] Numbers in surface sediments ranged from a high of 14×10^9 cells cm^{-3} in the late autumn (fall) and a low of around 5×10^9 cells m^{-3} in late summer. At 5 cm depth this pattern was repeated but with lower maximum and minimum numbers (7 and 3×10^9 cells cm^{-3}); at sediment depths of 10 and 20 cm the pattern did not reflect seasonality. In the Rhode River, Maryland, similar results were found among six stations including four vegetated marsh sites, a tidal creek site, and a subtidal mud flat (Figure 15).[799] A trend for lower numbers in the upper layers of creek and subtidal sediments as compared to intertidal sediments was evident. This trend paralleled a decrease in the organic carbon content of these sediments.[155,801] In contrast the highest numbers of bacteria were found in intertidal sediments with high (10 to 15%) organic carbon content. Dale[155] noted this relationship and Rublee[799] has plotted his data, together with that from six additional studies, and from the plots calculated a highly significant regression (Figure 16).

Dale[155] also showed that bacterial numbers, estimated by direct counting, which varied from 1×10^8 to almost 1×10^{10} g^{-1} of sediment, showed a highly significant negative correlation between numbers and mean grain size. This, for POM, as previously noted, suggested that the surface area of the grains controls density. Rublee[799] found that marsh interstitial water contained only 0.1 to 1% of the total sediment bacteria, and that when filtered water was flushed through short sediment cores it was demonstrated that only a small fraction (<1%) of the total sediment bacteria was washed out. These results suggested that the majority of the cells adhere to the sediment particles.

Total living microbial biomass may be estimated by using adenosine triphosphate (ATP) assays. Rublee[799] has listed data from five studies. In the Sapelo Island *Spartina* marsh studies (Table 5) ATP concentrations decreased with depth in all sediments[115,116] with concentrations in the tall *Spartina* decreasing less with depth than the short *Spartina*. However, the concentrations in the surface sediments were much higher in the short *Spartina*. This pattern was also found in a *Spartina* marsh at the Newport River Estuary, North Carolina,[800] and in mud flat sediments in the same estuary.[246]

Rublee[800] monitored the number and size of bacteria at four depths (0 to 1, 5 to 6, 10 to 11, and 20 to 21 cm) in the North Newport Estuary *Spartina* marsh for a period of 13 months. The number of bacteria reached a maximum of about 1.4×10^{10} cells cm^{-3} at the

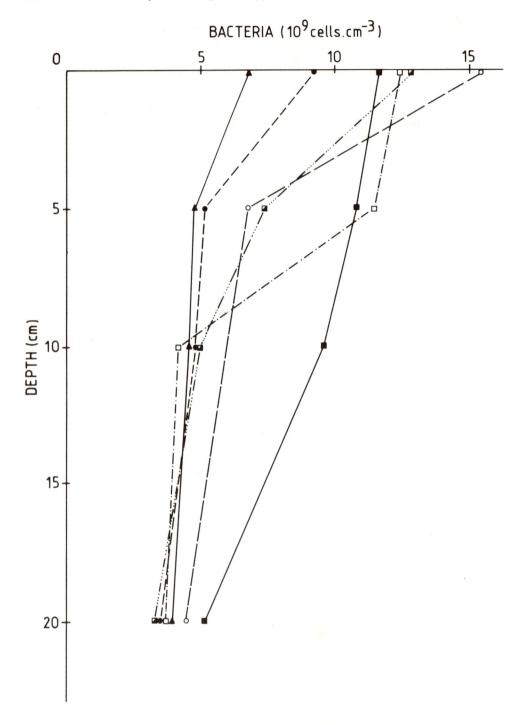

FIGURE 15.   Number of bacteria in sediments of Rhode River Estuary, Maryland. ○ = low marsh; ● = low marsh creek bed; ■ = high marsh *(Distichlis spicata/Spartina patens);* □ = high marsh *(Spartina cynosuroides);* ▨ = high marsh (mixed vegetation); and ▲ = subtidal mud flat. (From Rublee, P. A., *Estuarine Comparisons,* Kennedy, V. C., Ed., Academic Press, New York, 1982a, 167. With permission.)

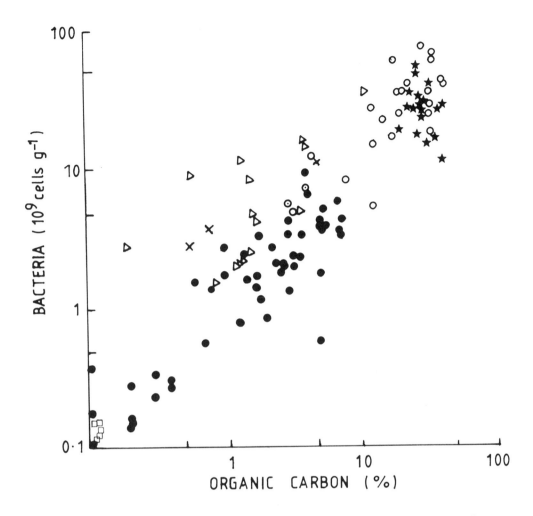

FIGURE 16.   Relationship between bacteria and organic carbon in intertidal and shallow subtidal sediments. ● = intertidal, Nova Scotia;[155] ○ = marsh intertidal, Maryland;[799] ☉ = subtidal, Maryland;[799] △ = marsh intertidal, North Carolina;[801] × = subtidal, North Carolina;[829] □ = subtidal North Sea coast;[1095] ★ = marsh intertidal, Massachusetts.[799] (From Rublee, P. A., *Estuarine Comparisons,* Kennedy, V. C., Ed., Academic Press, New York, 1982a, 168. With permission.)

Table 5
RECOVERABLE ATP FROM THE SEDIMENTS
OF THE SAPELO ISLAND SALT MARSH

	ATP (μg cm^{-3})	
Depth (cm)	Tall *Spartina alterniflora*	Short *Spartina alterniflora*
0—1	0.66	2.11
1—5	0.40	1.52
5—10	0.24	0.54
10—15	0.17	0.22
15—25	0.14	0.12
25—35	0.15	0.05

From Christian, R. R., Bancroft, K., and Wiebe, W. J., *Soil Sci.,* 119(1), 89, 1975. With permission.

Table 6
ESTIMATED CONTRIBUTION OF MICROBIAL AND BACTERIAL STANDING CROP TO TOTAL SEDIMENT ORGANIC CARBON AND NITROGEN IN A NORTH CAROLINA SALT MARSH

Depth (cm)	Microbial C (mg C cm^{-3})	Bacterial C (mg C cm^{-3})	Sediment C (mg C m^{-3})	Microbial N (mg N cm^{-3})	Bacterial N (mg N cm^{-3})	Sediment N (mg N cm^{-3})
0—1	1.013	0.147	25.40	0.170	0.044	2.19
5—6	0.080	0.088	31.90			
10—11	0.023	0.047	33.80			
20—21	0.010	0.014	13.20			

From Rublee, P. A., *Estuarine Coastal Shelf Sci.*, 15, 73, 1982b. With permission.

Table 7
AEROBIC MICROBIAL RESPIRATION OF CARBON IN THE DUPLIN RIVER (SAPELO ISLAND) MARSH

Microbial community	g C m^{-2} year^{-1}	Total respiration (%)	Aerobic primary production[a] (%)	Total primary production[b] (%)
Standing dead *Spartina*	180	53	22	12
Marsh sediment surface	90	26	11	6
Subtidal sediment surface	20	6	2	1
Water	50	15	6	3
Total	340	100	41	22

[a] 780 g C m^{-2} year^{-1} used as the estimate of total primary production.
[b] 1380 g C m^{-2} year^{-1} used as the estimate of total primary production.

From Christian, R. R., Hanson, R. B., Hall, J. R., and Wiebe, W. J., *The Ecology of a Salt Marsh*, Pomeroy, L. R. and Wiegert, R. G., Eds., Springer-Verlag, New York, 1981, 126. With permission.

sediment surface in October corresponding to the period of *Spartina* dieback. Cell numbers were lowest and most consistent throughout the year at the 20-cm depth. Cell volumes averaged 0.2 μm^3 at the marsh surface and decreased with depth. Table 6 gives estimates for microbial, bacterial, and sediment C and N at the various depths. Mean standing crop of bacteria to a depth of 20 cm in the sediment was about 14 g bacterial carbon m^{-2}. In the surface sediments bacteria contributed up to 15% and benthic microalgae up to 10% to the total living microbial biomass as estimated by ATP. Bacteria were the major biomass component at depths of 5, 10, and 20 cm. At all depths the microbial community contributed <4% total organic carbon and <8% of total nitrogen. The contribution of bacteria was estimated at about 0.6% of sediment organic carbon and less than 2% of nitrogen at the sediment surface. These values compare well with those reported for a Georgia salt marsh[116] and for shallow sediments[221,246] where the contribution of microbes to sediment organic carbon and sediment nitrogen was in the range of 0.05 to 2.98%. However, the apparently small contribution of the bacteria and total microbial community to organic carbon and nitrogen content of sediments does not give a true picture of their importance. Standing crop measures are not a good indicator of metabolic activity, especially for microorganisms which have high metabolic rates and turnover times.

Respiratory oxygen uptake in the waters of the Sapelo Island marsh area has been detailed above. In addition there is a substantial oxygen demand by both tidal and subtidal sediments. Aerobic respiration in the Duplin River area is summarized in Table 7. The decomposer

Table 8
GENERALIZED BIOMASS ESTIMATES FOR MICROBIAL GROUPS IN ESTUARINE MARSH SYSTEMS

Depth (cm)	Total microbial biomass	Bacterial biomass	Fungal biomass	Algal biomass	Protozoan biomass
0—1	775 (175—1400)	198 (92—354)	130 (63—228)	1170 (87—3420)	—
5—6	140 (80—258)	127 (57—226)	—	21	4—46[a]
10—11	56 (23—95)	80 (29—162)	—	3	—
20—21	25 (10—35)	49 (17—88)	—	3	—
0—20	3183 (1061—5678)	1975 (818—3670)	130 (63—228)	1125 (198—5756	4—46

Note: Ranges are given in parentheses, where multiple determinants were available 0 to 20 cm values for total biomass were determined by summation of linearly interpolated values from surface to 20 cm depth. Values are given in μg C m^{-3}.

[a] 0 to 5 cm estimate based on the range given by Fenchel,[228] assuming dry weight = 20% of wet weight, and carbon = 50% dry weight.

From Rublee, P. A., *Estuarine Comparisons*, Kennedy, V. C., Ed., Academic Press, New York, 1982a, 187. With permission.

microbial community of the standing dead *Spartina* accounted for one half of the total oxygen demand, a demand equivalent to 30% of the *Spartina* shoot production. The marsh sediment microorganisms accounted for only 26% of measured oxygen demand, and respiration of subtidal surface sediments totalled only 6% of the measured oxygen demand. Thus, it can be seen that the aerobic respiration of the sediments accounted for only somewhere between 7 and 13% of the total primary production, depending on how the latter is estimated. Anaerobic fermentation, methanogenesis, and the physical transport of plant detritus from depth to surface are other processes that account for the disappearance of carbon produced by the roots and rhizomes of *Spartina*. These processes will be discussed in the next section.

Rublee[799] has summarized estimates of the algal and microbial biomass in estuarine marsh systems (Table 8). These data suggest that algae are the dominant contributors to total microbial biomass in the surface sediments where they contribute 50% of the total. At greater depths, the bacteria generally comprise most of the biomass. Of the estimated total microbial biomass the bacteria comprise 60%. Rublee[799] points out that the estimates in Table 8 are generalizations at best, and there will be site-specific variations depending on the physical, chemical, and biological conditions.

3. Anaerobic Processes in the Sediments

In seagrass beds, marshes, and mangrove swamps most of the production occurs as the belowground production of roots and rhizomes in the sediment. Although these sediments accrete organic matter, net rate of accretion is only a small percentage of the belowground production. Most of this belowground production (which as we have seen in Chapter 3, Section II.C is considerable) is decomposed anoxically within the sediments. The anoxic sediments of the intertidal flats and subtidal bottoms, on the other hand, are dependent largely on the deposition and burial of organic matter for the fueling of the anaerobic decomposition processes.

The major anaerobic processes in estuarine sediments are fermentation, dissimilatory nitrogenous oxygen reduction (DNOR), dissimilatory sulfate reduction (sulfate reduction), and methanogenesis.[1001] Organic matter from primary producers and aerobic heterotrophs enters the anaerobic cycle through fermenters and nitrogenous oxide reducers (Figure 17). Sulfate reducers and methanogens utilize relatively few substrates, most of which are the

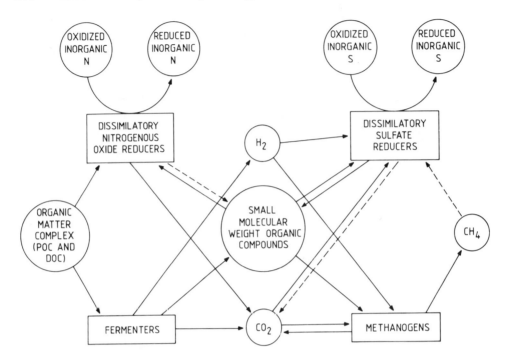

FIGURE 17. Conceptual model of the interactions of anaerobic microbial processes in salt-marsh sediments. Solid lines are confirmed fluxes; dashed lines are possible fluxes. (From Wiebe, W. J., Christian, R. R., Hansen, J. A., King, G., Sherr, B., and Skyring, G., *The Ecology of a Salt Marsh*, Pomeroy, L. R. and Wiegert, R. G., Eds., Springer-Verlag, New York, 1981, 138. With permission.)

end products of fermentation. Dissimilatory nitrogenous oxide reduction (DNOR) is a general term which encompasses a number of pathways, denitrification, dissimilatory reduction, and dissimilatory ammonia production. In the course of these processes, plant and animal matter is broken down into simpler substances, and eventually reduced to its mineral constituents.

The question has been posed as to whether the heterotrophic activity of marsh sediments is fueled primarily by the slow, constant, long-term breakdown of the large mass of roots and rhizomes, or by organic matter such as root exudates with a short turnover time. Christian et al.[117] concluded that long-term decomposition of roots and rhizomes was the main resource for microbes. However, Howarth and Hobbie[411] consider that rapid release of organic matter from living roots and rhizomes, and possibly other carbon sources as well, could play a part. These authors have attempted to estimate the magnitude of the various inputs of decomposable organic matter to sediments of short *Spartina alterniflora* in the Great Sippewissett Marsh (Table 9). These inputs are roughly divided into two pools: those believed to have a short turnover time and those likely to have a long turnover time. Valiela et al.[949] estimated belowground production of roots and rhizomes to be approximately 1680 g C m^{-2} year^{-1} and Valiela et al.[947] reported that about 20% of the weight of such roots and rhizomes is rapidly leached in buried litter bag studies. Therefore it is estimated that approximately 340 g C m^{-2} year^{-1} of readily disposable organic matter is leached from the roots and rhizomes as they die. Living roots and rhizomes may contribute additional organic matter by the sloughing of material or by excretion of DOC. These processes are virtually unstudied for marsh grasses but have shown to be important for other plants, amounting to a loss of material equivalent to as much as 50% or more of belowground structural root production.[378,550,826] According to Howarth and Hobbie[411] both sloughing of root material and excretion of organic matter may be considerable and represent net primary production not accounted for in investigations of changes in biomass of belowground components. Pro-

Table 9
RATES OF MICROBIAL HETEROTROPHY IN
STANDS OF SHORT *SPARTINA ALTERNIFLORA*
IN THE GREAT SIPPEWISSETT SALT MARSH,
MASSACHUSETTS, AND AT SAPELO ISLAND,
GEORGIA

Form of activity	Rate (g C m^{-2} year^{-1})[a]	
	Great Sippewissett	Sapelo Island
Oxygen respiration	60[VP 413]	180[P 201,913]
Denitrification	3[G 449]	10[F 1075]
Sulfate reduction	1800[G 412]	600[G 410]
Methanogenesis	16[F 412]	80[G 466]
Additional fermentative and decompositional activity not accounted for in the above measurements	712[D 411]	0[411]
Total	2591	870

[a] Adequacy of estimate: VP = very poor; P = poor; F = fair; G = good.

From Howarth, R. W. and Hobbie, J. E., *Estuarine Comparisons*, Kennedy, V. S., Ed., Academic Press, New York, 1982, 196. With permission.

duction of chemosynthetic bacteria represents another input of organic matter to marsh sediments. Howarth and Hobbie[411] then estimate that the inputs of readily decomposable carbon to the Great Sippewissett Marsh are in the range of 300 to 600 g C m^{-2} year^{-1} and possibly much more if excretions of ethanol and other end products of anaerobic plant respiration are large. Inputs of more slowly decomposable material are in the range of 1200 to 1500 g C m^{-2} year^{-1} or more. The total input is in the range of 1500 to 2100 g C m^{-2} year^{-1} or more, again possibly much more. This compares with an estimate of carbon consumption by heterotrophic activity of the order of 2600 g C m^{-2} year^{-1}.

4. Fermentation

Fermentation is the anaerobic dissimilation of organic matter in which the terminal electron acceptor is an organic compound. As a major consequence of fermentation, organic compounds of low molecular weight are produced and these can serve as substrates for the other anaerobic processes. Fermentation reactions are induced by a variety of both procaryotic and eucaryotic organisms which may be faculative or obligatory anaerobic; the former are capable of either aerobic respiration or fermentation, while the latter are inhibited or killed by oxygen.

Fermenters can utilize many substrates including simple sugars, as well as cellulose, starch, pectin, alcohols, amino acids, and purines. The end products include organic compounds, such as short-chain fatty acids and alcohols, as well as CO_2, H_2, and NH_3. Although a wide variety of organisms are capable of fermentation reactions, bacteria and yeasts are responsible for most of the activity in estuarine sediments. Some microfaunal and meiofaunal fermenters also occur in the anaerobic zones.[228,675]

Measurement of fermentation in nature is difficult because of the diversity of organisms and inorganic compounds. A number of investigators have added ^{14}C-glucose to samples of sediment and followed its incorporation into CO_2, particulate matter, and ether-soluble end products of fermentation.[120,121] The information from such experiments yields an estimate

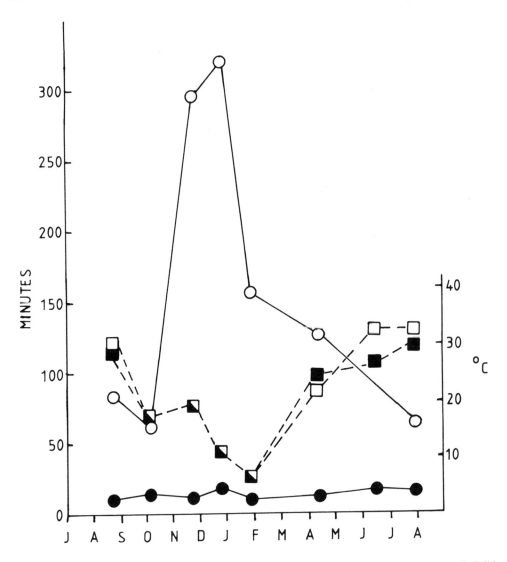

FIGURE 18.   Seasonal variations of soil temperature and the turnover time of glucose in salt-marsh soil. Solid symbols = tall *Spartina*; open symbols = short *Spartina*. Circles = turnover times; squares = soil temperatures. (From Wiebe, W. J., Christian, R. R., Hanson, J. A., King, G., Sherr, B., and Skyring, G., *The Ecology of a Salt Marsh*, Springer-Verlag, New York, 1981, 140. With permission.)

of *potential* rather than *actual* fermentation rates, since natural substrate concentrations are not determined. In the Sapelo Island marshes the effect of temperature on potential fermentation activity was studied by Hanson.[356] Turnover times of glucose uptake were always more rapid in tall *Spartina* (TS) than in short *Spartina* (SS) sediments (Figure 18). At both sites activity was highest at the surface and decreased with depth. Turnover times in TS sediments were generally minutes to approximately an hour, while in SS sediments they were generally several hours. The proportion of the ^{14}C label in the three fractions (CO_2, particulate matter, and ether-soluble end products) also differed. In SS sediments 30 to 70% of the radioactive carbon was found in the ether-soluble fraction, which contains the fermentation end products, while TS sediments yielded virtually no activity in this fraction. Christian and Wiebe[120] suggested rapid uptake of end products in TS sediments as the major difference between the two zones.

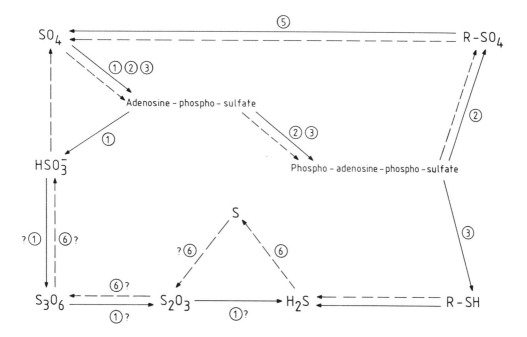

FIGURE 19. The sulfur cycle in estuarine sediments. Solid lines = reductive steps; dashed lines = oxidative steps; combined solid and dashed lines = organic-mediated, organic-organic, or organic-inorganic transformations; and a question mark = a postulated intermediate. Other pathways undoubtedly exist. They are (1) dissimilatory sulfate reduction; (2) assimilatory sulfate production; (3) assimilatory sulfate reduction; (4) sulfide mobilization; (5) sulfate mobilization; and (6) sulfide oxidation. (From Wiebe, W. J., Christian, R. R., Hansen, J. A., King, G., Sherr, B., and Skyring, G., *The Ecology of a Salt Marsh,* Springer-Verlag, New York, 1981, 142. With permission.)

5. Dissimilatory Sulfate Reduction

Dissimilatory sulfate reduction results from the anaerobic respiration of organic substrates, with sulfate being the terminal electron acceptor. This reaction is accomplished only by a small group of bacteria, called sulfate reducers, that release the sulfide produced into the environment. In some estuarine sediments, especially marsh ones, large quantities of sulfide are formed, the reaction being dependent on a plentiful supply of organic matter, mainly simple organic acids. For every mole of sulfate reduced, 2 to 6 mol of organic carbon are oxidized.

The pathway of dissimilatory sulfate reduction is shown in Figure 19. It is a complex cycle involving many possible steps. Many of the steps, those indicated by question marks, are postulated but not proven. Study is difficult because abiotic chemical conversions can take place at certain steps. Dissimilatory sulfate reduction, as carried out by *Desulfovibrio* sp. is

$$2 \text{ lactate } + \text{ SO}_4^{2-} \rightarrow 2 \text{ acetate } + 2\text{CO}_2 + \text{H}_2\text{O} + \text{S}^{2-}$$

Sulfate reducers possess cytochromes for ATP generation. The reaction takes place only under anoxic conditions but sulfate reduction can take place in oxidized sediments if anoxic microzones occur.[439]

Due to the abundance of SO_4^{2-} in sea water, sulfate reduction is a dominant process in estuarine sediments and the liberation of sulfide affects the chemical and biological environment in many ways. The presence of H_2S (or at the prevailing pH, HS^-) provides a sink for heavy metals, especially iron, which is precipitated as the black ferrous sulfide to which

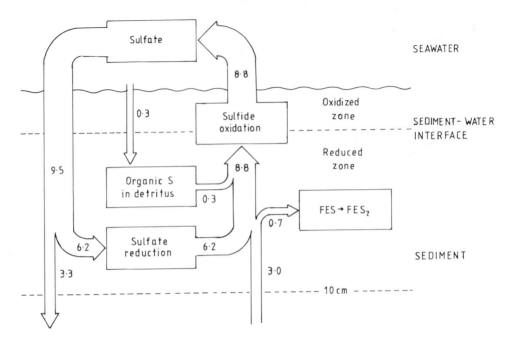

FIGURE 20.   Transfer rates of sulfur in a marine sediment. Weighted average for nine stations over 2 years. (From Jørgensen, C. B., *Limnol. Oceanogr.*, 22, 827, 1977a. With permission.)

reducing sediments owe their color. Under aerobic conditions phosphate tends to precipitate as ferric or calcium phosphate. In the presence of sulfide, phosphates are liberated from iron and again become available to the ecosystem.

The sulfur cycle has been well documented in shallow *Zostera* beds[245] and in a model system with *Zostera* as the sole carbon input.[441] The investigators measured oxygen consumption, rate of sulfate reduction and the concentrations of SO_4^{2-}, HS^-, FeS, and sulfur at different depths. Less than 4% of the H_2S derived from the system was found to be from heterotrophic metabolism; the remainder derived from SO_4. More than 50% of the total processing of added *Zostera* was due to sulfate reduction.

In an attempt to verify the results of the model system study discussed above in a natural habitat, Jørgensen[439] constructed a budget of the sulfur cycle at nine stations in a shallow coastal sediment in Denmark. He found (Figure 20) that oxygen uptake and sulfate reduction changed in parallel throughout the seasons. The weighted average daily rate of oxygen uptake for the area was 34 mmol O_2 m^{-2}, and the average daily rate of sulfate reduction was 9.5 mmol SO_4^{2-} m^{-2}, made up of 6.2 mmol in the upper 10 cm and 3.3 mmol in the deeper sediments. However, 1 mole of sulfate oxidizes twice as much organic carbon to CO_2 as does 1 mole of oxygen. Hence, we might conclude that a total of 34 mmol of organic carbon per m^2 was oxidized daily with $9.5 \times 2 = 19$ mmol being oxidized by sulfate-reducing bacteria and the remainder by processes in the aerobic zone. However, it was found that 10% of the sulfide produced remained in the sediments indicating a further 2 mmol of organic carbon per m^2 being oxidized and not leading to oxygen uptake. Hence, it was concluded that the proportion of carbon oxidation being catalyzed by sulfate-reducing bacteria was (19/36) $\times$ 100 = 53%. A budget for the sulfur cycle is given in Figure 20.

In the Sapelo Island marshes Oshrain[673] found that the patterns of sulfate distribution differed markedly between tall *Spartina* (TS) and short *Spartina* (SS) sediments. In TS sediments sulfate concentrations were nearly constant with depth, while in SS sediments they decreased with depth. Interstitial water salinity profiles also differed in the two zones; in TS sediments salinity was constant with depth, but in the SS sediments it increased with depth. The sulfide concentrations in the two sediments displayed no consistent pattern of change with depth. However, total sulfide concentrations were much greater in TS than SS sediments.

Sulfate reduction rates measured with $^{35}SO_4^{2-}$ provided further evidence of a distinction between TS and SS sediments. Active sulfate reduction in TS sediments occurred from the surface to the deepest layer measured (35 cm). However, activity in SS sediments was barely detectable below 10 to 15 cm.[847] Sulfate reduction was associated with the distribution of roots in the two sediments in a manner similar to the fermentation process and to ATP concentration[116] indicating that the processes of fermentation and sulfate reduction are linked. The differences between the two sediments are probably linked to the better drainage and lowered salinities in the TS sediments.

6. Dissimilatory Nitrogenous Oxide Reduction

Dissimilatory nitrogenous oxide reduction (DNOR) which encompasses four specific pathways (Figure 21) is (1) denitrification; (2) dissimilatory reduction (terminates at NO_2); (3) dissimilatory ammonia production; and (4) "nitrification" N_2O pathway = ammonia to nitrous oxide. Only bacteria have been shown to unequivocally possess this capability. It is an obligatory anaerobic process, since oxygen concentrations as low as 0.2 mℓ/ℓ inhibit enzyme activity.[298] The nitrogenous oxides serve as terminal electron acceptors during cytochrome-catalyzed oxidation of organic matter. When the end products are N_2O or N_2, the process is called denitrification. There have been few studies of DNOR in salt marshes, yet their sediments can be considered ideal habitats for DNO-reducing organisms; large portions of the soil are anaerobic with limited oxygen diffusion capacity, organic matter is abundant, the pH and Eh are optimal, and an extensive aerobic rhizosphere exists in which nitrate could be produced. These processes will not be discussed further here but will be considered in Volume II, Chapter 1.

7. Methanogenesis

Methane is produced under anaerobic conditions by bacteria using CO_2 or a methyl group as an electron acceptor. This is the least studied of the sediment anaerobic processes and the mechanisms are little understood in comparison with the other processes. In the Sapelo Island marshes methane release over a year varied significantly in the different marsh zones: 53.1 g m^{-2} for short *Spartina* (SS) sediments, 14 g m^{-2} for intermediate *Spartina* (MM) sediments, and 0.41 g m^{-2} for tall *Spartina* (TS) sediments.[466] This represented a loss of 8.8% of net carbon fixed in SS sediments, 0.002% in TS sediments, and about 5% for the entire marsh. It is probably that differences in water flow are responsible for the marked differences between methane production in TS and SS sediments. Recent work has also indicated that competition for substrates between sulfate reducers and methanogens is a major factor affecting methanogenesis.[466]

8. Controls and Interactions

Novitsky and Kepkay[633] measured heterotrophic activity through a vertical profile in Halifax Harbour, Canada, using sediments labeled with ^{14}C glucose, glutamate, and lactate as substrates. The gross uptake of the three substrates was characterized by a large peak of activity at the transition from aerobic to anaerobic sediments, the 40-cm level, (Figure 22). This is also the area where a peak in bacterial cell numbers (determined by acridine orange direct counts) occurs.[459] Kepkay et al.[459] and Kepkay and Novitsky[460] have shown that this transitional environment is also active with respect to the chemoautotrophic production of organic carbon. It was found that 49% of the organic carbon fixed by populations from the 40 cm horizon was present extracellularly. This means that there is a constant production of dissolved organic matter available for microbial populations. Novitsky and Kepkay[633]

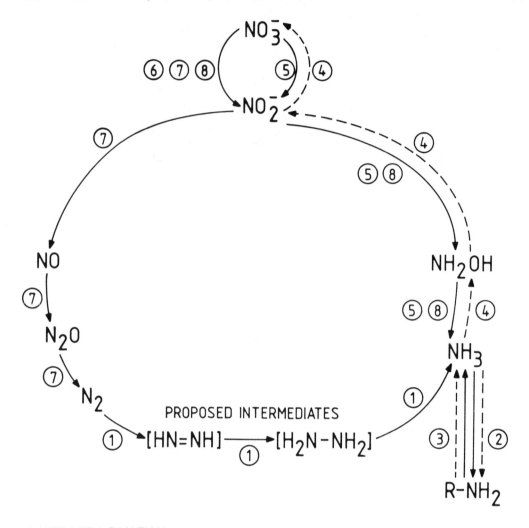

1. NITROGEN FIXATION
2. IMMOBILIZATION
3. AMMONIFICATION
4. NITRIFICATION
5. ASSIMILATORY REDUCTION
6. NITROGENOUS OXIDE REDUCTION - DISSIMILATORY REDUCTION
7.   "   "   "   - DENITRIFICATION
8.   "   "   "   - DISSIMILATORY NH_3 PRODUCTION

FIGURE 21. Diagram of nitrogen cycle. Solid lines = reductive steps; dashed lines = oxidative steps; and combined = neither oxidation nor reduction. (From Wiebe, W. J., *Ecological Processes in Coastal and Marine Systems*, Livingston, R. J., Ed., Plenum Press, New York, 1979, 471. With permission.)

consider that this is the major, if not the paramount, factor responsible for the elevated cell numbers and activity at this horizon.

Below the 40-cm horizon, uptake activity under simulated *in situ* anaerobic conditions for glucose and glutamate falls by two orders of magnitude, indicating that they are not

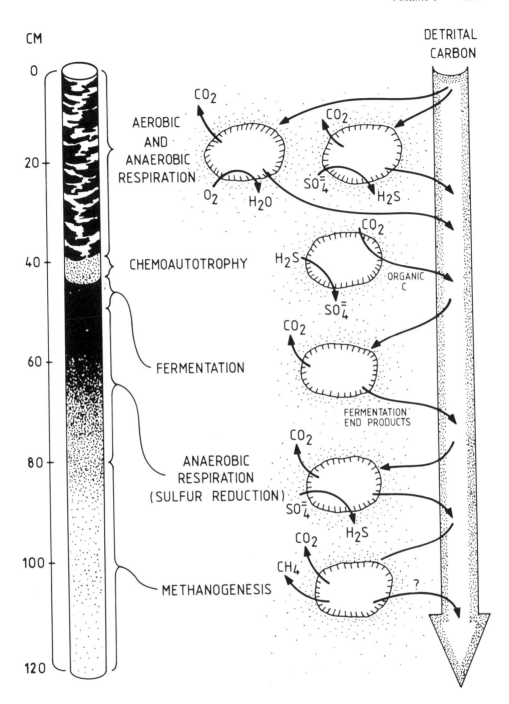

FIGURE 22.   Schematic diagram of microbial metabolism and carbon flow in Halifax Harbour sediments. Only the predominant type of metabolism is shown for the various horizons. This does not imply that they are exclusive or nonoverlapping. (From Novitsky, J. A. and Kepkay, P. E., *Mar. Ecol. Prog. Ser.*, 4, 5, 1981. With permission.)

normally encountered substrates *in vivo*. The horizons immediately below 40 cm were active with respect to sulfate reduction[460] and exhibited active uptake of lactate, a preferred substrate for sulfur-reducing bacteria. The data from the experiments carried out by Novitsky and Kepkay,[633] reflecting the fate of the organic substrate after uptake, demonstrate that a pattern of microbial metabolism exists that is related to the vertical environments encountered in vertical sections through the sediment. With the reduction in oxygen at the transition at 40 cm (RPD layer) it would be expected that fermentation would be the predominant type of metabolism. This conclusion was supported by the fact that the percent respired for all three substrates is low at the horizon below the transition.

A general succession of microbial metabolism in the sediments studied is shown in Figure 22. One of the primary factors governing this succession is the availability of oxygen and sulfate as terminal acceptors. The various groups of heterotrophs not only mineralize the organic matter but are also active in changing its composition through their metabolism. The constant production of organic carbon by chemoautotrophy at the transition between aerobic and anaerobic sediments fuels the active heterotrophic populations.[633] Novitsky and Kepkay[633] point out that the picture outlined above will vary according to the type of sediment, e.g., sand flats or silty marsh sediments. In the latter situation the RPD layer is only a few centimeters thick and in some cases this is reduced to a few millimeters.

Three broad categories of factors affect anaerobic processes: (1) abiotic factors including light, water movement, and sediment type; (2) aerobic biological processes such as root exudation, animal burrowing, and oxygen production by the benthic microflora; and (3) interactions among the anaerobic organisms, for example, substrate competition and production of stimulatory and inhibitory compounds.[1001] At Sapelo Island, sediment-water flow appeared to exert a major control on the type and magnitude of anaerobic processes. The texture of the sediment (percent of sand, clay, or silt), the slope of the sediment surface, and the rate of sedimentation all influenced the water flow characteristics.

Howarth and Hobbie[411] list the following as the likely controls of anaerobic decomposition in salt marshes:

1.   **Inherent structural components of the roots** — Roots and rhizomes contain much lignin- and ligno-cellulose which decompose slowly.
2.   **Supply of electron acceptors** — SO_4^{2-} is the major electron acceptor consumed in heterotrophic respiration. However, sulfate concentrations are apparently always high enough so that rates of sulfate reduction are not limited.
3.   **Supply of oxygen** — Molecular oxygen is apparently required for decomposition of some materials such as lignins, aromatic hydrocarbons, and perhaps some humic substances. Also, if oxygen were present, animal activity might increase the rate of decomposition and microbial activity by breaking up the roots and rhizomes.
4.   **Supply of labile organics from roots** — This may be large, and, if so, the rate of release of substrates would partially control the rate of microbial activity. Christian et al.[119] and Christian and Wiebe[120] considered root exudates to be relatively unimportant, but Howarth and Hobbie[411] believe that the experimental conditions used were such that the conclusion was not a valid one. There is a great need for careful measurements to resolve this impasse.
5.   **Metabolic buildup** — In salt marsh sediments, as in other sediments, microbes are present in high numbers but many are inactive. It is possible that a buildup of metabolites and toxins, including H_2S, limits their activity. However, this has not yet been proven.

In the Sapelo Island marshes the four anaerobic processes, fermentation, DNOR, sulfate reduction, and methanogenesis, are tightly coupled. The tall and short *Spartina* sediments operate quite differently. The major determinants of process rates are interstitial water flow, together with substrate availability.

V. MICROORGANISMS AS A FOOD RESOURCE

There is still considerable debate as to the productivity of bacteria in the water column in coastal marine areas and of their contribution to planktonic food webs. The question is whether DOM is being converted into biomass that is usable by large zooplankton. The overall efficiency with which marine systems produce fish biomass from primary production suggests that DOM is being utilized effectively rather than being respired by microorganisms.[576] As we have seen, Wiege and Smith[1003] have shown that in kinetic tracer experiments of about 10-hr duration, DOM was converted into bacterial biomass with an efficiency of 97 to 99%. Russian workers claim that bacterial production is often greater than primary production.[865,867,868] While these claims have been disputed by a number of workers,[434] there is support from a number of studies.[836,839] As Joint and Morris[434] point out, the arguments may be misleading because they relate bacterial production to phytoplankton production, implying that bacteria are competing directly with zooplankton for organic matter produced by phytoplankton. This need not be the case, and a significant proportion of the total primary production may be utilized by bacteria without depriving zooplankton of any food.

A. Consumption of Microorganisms by Zooplankton

Pomeroy,[730] as we have seen, has drawn attention to the possibility that Protista may be the major consumers of the bacteria. The major question, however, is whether bacteria or protozoa (individually or in clumps, or attached to organic matter) constitute an important food for zooplankton. Of interest in this connection are the findings of Robertson et al.[1088] on microbial synthesis of detritus-like particulates from dissolved organic carbon released by tropical seagrasses. Dried leaves of *Thalassia testudinum* and *Syringodium filiforme* released 12.6% and 19.4%, respectively, of their organic carbon (DOC) during 3 days of axenic leaching. When inoculated with microbes, the DOC was rapidly converted to bacterial aggregates of a size (from a few microns to a few millimeters) that could be ingested by macroconsumers. According to Robertson et al.[1088] macro- and microscopic aggregates are frequently present in south Florida waters during periods of low spring tides possibly as a result of pulse releases of seagrass DOC. Large populations of ciliates and flagellates also developed in the cultures presumably feeding on the unaggregated bacteria. The protozoan community was dominated by free-swimming flagellates. In treatments containing the residual macroparticulate organic carbon (MPOC), 75 to 95% of the bacteria were present attached to the leaves and suspended aggregates were not observed.

Robertson et al.[1088] present a model of dissolved and particulate carbon flow through the initial stages of detritus-based food chains (Figure 23). This model distinguishes between utilization of dissolved and particulate plant material. The contribution of particulate matter to primary consumers involves colonization, decomposition, and disintegration by microbes. Utilization of the dissolved fraction can occur either through flocculation and subsequent microbial colonization, or through the uptake of DOC by bacteria, followed either by ingestion of the cells into intermediate consumers such as Protozoa, or by the formation of bacterial aggregates. The utilization of bacteria as food is precluded for many animals because of their inability to trap individual cells.[235] Therefore the phenomenon of protozoan grazing, as emphasized by Pomeroy,[730] may increase the contribution of DOC to secondary productivity through the formation of food particles of more suitable size for a variety of consumers.[548,824] As Robertson et al.[1088] point out, the importance of each of the pathways depicted in Figure 23 has received little study and certainly must vary from location to location, depending on considerations such as physical energy and DOC concentrations and release rates. There is a need for future research directed at the quantification of the pathways.

In a recent paper Hobbie and Lee[399] have advanced the view that extracellular mucopolysaccharides of microorganisms are more abundant than the microorganisms themselves,

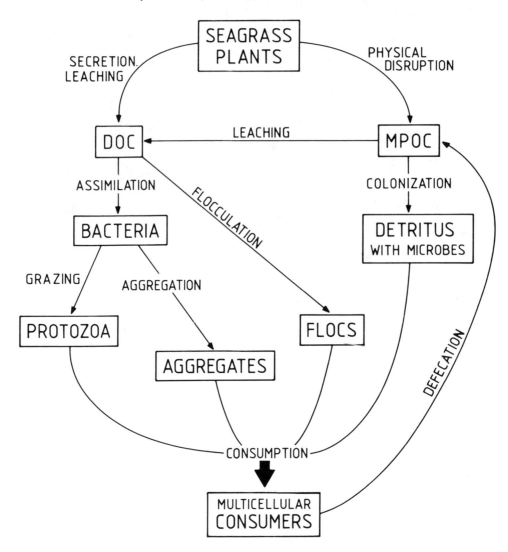

FIGURE 23.   Pathways of dissolved and particulate carbon flow through the initial stages of detritus-based food chains. (From Robertson, M. L., Mills, A. L., and Zieman, J. C., *Mar. Ecol. Prog. Ser.*, 7, 279, 1982. With permission.)

and that they may provide a major source of food for consumers. Bacteria produce extra-cellular polymer fibers with which they attach to surfaces.[126,682] Geesey et al.[280] reported that the mucilage-producing algae on rock surfaces in streams may be as important as the bacteria in producing polysaccharide material; the mass of extracellular material may be four to five times greater than the living biomass. It is also possible that bacteria remove dissolved organic matter from the water and produce large quantities of particulate material by extracellular secretion and death.[683] If this hypothesis is true it would explain some of the discrepancies between energy needed by detritivores and the microbial food available.

One of the best documented studies on the utilization of detritus by zooplankton has been carried out in the Patuxent River estuary, Maryland.[385,387] A carbon budget was constructed for a population of the copepod *Eurytemora affinis,* and it was found that during March, April, and May the carbon demands of the copepods greatly exceeded primary production. During April, the carbon demands exceeded supply by factors of 5 to 75 (Figure 24). Since

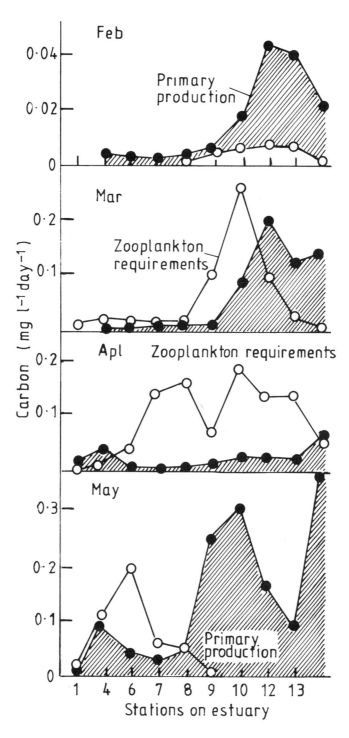

FIGURE 24.   Primary production and estimated carbon requirements of the zooplankton at various stations in the Patuxent River estuary. (From Heinle, D. R. and Flemer, D. A., *Mar. Biol.*, 31, 235, 1975. With permission.)

during this period the amount of detritus greatly exceeded that of the phytoplankton, it was concluded that the detritus must be a food supply for the copepods. The hypothesis was tested by experiments rearing copepods on standardized diets of detritus, with and without microorganisms, using copepods grown on algal diets as controls. The conclusion was that copepods could grow and produce eggs on a diet of detritus when abundant microorganisms were present, or on a mixed diet of algae and detritus, but they did not thrive on a diet of detritus that had been autoclaved to control the microbial flora. The results also suggested that copepods could do very well on a diet of ciliates. This conclusion is supported by Sorokin's[865] work on the consumption of bacteria by zooplankton. Heinle et al.[387] investigated the role of detritus as food for a number of estuarine copepods. A variety of detrital foods derived from *Spartina* and other marsh plants were fed to the copepods *Eurytemora affinis* and *Scottocalanus canadensis*. They found that when a rich and abundant microbiota was present the copepods did well. Ciliated protozoans appeared to be particularly important in the transfer of detrital energy to copepods. When fed on a protozoan infusion, egg production was as high if not higher than for copepods fed on algal cultures. As found by Robertson et al.,[1088] protozoa were particularly numerous when cultured in an infusion of naturally dried *Spartina*.

In the mesohaline zone of the Westerschelde estuary in the Netherlands, the dominant copepod *Eurytemora affinis* is abundant throughout the year. In the nearby mesohaline of Lake Veere, *Arcatia tonsa* is the dominant copepod, but its abundance is limited to the summer. Bakker et al.[31] suggests that the continuous supply of suspended detritus in the Westerschelde explains the high biomass *Eurytemora*, while the seasonal availability of phytoplankton limits the summer abundance of *Acartia*.

In the Cochin Backwater in India, (Figure 25) phytoplankton maximum occurred during the midsummer, but there was no corresponding zooplankton maximum.[984] There was a clear peak of Copepoda in the autumn and minor peaks of other zooplankton in the spring. It would appear that the zooplankton must feed on sources other than endogenous phytoplankton. Abundant organic matter is supplied by the rivers particularly during the monsoon season and this, with its associated microbes, is probably their main food.

As we have seen in Section IV. A.2, Pomeroy[730] was the first to emphasize the role of microheterotrophs in pelagic food webs and to integrate them into a dynamic model of the coastal ecosystem. The study of Newell and Linley[1084] which established that heterotrophic microflagellates comprised a significant proportion of the heterotrophic community of protozoans and bacteria in waters off the western approaches to the English Channel, has also been discussed. Newell and Linley[1084] estimated that the flagellate consumption requirements accounted for 81 to 97% of the production of their bacterial prey (e.g., at a mixed water station bacterial production was estimated at 625.3 mg C m^{-2} day^{-1}, while microflagellate consumption was estimated at 606.9 mg C m^{-2} day^{-1}). Similar microflagellates are an important component of estuarine plankton.

In addition to the flagellates, ciliates are also an important component of coastal and estuarine plankton. Among the larger ciliates, a dominant group is the loricate ciliates belonging to the suborder Tintinnida. These ciliates graze most effectively on particles in the nannoplankton size range.[52,383,384] As mentioned previously, nannoplankton can seasonally dominate phytoplankton biomass and productivity in estuaries.[205,206,548,956] It is therefore likely that under such conditions a food-chain comprising nannoplankton-tintinnids-metazoan zooplankton could play an important role in the pelagic food web dynamics. Robertson[795] carried out a series of laboratory experiments to determine the impact of crab *(Uca)* zoea and the estuarine copepods *Acartia tonsa* and *Tortanus setacaudatus* upon the co-occurring tintinnids *Favella panamensis* and *Tintinnopsis turbulosa*. Robertson[795] found that both *Uca* zoea and *Acartia* ingested both tintinnid species, while *Tortanus* ingested *Favella* but not *Tintinnopsis*. He concluded that when the phytoplankton was dominated by small (diameters

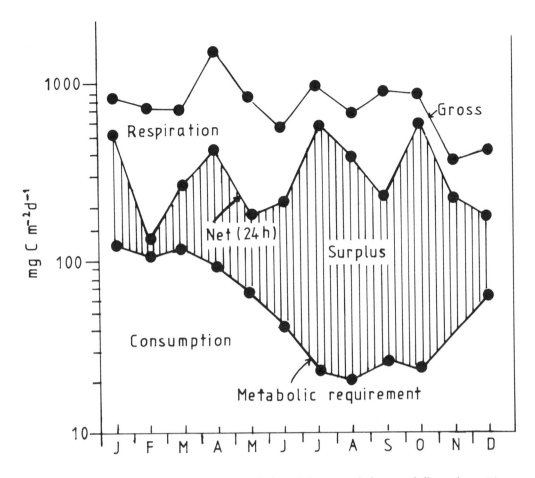

FIGURE 25. Primary carbon production and respiration relative to zooplankton metabolic requirement (consumption). (From Smayda, T. J., *Estuaries and Enclosed Seas,* Ketchum, B. H., Ed., Elsevier, Amsterdam, 1983, 65. With permission; modified from Reference 750.)

<10 μm) species, tintinnids in concentrations exceeding 10^3 organisms ℓ^{-1} can be important items in the diet of *Acartia*. At lower tintinnid concentrations, or when algal species with diameters >10 μm are present in significant concentrations, tintinnids merely supplement algae in the diet of *Acartia*.

The possible pathways which may be involved in the utilization of organic detritus and associated microorganisms are shown in Figure 26. Direct consumption of detritus (7) is relatively unimportant. As we have seen, it is acceptable providing it is combined with algal cells (8). With *Eurytemora affinis,* Heinle et al.[387] have shown that a two-stage route (1 and 4) via bacteria can sustain growth. Protozoa are also a good food source which implies either a two-stage (5 and 3) or a three-stage route (1, 2, and 3). It is probably that many of the routes shown in Figure 26 are important and that combinations of pathways provide an important stabilizing effect on the pelagic food webs in estuaries. As we have seen, the pulsed seasonal production of detrital carbon from tidal marshes results in similarly pulsed production of zooplankton in some estuaries without a time lag.[386] Thus, it is clear that bacteria can provide a food source for zooplankton, but the effectiveness of this and what proportion of food intake is involved has yet to be determined in detail.

B. Consumption of Microorganisms by Benthos

Odum and Heald[662] divided benthic detritivores into shredders and grinders, such as

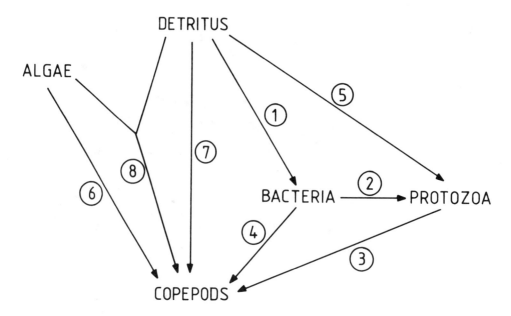

FIGURE 26.   Possible pathways involved in the utilization of organic detritus and associated microorganisms by copepods. (From Heinle, D. R., Harris, R. P., Ustach, J. F., and Flemer, D. A., *Mar. Biol.*, 40(4), 351, 1977. With permission.)

amphipods[229] which chew detritus particles, deposit feeders which select particles from the sediments, and filter feeders. There are three factors affecting detrital utilization:[924] (1) mechanical breakdown of particles, (2) orthochemical processes (i.e., leaching and sorption), and (3) nutritional composition, trophic efficiencies, and trophic transfer. The mechanical breakdown of large recognizable particles of plant materials or fecal pellets into amorphous material is a function of the mechanical activity of physical processes such as wave action and sediment abrasion, bioturbation by meio- and macrofauna, and the action of shredders and grinders. The end result of these processes is the reduction of the particle size of detritus derived from dead plant material, an increase in the surface-to-volume ratio, and the ratio of bacterial biomass to detrital biomass. Data from Fenchel[229] showed that as particle size decreases both the oxygen uptake and the number of microorganisms per unit weight increased markedly. One consequence of the increase in microbial biomass, as we have seen in Section II.C, is that the protein content of the detritus also increases with decreasing particle size. This nitrogen must be taken up from the water by the microorganisms.[237]

The relationship between particle size and microbial densities in estuarine sediments suggests that potential food resources on particles of organic and inorganic origin increase logarithmically in the fine silts and muds near the sites of plant production. Although it is believed that the biomass of bacteria in sediments is insufficient to provide more than a small fraction of the food requirements of detritivores, it is important to take into account the high rates of production of bacteria which may achieve with multiplication rates of up to six times per day.[364] As Newell[624] points out, the significance of microorganisms in the nutrition of deposit feeders is demonstrated by two lines of evidence. First, by correlations between the biomass of certain selected deposit feeders and the organic carbon and nitrogen content of deposits,[533,623] and more recently with bacterial numbers in the sediments.[939] Second, direct measurements of the assimilation of different components of the food of a variety of animals from both marine and fresh water habitats suggests that microorganisms may be primarily digested, while the component of organic detritus is voided as indigestible feces which are later colonized by heterotrophs and autotrophs.[364,365,623,639,835]

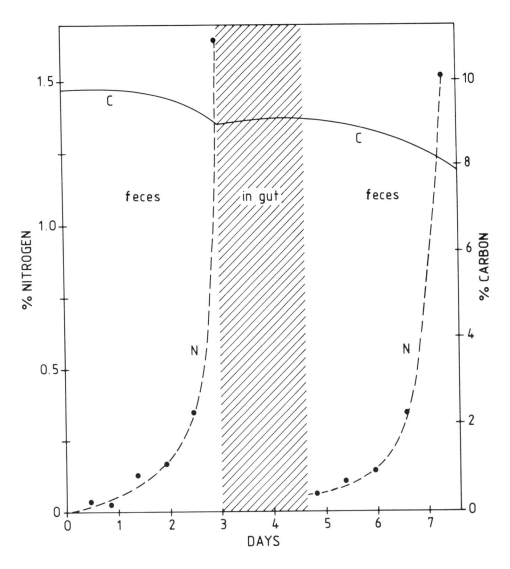

FIGURE 27.   Graph showing the percentage of organic carbon and nitrogen in feces of *Hydrobia ulvae* at 18°C in sea water. (A) Before and (B) after feeding to *Hydrobia ulvae*. Broken line indicates organic carbon; continuous line indicates organic nitrogen. (From Newell, R. C., *Proc. Zool. Soc. London*, 144, 33, 1965. With permission.)

Newell[623] attempted to demonstrate experimentally that the mud snail *Hydrobia ulvae* and the bivalve *Macoma balthica* digest protein in the form of microorganisms from the sediments in which they live. A large number of *Hydrobia* were placed in filtered sea water and their fecal material collected, a subsample being analyzed for organic carbon and nitrogen. The organic carbon was found to be high while the organic nitrogen was low (Figure 27). The remainder of the collected feces was cultured at 18°C and subsampled over a period of 3 days. From Figure 27 it can be seen that the organic nitrogen component increased from 0.025 to 1.75% of the dried fecal material. At the same time the organic carbon content fell from about 10 to 8% of the dried weight. This was taken to indicate that a population of microorganisms had built up on the voided feces and had broken down a proportion of the organic carbon in the process. Stored *Hydrobia* were then fed the cultured feces and after 1.5 days the freshly voided fecal pellets were collected and analyzed. It was found that the organic nitrogen component was low (c. 0.075%), while the organic carbon com-

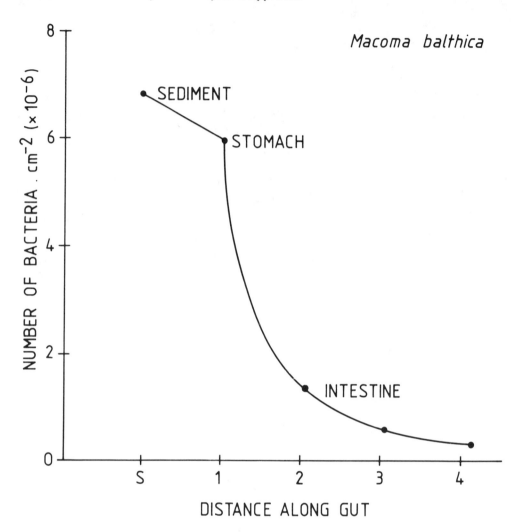

FIGURE 28.   Graph showing the numbers of bacteria cm^{-2} in sediment prior to ingestion S, (1) in the stomach; (2) in the anterior; (3) in the mid-gut; and (4) in the hind-gut of *Macoma balthica*. (From Newell, R. C., *The Biology of Intertidal Animals*, 3rd ed., Marine Ecological Surveys, Faversham, Kent, 1979, 781 pp. With permission; based on Reference 231.)

ponent was similar to that in material fed to the animals. Subsequent culturing of the feces resulted in an increase in the organic nitrogen content.

This suggested that fecal material is subject to a cycle which is repeated many times until the organic carbon is used up. Similar experiments with the lugworm *Arenicola marina* and the bivalve *Macoma balthica* confirmed the digestion of the microorganisms in the sediments and fecal pellets they ingested during passage through the gut. Again *Macoma balthica* has been shown to remove 95 to 99% of the bacteria, all the Protozoa, and 50 to 75% of the diatoms from the sediment during passage through the gut[231] (Figure 28).

Studies on the microbial colonization of fecal pellets and their utilization as food by the same or other species (coprophagy) has been carried out on a range of species including crabs, shrimps, polychaete worms, bivalves, hydrobiid snails, and mullets. Frankenberg et al.[259] and Frankenberg and Smith[260] have reviewed the utilization and trophic significance of fecal pellets in a variety of organisms. Frankenberg et al.[259] showed that the burrowing shrimp *Callianassa major* can produce as much as 456 ± 118 fecal pellets per burrow per day on Sapelo Island, Georgia. This amounts to between 175 ± 46 and 2600 ± 670 pellets m² day⁻¹ depending on the population density. These pellets are clearly a major component of the surface sediments. Feeding experiments indicated that several members of the local communities such as the blue crab *Callinectes sapidus,* the hermit crab *Pagurus,* the bivalve *Mulinia lateralis,* and the amphipods *Parahaustorius longimerus* and *Neohaustorius schmidti* consumed these pellets. Frankenberg and Smith[260] subsequently reviewed the evidence of ingestion of fecal material by a variety of marine animals and showed that consumption rates varied from 0 to 83% of the body weight in fecal pellets during 48 hr. There are major differences in the rates at which fecal pellets from different animals are ingested. In some cases the ingestion of fecal material was sufficient to meet the total metabolic requirements of the animals.

The studies outlined above gave rise to the concept of enrichment of detritus as a food resource by colonization by microorganisms. Tenore[917] has shown that the incorporation of detritus from eelgrass by the polychaete *Capitella capitata* increases with the age of the detritus and that this probably represented the increasing microheterotrophic community living on the surface of the detritus. Further, the worms appeared to initially select the fine particles on which a microbiota first develops. Tenore et al.[925] have subsequently shown that the presence of meiofauna dominated by nematodes has an important effect on increasing the rate of incorporation of detritus from the eelgrass *Zostera marina* into the polychaete *Nepthys incisa.*

According to Tenore and Rice,[924] emphasis on the concept of enrichment seems an oversimplification of the role of aging and microbial decomposition of detritus. They point out that the decompositional role of microbes (bacteria and fungi) that depolymerize complex organic materials into caloric substrates utilizable by macroconsumers should not be deemphasized.[592,923] Also, many types of detritus, especially macroalgal, contain organic substances that are themselves readily available to macroconsumers.[921,923] As mentioned above, there is some evidence that nitrogen in aging detritus may be due to the accumulation of extracellular products. Suberkropp et al.[893] suggested that part of the nitrogen built-up in leaf litter results from the condensation products of microbial proteins and plant phenolics. Rice,[787] in studies of decomposition of detritus derived from *Spartina, Rhizophora,* and the brown macroalga *Spathoglossum,* found that microbially produced extracellular proteins combine with highly reactive phenols and are exported from or retained in the detritus, depending on the mobility of the phenols.

Macrodetritivores such as *Capitella capitata* incorporate carbon from many different detrital sources that are in various stages of decomposition.[921] This detrital pool may range from high nitrogen easily decomposable seaweed detritus, to low nitrogen decay-resistant vascular plant detritus. In the latter, most of the energy remains unavailable for immediate assimilation, while in the former more calories are available sooner. Tenore[921] showed that available calories and nitrogen content, but not total calories, of the detritus were significant contributors to the biomass (carrying capacity) of *Capitella.* Hanson[358] has studied the effects of organic nitrogen and total caloric and/or available caloric-content of detritus on the regulation of microbial dynamics using 23 types of detrital ration. The results of the series of experiments suggested the rate of organic enrichment to benthic systems is an important variable to microbial dynamics, as is the organic nitrogen and/or available caloric content of the ratio, but not the total caloric content. It is clear that microbial nitrogen is not the only important variable that influences the food value of detritus. It seems that microbial extracellular products, biomass turnover, and chemical adsorption or organic nitrogen are important.[358,399,787] It is clear that studies of the nutritional aging chemistry of detritus and new research directions are needed to fully understand how microbes influence detritus decomposition and contribute to the nutrition of consumers, and how detritus influences microbial processes in decomposition.

1. Factors Affecting Detrital Utilization by Macroconsumers

Tenore and Rice[924] have recently reviewed the factors affecting detrital utilization. As they point out there is a growing body of information gained from laboratory studies on the effect of quantity and quality (biochemical composition) of detritus on detrital availability. Detrital source, amount, and state of decomposition can affect nutritional quality and its utilization by benthic deposit feeders.[918,919,921,923,925]

According to Tenore and Rice[924] the biochemical composition of detritus changes is due to: (1) leaching of substances that inhibit microbial growth or macrobenthic feeding and (2) nutritional enrichment of detritus by either microbial biomass itself, or transformation products produced by microbes. Detritus ages at different rates depending on its biochemical composition and resistance to decay. Thus, parts of a detrital pool will become available to macroconsumers at different times. This means that food availability to macroconsumers is spread over time.

There are limited data on the exact role of micro- and meiobenthos on detrital availability to macroconsumers. The presence of meiofauna increases both detrital oxidation and utilization by the polychaetes *Nepthys incisa*, *Nereis succinea*, and *Capitella capitata*,[73,925] but the mechanisms of the interactions are undocumented. When dealing with detrital-based food chains, as Tenore and Rice[924] emphasized, we are dealing with a "detrital trophic complex" comprising microbial populations and/or meiobenthos that exhibit an array of trophisms (i.e., nutritional sources) and feeding interactions. These authors use a ratio P:O, i.e., net production of the macroconsumer (P) to detrital oxidation by the detrital trophic complex (O) (i.e., the total metabolic expenditure of bacteria, microfauna, and meiofauna), to measure the cost of net production of detritivores. This ratio reflects the relative amounts of the resource that are conserved and expended in food chain transfer from detritus to macroconsumers.

Experimental studies of detrital utilization by deposit-feeding polychaetes in laboratory microcosms show that P:O is affected by: (1) the presence of ciliates and meiobenthos; (2) detrital source and amount; and (3) length of detrital aging.[73,925,1091] In general, the presence of ciliates and meiofauna results in an increase in the P:O ratio. Results of utilization rates of aging detritus derived from different sources (periphyton, *Gracilaria*, *Spartina*) suggests that: (a) overall the more rapidly degradable the detritus, the higher the P:O, and (b) with time of aging, P:O increased for all three detritus types.[1091]

Tenore and Rice[924] have developed a conceptual model of the factors affecting secondary production in detrital-based food chains (Figure 29). In the model, lower case letters are specific transfer rates from one storage to another, and the subscript indicates the storage to which energy is transferred. D_u and D_a are gross rates of unavailable and available detritus input from outside the detritus-based system. Only available detritus (A) enters the macroconsumer level, either directly or indirectly as microbial biomass. Unavailable substrates (u) must be microbially processed, the specific rate of conversion to available material (U_a) being enhanced by inorganic N and by meio- and macrofaunal activity. Bacteria are consumed by grazing meiofauna (M) which in turn may (or may not) be consumed by the detritivores. Nonliving organic material from these organisms may be returned to the detrital pool (U or A). Because either N content or available caloric content of the detritus may be the important limiting factor, a factor α (indicating whether detritus organic N or available caloric content is the limiting factor to macroconsumer production) has been incorporated by Tenore and Rice[924] into the rate equation for the secondary production of the macroconsumers.

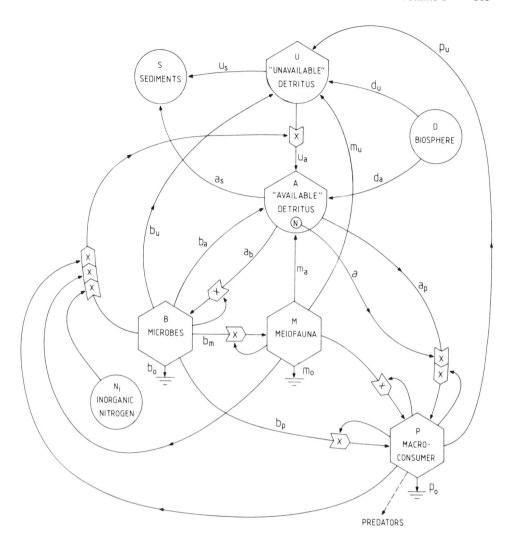

FIGURE 29. A model of trophic factors affecting production of macroconsumers in detrital food chains. U = "unavailable detritus"; A = "available detritus"; B = microbes; M = meiofauna; P = secondary production; N_i = inorganic nitrogen; and S = losses from detritus ecosystem. (From Tenore, K. R. and Rice, D. L., *Marine Benthic Dynamics,* Tenore, K. R. and Coull, B. C., Eds., University of South Carolina Press, Columbia, 1980, 334. With permission.)

Chapter 5

CONSUMERS

I. INTRODUCTION

The components and distribution patterns of the estuarine fauna have been dealt with in Chapter 2. Here we will be concerned in more detail with the functional roles they perform. There are many approaches that could be used in a discussion of the roles of the consumer populations. They can be subdivided in habitat terms into zooplankton, nekton, epifauna, and infauna. The latter two categories comprise the benthic fauna and this can be further subdivided on a size basis into micro-, meio-, and macrobenthos. An alternative approach is to subdivide the populations on the basis of their functional roles in the estuarine food web into primary and secondary consumers. The former category on the basis of their primary food source can be subdivided into herbivores, detritivores, and omnivores. As we shall see, such a classification is not particularly useful in a consideration of estuarine ecosystems and will not be followed. Instead various groups of organisms will be selected and approached from the viewpoint of their functional role in the ecosystem. There is a vast amount of literature on estuarine fauna and only a very small portion of it can be discussed in this section. For more detailed discussions readers are referred to Green[330] Perkins,[710] Day,[177] Gray,[329] and McLusky.[559]

II. ZOOPLANKTON

The zooplankton of estuaries in contrast to that of the ocean is limited by two features. First, the turbidity, can limit phytoplankton production and thus limit the food available for zooplankton, although as we shall see many estuarine zooplankters have been shown to feed on bacterial aggregates and detritus, and thus phytoplankton production may not be as limiting as has been assumed. Secondly and often more importantly, currents, particularly in small shallow estuaries or those dominated by high river flow, can carry members of the zooplankton out to sea. A proportion can return on the incoming tide but a varying proportion will be dispersed in coastal waters. Grinley[338] has recently reviewed the zooplankton of estuaries with special reference to South African estuarine systems, while Miller[594] has compiled a general review of estuarine zooplankton.

A. Composition and Distribution
1. Composition
As in the adjacent coastal waters, the zooplankton of estuaries can be subdivided into the *holoplankton* species, which are planktonic throughout their life, and the *meroplankton* species, which are planktonic part of the time — for example at night only. The meroplankton also includes the larval stages of benthic invertebrates which spend part of their life cycle in the plankton.

The holoplanktonic fauna is dominated by small species of copepods, although most of the taxonomic groups present in neritic seas may be found in estuaries, particularly if marine dominated with high salinities. Chaetognaths, ctenophores, and ciliates including Tintinnidae may be common, and large rhizostomid jellyfish are seasonally abundant, but planktonic foraminifera, hydroid medusae, euphausiids, salps, and larvaceans are usually scarce. Meroplanktonic mysids, cumaceans, tanaidaceans, amphipods, and isopods are abundant at night but spend the daytime on the bottom. Invertebrate larvae are seasonally abundant, particularly the larvae of barnacles, crabs, and polychaete worms. Pelagic eggs, larvae, and juveniles of fishes are seasonally common.

Several genera of copepods are characteristic of estuaries in different parts of the world. Species of *Eurytemora, Acartia, Pseudodiaptomus,* and *Tortanus* are particularly well represented. The mouth area is usually occupied by euryhaline marine species of *Paracalanus, Centrophages, Oithona, Pseudocalanus, Temora,* and many harpacticoids including *Enterpina* and *Harpacticus.* Grinley[338] lists 31 species of copepods that are widespread in South African estuaries. Prominent genera are *Acartia, Centrophages, Oithona, Paracalanus,* and especially *Pseudodiaptomus.* Australian estuaries are often dominated by other genera such as *Boeckella, Gladioferens,* and *Sulcanus.*

A dominant copepod in many estuaries is *Acartia tonsa,* a euryhaline (0.3 to 30 + ‰) and eurythermal (5 to 35°C) species.[159,292] In studies of the zooplankton of Barataria Bay, Louisiana, *A. tonsa* accounted for 60% of the organisms counted by Gillespie[292] and 83% of those counted by Cuzon du Rest.[154] In Biscayne Bay and Card Sound, Florida, *A. tonsa* and *Paracalanus parvus* were the dominant species.[763] Mean annual numbers in Card Sound and south and central Biscayne Bay were 300, 187, and 2833 m^3 for *A. tonsa,* and 524, 87, and 516 for *P. parvus,* respectively. In the Damariscotta River estuary, Maine, Lee and McAlice[503] estimated that there were 28,882 copepods m^{-3}, with 8571 *A. tonsa* m^{-3}, 7360 *Eurytemora herdmanni* m^{-3}, and 3753 *Acartia clausi* m^{-3}.

Woolridge[1043] has studied the zooplankton of the mangrove-fringed Mngazana estuary in Transkei, South Africa. He found Mysidacea to be more important than in nonmangrove estuaries making up more than 75% of the zooplankton. *Gastrosaccus brevifussura* and *Mesopodopsis africana* were common, the latter reaching a density of over 10,000 m^{-3}. Some species of copepod also become abundant at times. *Pseudodiaptomus hessei* reached 42,700 m^{-3} in Richards Bay, South Africa, a shallow estuary on the east coast,[339] while *Acartia natalensis* reached 10,600 m^{-3} and *Oithona similis* reached 123,000 m^{-3} in the Mngazana estuary.[1043]

2. Seasonal Changes

Zooplankton in estuaries usually reaches its peak in the spring and summer. In the salt marshes east of the Mississippi River, peak numbers and biomass 8 g dry wt m^2 occurred as a sharp peak in mid-April[154] (Figure 1). However, in Card Sound and Biscayne Bay the general pattern of seasonal variations of the zooplankton was one of rapid fluctuations throughout the year,[763] with a low point in midsummer (Figure 2). Possible explanations include the high maximum summer temperatures (30°C +), shortage of food over the summer, or seasonal weather patterns. Basic seasonal patterns can frequently be disrupted by flooding or by opening and closing of the estuary mouth.

3. Diversity and Distribution Patterns

In almost all estuaries, the greatest diversity occurs near the mouth where a wide range of neritic species occur. In the lower reaches, various marine species penetrate to different degrees. If species diversity is plotted against salinity for a particular estuary, it is found that peak diversity occurs near 35‰ and diversity decreases with greater variations in salinity. Below 2‰ an increase in diversity may occur as a result of the penetration of fresh water species (Figure 3). On the basis of the observed patterns, the zooplankton of estuaries may be divided into four components on the basis of salinity tolerance: (1) a stenohaline marine component penetrating only into the mouth, (2) a euryhaline marine component penetrating further up the estuary, (3) a true estuarine component comprising species confined to estuaries, and (4) a fresh water component comprising species normally found in fresh water (Figure 4).

4. Vertical Migration

In many estuaries, there is hardly any zooplankton in the water during hours of daylight.

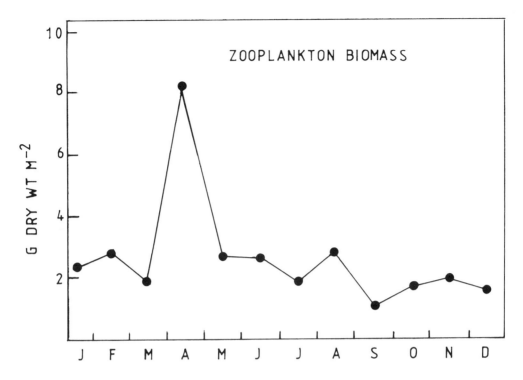

FIGURE 1.   Annual cycle of zooplankton from salt marshes east of the Mississippi River. (From Cuzon du Rest, R. P., *Publ. Inst. Mar. Sci. Univ. Tex.*, 9, 132, 1963. With permission.)

The settled volumes of plankton obtained in surface hauls is commonly less than one tenth of that found at night. Owing to the differences in the vertical migration behavior of different species, their relative numbers may change markedly.

Grindley[334] has carried out studies of the diurnal fluctuations of zooplankton at the surface in six different estuaries and lagoons in South Africa. Vertical migration appeared to occur in almost all of the species of zooplankton observed under a variety of conditions. The patterns of vertical migration differed from the usual type found in oceanic plankton. The plankton rose *after* dark, reached peak concentrations at the surface near the middle of the night, and descended *before* dawn. This pattern of behavior which is best described as an endogenous rhythm has been confirmed by laboratory experiments in the case of the copepod *Pseudodiaptomus hessei*.[334] It was found that *Pseudodiaptomus* migration was largely inhibited by water of lowered salinity on the surface.[333]

According to Grindley[338] the survival value to estuarine zooplankton of inhibition by low surface salinity is probably in helping them to maintain their position in the estuary. In estuaries there is usually a net surface outflow of low salinity water, and a compensating influx of saline waters along the bottom which maintains the salinity gradient. By vertical migration, the planktonic animals drift alternately upstream and downstream tending to maintain their position in the estuary. When floods occur, however, the planktonic animals would be in danger of being swept out to sea if they rose to the surface. The observed inhibition of migration produced by a strong salinity discontinuity would prevent this. This behavioral mechanism would reduce the coefficient of reproduction required to maintain a zooplankton population in an estuary for a given exchange ratio of water escaping seaward at each tidal cycle.[333,334]

During a flood much of the plankton is swept out of the estuary, but it reappears surprisingly quickly.[338] Conversely, if the flow is sluggish and the residence time of the estuarine water is long, planktonic populations can build up high densities.

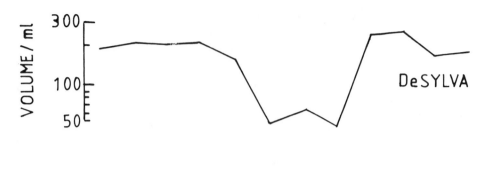

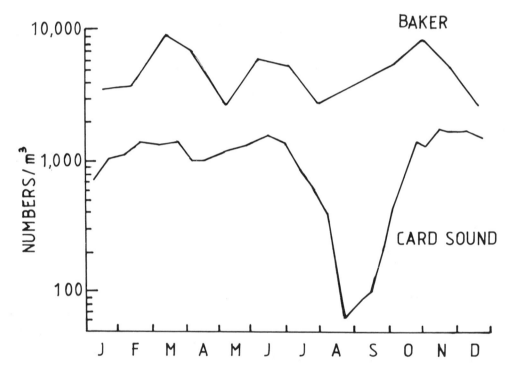

FIGURE 2.   Seasonal changes in total numbers of macrozooplankton from a 200-μm net in Card Sound[764] and Central Biscayne Bay,[1064] and settled volume from a 500-μm net.[194] (From Reeve, M. R., *Estuarine Research,* Vol. 1, Cronin, L. E., Ed., Academic Press, New York, 1975, 363. With permission.)

B. Biomass and Production

The biomass of zooplankton is often significantly higher in the upper reaches of estuaries where diversity is low and estuarine species predominate.[338] Grindley and Woolridge[339] have recorded densities for the copepod *Pseudodiaptomus stuhlmanni* up to 42,700 m^{-3}.

Zooplankton biomass is usually estimated from the settled volume and expressed as dry wt m^{-3}, or g C m^{-3}. A significant correlation coefficient has been shown between settled volume and biomass (dry wt). Zooplankton biomass can vary greatly from time to time and place to place within an estuary and consequently single figures are of little value.

For estuaries around the South Africa coasts, Grindley[338] records the following range of values:

1.   East coast estuaries: 1.2 to 1200 mg m^{-3}
2.   South coast estuaries: 1.0 to 112.7 mg m^{-3}
3.   West coast estuaries: 1.4 to 1014.5 mg m^{-3}

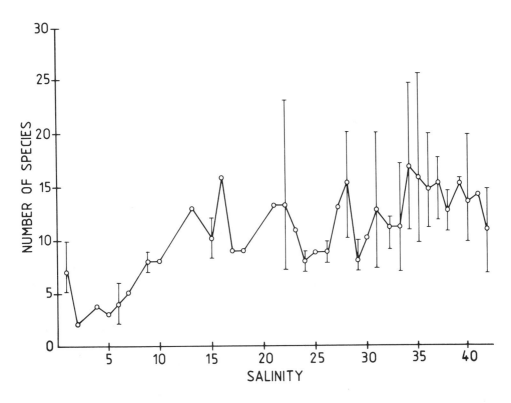

FIGURE 3. Zooplankton diversity (species richness) of samples taken in the Swartkops estuary in 1967 to 1968. Means and ranges are shown for each salinity. (From Grindley, J. R., *Estuarine Ecology with Particular Reference to Southern Africa*, Day, J. H., Ed., A.A. Balkema, Rotterdam, 1981, 127. With permission.)

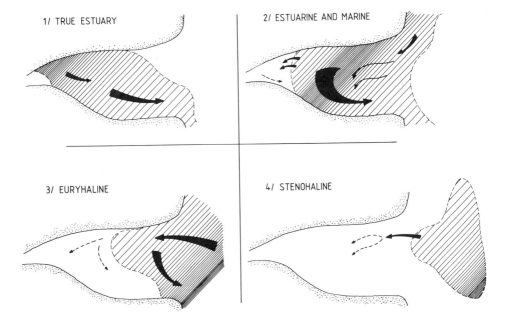

FIGURE 4. Components of the estuarine zooplankton.

However, the majority of biomass figures fall between 10 and 100 mg m^{-3}. In the Damariscotta River estuary, Maine, Lee and McAlice[503] record a mean biomass of 86 mg m^{-3}. In Barataria Bay, Louisiana, when *Acartia tonsa* constituted an average of 80% of the zooplankton, Day et al.[184] estimated a dry weight biomass of 2.5 g m^{-2}. Other values recorded in the literature include 3.45, 5.28, and 2.02 mg m^{-3} for the St. Andrews Bay system in northwest Florida,[406] and the 44.8 mg m^{-3} for Crystal River in South Florida.[586]

Values for annual production like those for biomass vary widely. For Richards Bay, South Africa, Grindley and Woolridge[339] estimated net secondary zooplankton production at 12 mg (dry wt) m^{-3} day^{-1} or approximately 4.4 g m^{-3} year^{-1}. This gave a P:B ratio of 0.04/24 hr or a P:B ratio of 13/year. Reeve,[763] for central Biscayne Bay and Card Sound, Florida, calculated a mean daily production of 92.7 mg m^{-3} (dry wt) and 9.2 mg m^{-3}, respectively, corresponding to 46.1 and 4.6 g C m^{-3} (carbon as 50% of organic dry wt). This may be compared with values provided by Riley[791] for zooplankton production in Long Island Sound, which he described as a "somewhat estuarine environment of moderately high productivity" of 27 mg m^{-2} day^{-1} in a 20-m-depth water column. In the case of Long Island Sound the P:B ratio was only 0.027 (i.e., one tenth of that for Biscayne Bay). On the other hand in Barataria Bay, Day et al.[184] calculated a net annual production of 25 g dry wt m^{-2} year^{-1} (12.5 g C m^{-2} year^{-1}).

The production data detailed above indicate that there are wide variations in estuarine zooplankton production. One of the reasons for this is that estuarine zooplankton, as we have seen in Chapter 4, play an important role in the utilization of heterotrophic secondary production. This is based on the breakdown of detritus and feeding on bacterial aggregates, detritus with its associated microbial community, and protozoa. As Grindley[338] points out, studies of zooplankton are only beginning to provide some understanding of secondary productivity of estuaries. He points out that as yet little is known of the relative importance of bacteria, protozoa, nannoplankton, phtyoplankton, zooplankton, and detritus in estuarine planktonic food webs.

C. Factors Influencing Distribution and Production

A number of the factors influencing the distribution and production of estuarine zooplankton has already been touched upon above. They include: the type of estuary, the volume of fresh water inflow, salinity tolerance of the species, temperature, predation, and detrital input.

1. The Type of Estuary

Especially important is the ratio of estuarine water volume:tidal prism volume, and the consequent residence time of the water in the estuary. If this is high, zooplankton have a longer period to develop and grow within the estuary and less risk of being flushed out of the system. In most estuaries there is a point above which the resistence time of the water is sufficient for estuarine plankton to survive, while nearer the sea the rapid tidal exchange results in the area being dominated by neritic plankton. For each species there is a point in the estuary where there is a balance between the exchange ratio of water and the coefficient of reproduction required to maintain the population. Population increase can occur upstream, while nearer the mouth tidal dispersion will produce a decrease.[336]

2. The Volume of Fresh Water Inflow

The volume of fresh water inflow is related to the tidal prism and estuarine water volume.

3. Salinity Tolerance of the Species

Salinity tolerance is one of the most important factors limiting the distribution of estuarine plankton. Grindley[338] has carried out experiments on the salinity tolerance of species of

Pseudodiaptomus which are well known for their ability to inhabit a wide range of salinities. *P. hessei* has been found in water from less than 1 to 74‰, while *P. stuhlmanni* occupies a similar range from less than 1 to 75‰. Survival experiments indicate for both peak survival at 35‰ but a wide tolerance reaching above 70‰. However, like the Australian species *Gladioferens imparipes*, which also has a wide salinity tolerance,[401] their distribution is limited by other factors since they dominate in the hypo- and hypersaline parts of the estuary. In the more normal salinities near the mouths of estuaries they cannot compete with marine neritic species. In the case of *Gladioferens* it is seldom present where either of the two less-euryhaline species, *Sulcanus conflictus* or *Acartia clausi* are abundant. These two species prey on the larvae of *Gladioferens*. Grindly[338] lists the recorded salinity ranges for 31 estuarine copepods from South Africa estuaries. Of these, five belonging to the genera *Acartia, Halicyclops, Oithona,* and *Pseudodiaptomus* have wide salinity tolerance. Six other species have tolerances ranging from about 20 to 35‰, while the rest are restricted to waters around 30 to 35‰.

4. Temperature

Temperature has been shown to be a major factor affecting growth and development rates of copepods.[139] For example, in Arctic estuaries, copepods as a rule have a single generation in a year and, as an exception, 2 generations,[223,322] whereas in Biscayne Bay *Acartia tonsa* passes through about 11 generations a year.[1038]

5. Predation

Zooplankton predators may appear in large numbers when their prey are abundant. Among the best documented instances are the interactions between copepods and ctenophores.

Kremer[492] has modeled the seasonal cycle of abundance of the ctenophore *Mnemiopsis leidyi* in Narragansett Bay, Rhode Island. This cycle is characterized by a dramatic biomass increase of several orders of magnitude, reaching a peak of 15 to 60 g wet wt m^{-3} in late summer, followed by a rapid autumn decline. The results of studies of fecundity, feeding, and metabolism of *M. leidyi* were incorporated in a population model (Figure 5). Zooplankton concentration and temperature (on the left-hand side) are the forcing functions of the model. An assimilation efficiency of 75% has been determined for another *Mnemiopsis* sp., *M. mccradyi*,[970] and is consistent with the generally high assimilations found by many other investigators for a wide variety of zooplankton.[138,140,498] Of the assimilated food there are losses due to excretion, respiration, and predation. The remainder is channeled either into growth or reproduction.

Simulations of the model with a biomass of 0.001 mg C m^{-3} (the approximate ctenophore density observed during the winter-spring sampling) were initiated on June 1. In the runs the assumed size of the animals was varied from 1 to 10 g wet wt. The difference in the initial size altered the results producing a "simulation envelope" (Figure 6). In general, the simulation agreed well with the average abundance observed in field sampling. After the 2nd week in August, when food availability became low, the model predicted a cutoff in reproduction and an overwhelming dominance of small animals (less than 1 g wet wt) during late August through October. Field observations confirmed this trend although it was less dramatic than in the model. In the autumn (fall), the model failed to predict the rapid decline which was observed in field sampling. It is likely that this discrepancy is due to the lack of predation on adults in the model. In Narragansett Bay, the butterfish *Peprilus tricanthus* is seasonally abundant during the late summer and fall and has been observed to feed voraciously on *Mnemiopsis*.[677] Starvation and parasitism are other factors which may contribute to population decline. Other studies of ctenophore population dynamics[397] have concluded that predation is the primary mechanism of population regulation. In contrast, Kremer's[492] study and the population model indicated that in *M. leidyi* population density

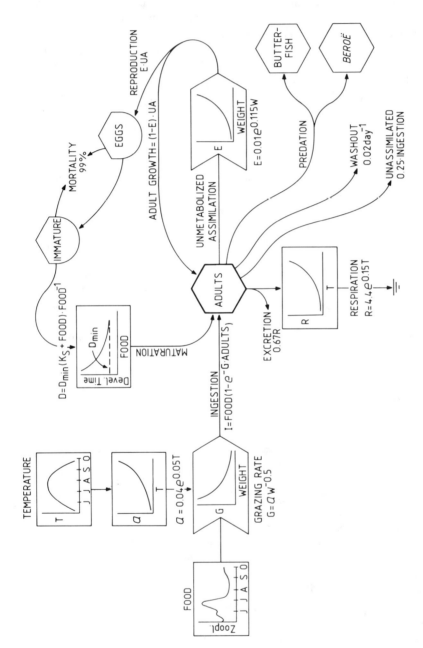

FIGURE 5.   Schematic representation of the ctenophore population in Narrangansett Bay, R. I. All major flows and functional relationships considered in the ctenophore population model are included with the appropriate equations and coefficients. (From Kremer, P., *Estuarine Processes*, Vol. 1, Wiley, M., Ed., Academic Press, New York, 1976, 211. With permission.)

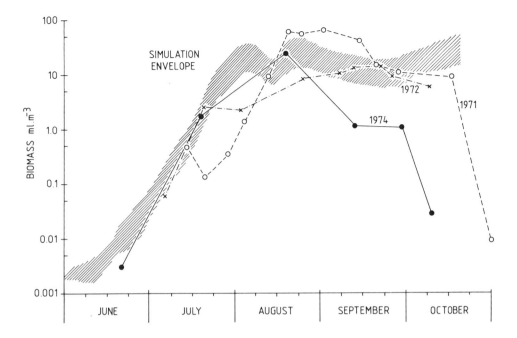

FIGURE 6. Biomass of *Mnemiopsis leidyi* as predicted from the ctenophore population model plotted with average field biomass estimates for 3 years. The simulation envelope results from assuming different size distributions for the initial biomass. (From Kremer, P., *Estuarine Processes*, Vol. 1, Wiley, M., Ed., Academic Press, New York, 1976, 212. With permission.)

is primarily a function of food supply and that the role of predation is mainly in bringing about the fall biomass decline, but not in limiting the biomass maximum. In Narragansett Bay *M. leidyi* seems to exploit its environment fully, and through its ability to increase its population very rapidly it is well adapted to the large seasonal fluctuations in food and temperature in a northern estuary.

Kremer[492] points out that the simulation model has provided a synthesis and rigorous evaluation of field and laboratory observations of the major ecological processes of an important estuarine carnivore. The model's prediction of population increase and peak biomass agreed quite well with field observations. Kremer[492] concluded that mechanistic simulation models can be effective tools in the synthesis and analysis of information and hypotheses about component parts of ecosystems.

Greve and Reiners[332] studied the population dynamics of small copepods and the ctenophores *Pleurobrachia pileus* and *Beroe gracilis* in the Jude, Weser and Elbe estuaries, and the nearshore Heligoland Bight North Sea coast of the Federal Republic of Germany. The early growth phases of the copepods and ctenophores occur in the estuaries proper, from which they pass out into the inner Bight which can be regarded as part of the estuarine system having a salinity regime ranging from 15‰ immediately adjacent to the shore, to 30‰ some distance offshore. The spreading out of these populations is described by Greve and Reiners[332] as "prey-predator waves". Figure 7 depicts the population changes that occur in small copepods, nauplii, *Pleurobrachia pileus, Beroe gracilis*, and fish larvae. The predatory effect of *P. pileus* on small copepods and nauplii is obvious, as is the predatory effect of *B. gracilis* on *P. pileus*. As the predation impact of *B. gracilis* reduces the *P. pileus* population, the copepod populations recover.

The figure shows a development of two successive prey-predator waves which are characterized by a time lag of about 3 weeks from the estuaries into the waters of the Bight. The population of *P. pileus* is first depleted in the outer estuary with the population decline

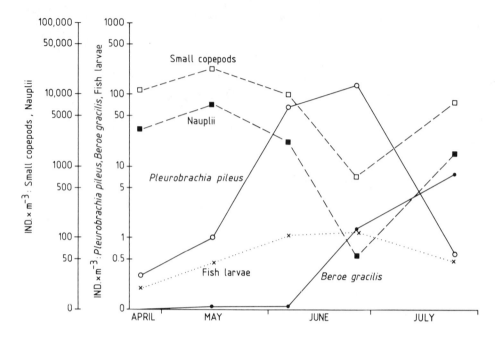

FIGURE 7.   Dynamics of small copepods, nauplii, *Pleurobranchia pileus, Beroe gracilis,* and fish larvae in Helgoland Bight. Mean abundance from 30 stations. (From Greve, W. and Reiners, F., *Estuarine Perspectives,* Kennedy, V. S., Ed., Academic Press, New York, 1980, 419. With permission.)

then moving out into the Bight. Greve and Reiners[332] consider that any managerial impact on the system that changes the size of the estuaries, the availability of detritus, or the temperature/salinity regime as such will have an effect on the ctenophore populations. This in turn would affect the trophic basis and possible survival of fish.

6. Detrital Input

As we have seen, estuarine zooplankton secondary production is to a large extent dependent on heterotrophic activity associated with detrital breakdown.

III. NEKTON

In contrast to the zooplankton, those larger animals which control the direction and speed of their own movements rather than drifting with the water constitute the *nekton*. In the open sea nektonic animals are largely independent of the bottom, whereas in estuaries few nektonic species can be considered truly independent of the bottom. Most estuarine nektonic species live near or just off the bottom, but periodically feed or swim in the water column just above the bottom. Estuarine nekton comprises primarily fishes, but squids, scallops, and natant crustacea (including crabs, lobsters, and shrimp) can be considered part of the nekton. In tropical estuaries in particular, crocodiles, turtles, and dugongs are an element that have a role parallel to that of the true nekton.

A. Estuarine Fishes

Day,[173] McHugh,[551] and Perkins[710] have divided estuarine fishes according to their breeding and migratory habits.

1.   The largest group may be termed *marine migrants*. The great majority of fishes found in estuaries are the juveniles of species which breed in the sea and use the estuary as

a nursery area for food and shelter until their gonads begin to develop. Common examples are mullets, croakers, anchovies, menhaden, flatfishes, and soles. Species in this category increase dramatically from high latitude to tropical estuaries.

2. The *anadromous fishes* include various species of salmon (*Salmo* and *Oncorhynchus*), sturgeons (*Acipenser*), the northern smelts (Osmeridae), the shad *(Alosa),* the white perch *(Roccus americanus),* the striped bass *(R. saxatilis),* and the lampreys (Petromyzontidae). In their migrations to and from fresh water these fishes dominate the northern cold estuaries. According to Korringa,[486] anadromous fish were equally important in Europe at one time but now, due to pollution and dam construction, they are scarce or absent except in Norway and the western coasts of Scotland and Ireland. There are no endemic anadromous fish in the southern hemisphere apart from the lamprey *Geotria australis* in New Zealand, Australia, and South America.

3. The *catadromous fishes* include the fresh water eels (*Anguilla* spp.) which breed in the deep ocean and return to the estuaries as elvers before entering fresh water.

4. *Estuarine residents* are species which pass through their entire life cycle in the estuary. The majority of estuarine residents are small species such as gobies, sygnathids, ambassids, antherinids, stolephorids, and some clupeids. Of the 58 common species listed for South African estuaries, 12 breed there.[183]

5. Finally there is an anomalous group whose breeding habits and migrations do not fit any well-defined group. The American garfish *Lepidosterus spatula* and *Dorosoma cepidianum* breed in fresh water but live mainly in estuaries. The menhaden (*Brevoortia* sp.), which is one of the most important commercial fishes in the Atlantic and Gulf coasts of the U.S., differs from one species to another. *B. tyrannus* on the Atlantic coast actually breeds in the sea but migrates at an early stage into low salinities or even fresh water.[343] The Gulf species *B. patronus* and *B. gunteri* both breed in the upper reaches of estuaries, but the adults migrate to the sea.

1. Feeding Habits of Estuarine Fishes

There have been a number of recent studies on estuarine fish communities which have emphasized the importance of reproduction, food habits, and interspecific interactions in structuring the community.[276,529,678,830,1073] Various studies have described trophic interactions in estuarine and coastal fishes with particular emphasis on spatial/temporal abundance patterns, inter- and intraspecific competition, and resource partitioning. Oviatt and Nixon[678] found little overlap in feeding habits of dominant demersal flatfishes in a Rhode Island estuary, and trophic resource partitioning was viewed as a mechanism to reduce competition. Haedrick and Haedrick[1073] found trophic partitioning among dominant resident fishes in a Massachusetts estuary. Stickney et al.[888] found that estuarine flatfishes of Georgia showed temporal partitioning of trophic resources with morphological adaptation (relative mouth size) as a functional determinant of feeding relationships. Stickney et al.[889] investigated sciaenid trophic patterns and found a direct relationship between prey size and predator size.

Sheridan and Livingston[830] have studied over a 6-year period the community structure and feeding relationships of the dominant fishes in the Apalachicola estuary of north Florida. Six species, in decreasing order of numerical abundance, the bay anchovy *(Anchoa mitchilli),* the Atlantic croaker *(Micropogonias undulatus),* the spot *(Leiostomus xanthurus),* the sand sea trout *(Cynoscion arenarius),* the Gulf menhaden *(Brevoortia patronus),* and the silver perch *(Bairdiella chrysura)* comprise 85% of the trawl susceptible fishes. Comparison of the four dominant species *Anchoa, Leiostomus, Micropogonias,* and *Cynoscion* demonstrated that although there were overlaps in diet components, the individual feeding patterns were different. Feeding differences were related to food particle size spectra as well as the food type (Figure 8).

The two benthic omnivores (*Micropogonias* and *Leiostomus*) exhibited high spatial and

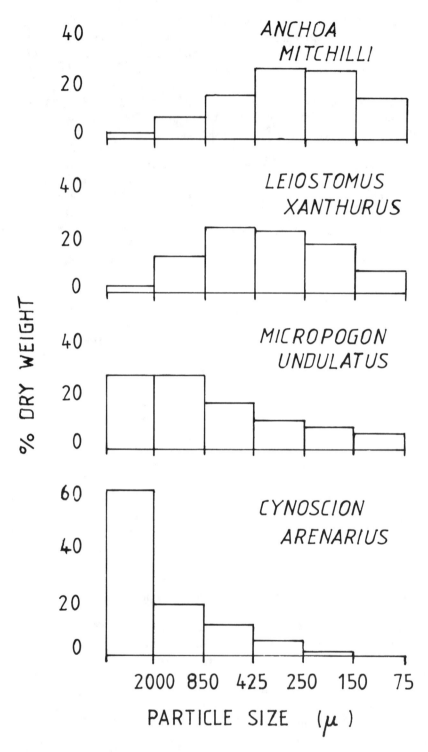

FIGURE 8.   Food particle size spectra of the four dominant fishes in the Apalachicola estuary in north Florida *(Anchoa mitchilli, Cynoscion arenerius, Leiostomus xanthurus,* and *Micropogonias undulatus)* expressed as mean percent dry weight in each of six particle size fractions. Data averaged over all size classes for each species. (From Sheridan, P. F. and Livingston, R. J., *Ecological Processes in Coastal and Marine Systems,* Livingston, R. J., Ed., Plenum Press, New York, 1979, 156. With permission.)

temporal overlap but differed in prey type and size (Figure 8). The two species utilized the estuary subsequent to high river discharge/detritus input, and concurrent with maximum benthic standing crops. Two epibenthic carnivores (*Cynoscion* and *Bairdiella*) also used the estuary but differed in times of peak abundance and prey types. *Cynoscion* preys mainly on mysids, but becomes increasingly piscivorous with growth, while *Bairdiella* prey on co-pepods, amphipods, and mysids with the larger individuals eating fish and shrimp. The two planktivores (*Brevoortia* and *Anchoa*) also frequent the estuary but at different seasons (spring and fall, respectively), yet neither concurs with the summer maximum zooplankton standing crop. *Anchoa* is prevented from doing so by the piscivorous *Cynoscion* population. Thus, regular seasonal progressions of dominant fishes are linked to available trophic resources, competition, and predation, which in turn are dependent upon factors such as river flow, detrital input, plankton production, and offshore processes.

Mullets are dominant estuarine fishes, especially in tropical regions. Ten species of mullets occur in the St. Lucia lakes in Zululand, South Africa, and Blaber[51] has analyzed their feeding habits. Many of the mullets feed on the small gastropod *Assiminea bifasciata*, but it is abundant and competition only becomes apparent when it is in short supply. However, the various species of mullet also feed on foraminifera, the large centric diatom *Aptinoptychus splendens*, small centric and pennate diatoms, filamentous algae, blue-green algae, plant fragments in varying quantities, and sediment particles of different sizes. Blaber[51] found that the different species of mullet ingest different size ranges of particles and suggested "that interspecific competition is reduced by substrate particle selection, and perhaps by differences in feeding periodicity."

There is a paucity of information on the biomass of fish within estuaries and the seasonal changes due to immigration, growth, mortality, or emigration. This is primarily due to sampling problems.[551] In Chesapeake Bay in 1962 the total catch was estimated at 180 kg ha^{-1} of which 173 kg ha^{-1} was caught by commercial fishermen.[551] Korringa[486] reported that before the Zuider Zee was reclaimed, the annual catch was 135 kg ha^{-1}. Milne and Dunnet[598] include an account of the seasonal changes in the populations of gobies and flounders, the two principal fish species of the Ythan Estuary, Scotland. The seasonal changes were as follows:

Biomass (kg ha^{-1})	Winter	Spring	Summer	Autumn
Gobies	3.5	1.6	2.4	6.9
Flounders	120	122	180	55
Total	123.5	123.6	182.4	61.9

The main contribution to the biomass is due to the flounders. The juveniles arrive in the spring, feed rapidly in the summer, and many return to the sea in the autumn.

B. Nektonic Food Webs in Estuaries

de Sylva[195] has reviewed nektonic food webs and the environmental variables affecting them. The major environmental variables are salinity, temperature, water transparency, tidal streams, and food availability. Generalized food webs (Table 1) are basically fueled by either phytoplankton or detritus, and its associated microbial community as the *energy* source. Secondary trophic levels are benthic invertebrates (either infaunal or epifaunal) or zooplankton and micronekton, or both, and include primary carnivores, omnivores, or benthic herbivores.[657] Middle carnivores include planktivores, benthophagous fishes and invertebrates, pelagic fishes, and invertebrates such as squids.

Day et al.[185] have recently analyzed the environmental factors regulating community metabolism with special reference to fisheries production in a Louisiana estuarine system,

Table 1
CLASSIFICATION OF NEKTONIC FOOD WEBS IN ESTUARIES

Phytoplankton-generated food webs

Phytoplankton → zooplankton → planktivorous pelagic and benthopelagic fish
Phytoplankton → zooplankton → planktivorous fish → large fish predators
Phytoplankton → phytoplanktonic fishes such as menhaden *(Brevoortia):* summer
Phytoplankton → zooplankton → menhaden: winter
Phytoplankton → zooplankton → large carnivores *(Manta)*
Phytoplankton (dinoflagellates) → mullet (Mugilidae): alteration of usual feeding habits

Detritus-generated food webs

Detritus → benthos (epifauna) → benthophagous fishes
Detritus → benthos (infauna) → benthophagous fishes
Detritus → benthos → benthophagous fishes → large fish predators (sharks)
Detritus → small benthos → larger invertebrates and small benthic fishes → large fishes
Detritus → large detritivorous fishes (mullet): ''telescoping'' of food chain
Detritus → benthos → large predators
Detritus → micronekton → intermediate predators such as snappers (Lutjanidae) and croaker (Sciaenidae)
Detritus → zooplankton → small fishes and invertebrates
Detritus → zooplankton → small fishes and invertebrates → larger fishes

From De Sylva, D., *Estuarine Research,* Vol. 1, Cronin, L. E., Ed., Academic Press, New York, 1975, 420.
With permission.

the Barataria Basin. This is a large (> 400,000 ha) interdistributory estuarine-wetland system located between the natural levees of the Mississippi River and Bayou Lafourche, an abandoned river channel. The physical characteristics of the species of interconnected water bodies of the Basin are listed in Table 2. Gross aquatic primary production (APP) ranged from 1058 to 3286 g O_2 m^{-2} year^{-1}, with the highest values in lakes with direct upland runoff, suggesting that nutrient inputs were important regulators of APP. All the aquatic habitats are strongly dependent on allochthonous organic inputs from adjacent watersheds, and upstream habitats are significant sources of organic matter for downstream habitats. The proportion of wetland primary production exported to adjacent water bodies was lowest in the fresh-water swamp (~2%) and greatest in the salt march (30%).

Louisiana has the greatest area of coastal wetlands[942] and the largest commercial fishery in the U.S. It is a commonly held belief that the coastal wetlands play an important role in supporting the fisheries.[518] Day et al.[185] believe that there is strong evidence showing coupling between fisheries and the marsh estuarine system, and they review several lines of evidence from Barataria Bay.

Chambers[103] has reviewed some 20 studies of nekton community composition, biomass distribution, and migratory patterns in the Barataria Basin system. Seven to eight species comprise 80 to 95% of the total numbers and biomass:

1. *Anchoa mitchilli* (bay anchovy)
2. *Micropogon undulatus* (croaker)
3. *Arius felis* (sea catfish)
4. *Mugil cephalus* (striped mullet)
5. *Leiostromus xanthurus* (spot)
6. *Brevoortia patronus* (menhaden)
7. *Menidia beryllina* (silverside)
8. *Penaeus* sp. (shrimp)

Table 2
PHYSICAL CHARACTERISTICS OF WATER BODIES IN THE BARATARIA BASIN

Ecological zone	Example	Turnover per year	Average depth (m)[1024]	Salinity (%o)[1024]	Secchi depth (cm)[1024]	Tidal range (cm)	Upland to wetland + water ratio[983]
Upper basin	Lac des Allemands	4.6[150]	2.0	0	33	3.2[90]	1:2.3
Middle basin	Lake Cataouatche	1.5[150]	2.0	0—3	30	NA[a]	1:6.7
	Lake Salvador	1.0[150]	2.5	0—6	72	8.5[b]	
Lower basin	Little Lake	NA[a]	1.5	0—15	72	12[579]	
	Lower Barataria Bay	14.6[361]	2.0	10—35	68	30[579]	1:33.3

[a] NA = not available.
[b] Interpolated between Barataria and des Allemands stations.

From Day, J. W., Jr., Hopkinson, C. S., and Conner, W. H., *Estuarine Comparisons*, Kennedy, V. S., Ed., Academic Press, New York, 1982, 123. With permission.

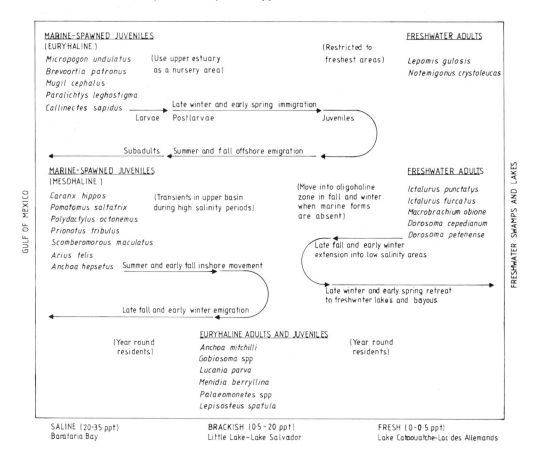

FIGURE 9. Patterns of estuarine use by nektonic organisms in the Barataria Basin. (From Day, J. W., Jr., Hopkinson, C. S., and Conner, W. H., *Estuarine Comparisons*, Kennedy, V. S., Ed., Academic Press, New York, 1982, 132. With permission.)

Anchoa is considered to be an estuarine resident that normally completes its life cycle within the estuary. The other species spawn offshore and use the estuary as a nursery and feeding area.

Chambers[103] presented a diagram of the pattern of use of the Barataria Basin by four different nekton groups (Figure 9): (1) euryhaline larvae, postlarvae, and juveniles of marine nekton which spawn offshore, migrate far up the Basin in the late winter and spring, and then gradually move downbay as they grow, eventually emigrating to the Gulf in late summer and fall; (2) juvenile and adult fresh water species which move southwards in the fall into oligohaline areas as they become fresher and replace the emigrating marine species, returning to fresh water areas in the late winter; (3) during the warmer months, mesohaline juveniles of certain marine species move up to the mid-Basin during periods of high salinity, and later return to the lower bays and Gulf in the late fall and winter as salinities decrease; and (4) some euryhaline species spend their entire life cycles in the estuary and often may be found anywhere from the fresh water swamps to the lower bays and barrier islands bordering the Gulf. The data of both Wagner[964] and Chambers[103] suggest that euryhaline marine-spawned individuals preferentially migrate into waters with low salinity and slowly move into waters of higher salinity as they grow.

The available data suggest that nekton species, especially larval and juvenile forms, preferentially seek out shallow water adjacent to wetlands, such as marsh ponds, tidal creeks, and the marsh edge in general. Data from Barataria Basin and Lake Pontchartrain show that

Table 3
COMPARISON OF ESTIMATED NEKTON
STANDING CROPS IN SHALLOW, MARSH
AREAS, AND OPEN-WATER AREAS

Area	Standing crop (g wet wt m^{-2})	Ref.
Upper Barataria Basin, shallow water	3.41	103
Upper Barataria Basin, open water	0.50	
Caminada Bay, shallow marsh ponds	13.8—46.1*	964
Caminada Bay, open water	1.19	
Lake Pontchartrain, shallow water	2.57	
Lake Pontchartrain, open water	0.32	

Note: All fish collected by otter trawl, except * = Antimycin.

From Day, J. W., Jr., Hopkinson, C. S., and Conner, W. H.,
Estuarine Comparisons, Kennedy, V. S., Ed., Academic Press,
New York, 1982, 123. With permission.

nekton biomass is 7 to 12 times higher in shallow water marsh areas as compared with open waters (Table 3). It is a possibility that the shallow nature of these waters attracts nekton seeking either food or refuge from predators. These shallow water areas seem to satisfy the three requirements outlined by Joseph[443] for a nursery area: (1) physiologically suitable in temperature, salinity, and other physico-chemical parameters; (2) abundant suitable food with a minimum of competition at critical trophic levels; and (3) a degree of protection from predators.

A number of workers have shown correlations among estuaries, wetlands, and fisheries. Turner[940] correlated shrimp yield (kg ha^{-1}) and intertidal wetlands on a world-wide basis. In the northern Gulf of Mexico he found that yields of inshore shrimp were directly related to the area of estuarine vegetation, whereas they were not correlated with area, average depth, or volume of estuarine water. Moore et al.,[603] in their analysis of the distribution of demersal fish off Louisiana and Texas, suggested that the greatest fish populations occur offshore from extensive wetlands with a high fresh water input. This is in agreement with the evidence from the Barataria Basin which suggests that wetlands enhance fisheries productivity.

IV. BENTHIC FAUNA

A. Introduction

There is vast literature devoted to the description of species composition in benthic communities. In such studies different communities were characterized by so-called "dominant" species, or "indicator" species. From this, Thorson[935,936] developed the concept of "parallel communities", which implied that similar animals would be found in association wherever similar environmental conditions exist. Other studies[811] examined the species structure of communities and used diversity indices to compare different communities. From this there developed concepts of latitudinal and depth gradients in community diversity. Other studies have concentrated on relating patterns of distribution to environmental factors such as grain size composition, organic content of the sediments, and salinity distributions.

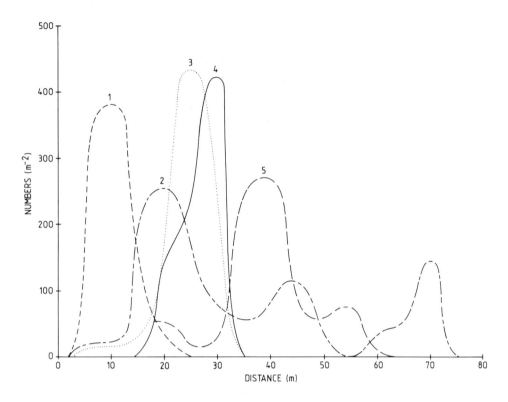

FIGURE 10.   Distribution curves of macrofaunal species along an oyster bed in Canada: (1) *Mya arenaria* (× 2);
(2) *Nereis vivens* (× 10); (3) *Nassarius obsoletus;* (4) *Zostera marina* (× 10); and (5) *Neopanope texana* (× 30).
(From Hughes, R. G. and Thomas, M. L., *J. Exp. Mar. Biol. Ecol.,* 7, 21, 1971. With permission.)

In spite of the considerable literature on community structure and organization in benthic
communities, the validity of the community concept has been challenged, particularly during
the last 15 years. The traditional view held that it was possible to classify assemblages of
species into communities, and that there were sharp boundaries between adjacent communities corresponding to discontinuities in the habitat. On the other hand, there were those
who argued that rather than occurring in discrete groups with sharp boundaries, species
occurred along gradients of environmental factors with each species having an optimum
somewhere along the gradient, i.e., species are distributed in the form of continua. A good
example of this is provided by the work of Hughes and Thomas[415] from an intertidal beach
in Canada. Figure 10 illustrates the distribution patterns they found with the species distributed in approximately normal distributions along a gradient of distance (probably reflecting sediment grain size). Similar distribution patterns have been shown for estuarine
benthos along a salinity gradient in the Chesapeake Bay region.[60] Thus, species seem to be
distributed in log-normal curves of abundance along environmental gradients and do not
form discrete communities. From these findings stems a modern definition of a benthic
community[597] which is widely accepted: "Community means a group of organisms occurring
in a particular environment, presumably interacting with each other and with the environment,
and separable by means of ecological survey from other groups."

Recent attention to the seasonal dynamics of shallow-water macrobenthic
communities[58,212,979] has shown the existence of strongly seasonal patterns of recruitment
and mortality, which lead to markedly changing community structure. Boesch et al.[61] have
extended such studies to the examination of long-term fluctuations in the composition of
macrobenthic communities at a polyhaline mud-bottom site and an oligohaline environment

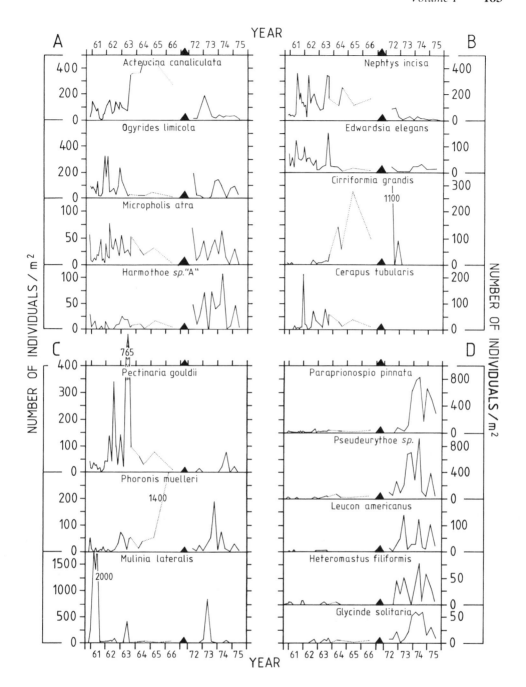

FIGURE 11.   Population fluctuations in species at a polyhaline site in the York River estuary, Chesapeake Bay, over the period 1960 to 1966 and 1972 to 1975: (A) species which were more or less equally abundant; (B) species that were very much reduced in abundance in the 1972 to 1975 period; (C) irruptive species; and (D) species which have become much more abundant since 1972. (From Boesch, D. F., Wass, M. L., and Virnstein, R. W., *Estuarine Processes*, Vol. 1, Wiley, M., Ed., Academic Press, Ney York, 1976, 180. With permission.)

in southern Chesapeake Bay. The sites were sampled frequently from November 1960 to July 1963, twice in 1964, once each in 1965 and 1966, and quarterly from May 1972 to July 1975. Figure 11 presents data from the study. The species could be divided into four groups: (1) species such as the brittlestar *Micropholus atra* which fluctuated about a mean population level and was more or less equally abundant over the sampling period; (2) species

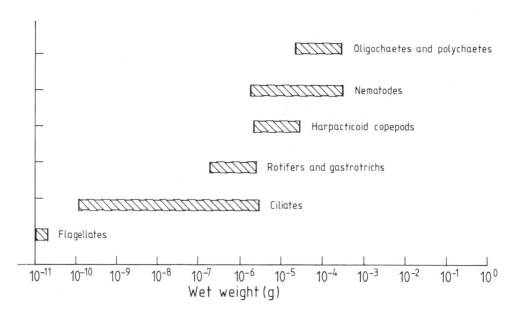

FIGURE 12.   Approximate weight ranges of meiobenthic faunal groups. (From Mann, K. H., *The Ecology of Coastal Waters*, Blackwell Scientific, Oxford, 1981, 186. With permission.)

such as the polychaete *Nepthys incisa* which was a regular dominant during 1960 to 1966, but which declined markedly in the 1972 to 1975 period; (3) species such as the bivalve *Mulinia lateralis* which exhibited marked, somewhat ephemeral population eruptions. *Mulinia* is very fecund, grows rapidly, matures quickly, and as such is adapted for opportunistic exploitation of resources; and (4) species such as the cumacean *Leucon americanus* which increased markedly in abundance in the 1972 to 1975 period.

From this study and others it is clear that both the species composition and the populations of individual species in estuarine benthic communities fluctuate markedly over time. Ziegelmeier[1055,1056] presented results of semiannual sampling over 17 years in the German Bight which showed population eruptions (e.g., the polychaete *Spiophanes bombax*), temporary extinctions (e.g., the bivalve *Tellina fabula*), and longer term declines (e.g., the bivalve *Nucula nitida*). Many complex factors are responsible for short- and long-term variation in estuarine benthic populations. Physico-chemical variables include salinity changes,[1027] temperature fluctuations, changes in the nature of the substrate,[212] and changes in dissolved oxygen concentrations.[979] Man's activities through the introduction of domestic and industrial wastes[797] and toxins[844] can profoundly affect the structure of estuarine communities. Not less important, but often less obvious, are the many biological factors which affect reproduction, recruitment, and survival. These factors will be further discussed in Volume II, Chapter 2, Section VIII.

It is often convenient to separate the benthos according to size into macrobenthos, meiobenthos, and microbenthos with dividing lines at about 1.0 mm and 0.1 mm and 10^{-4} and 10^{-10} g wet wt (Figure 12). The meiofauna usually consists of nematodes, harpacticoid copepods, turbellarians, small annelids, and a phylum unique to the meiofauna, the Gastotricha. The meiofauna can be further subdivided into the *temporary* meiofauna comprising the larval stages of the benthic macrofauna, and the *permanent* meiofauna, those species that pass through their complete life cycle as members of the interstitial fauna living in the interstices between sand grains. There is an overlap at the lower end of the meiofaunal size range with the protozoa of the microfauna.

There are a limited number of studies in which all components, macro-, meio-, and

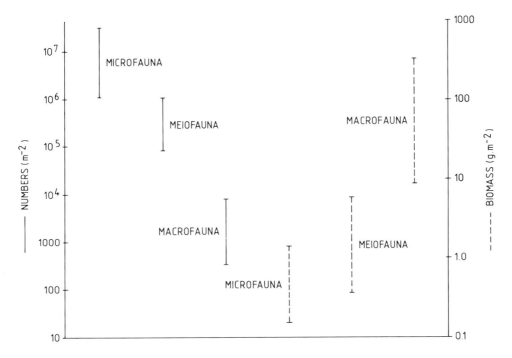

FIGURE 13.   Abundance and biomass ranges of macro-, meio-, and microfauna from sublittoral, sandy sediments. (From Fenchel, T., *Annu. Rev. Ecol. Syst.*, 9, 99, 1978. With permission.)

microfauna have been measured. Figure 13 shows data from a sand beach where the smaller animals (the microfauna) dominate numerically, but the macrofauna dominate in terms of biomass.[237] The actual ratios between the three groups depend largely on the sediment type, with for example, microfauna being very common in fine sand but scarcer in muddy sediments.

B. Meiofauna

Swedmark[904] McIntyre,[553] Gerlach,[281,282] Gray,[326] Coull,[147] Fenchel,[237] and Coull and Bell[148] have reviewed various aspects of meiofaunal ecology. Meiofauna include a wide assortment of metazoan animals with small, elongated, often flattened bodies, adapted to an interstitial mode of life. Fenchel and Reidl[245] have summarized the information on the occurrence and interrelationships of what they have termed the "sulfide system", the microorganisms living below the oxidized surface layer of marine sediments. From this it is clear that both the micro- and meiobenthos form part of a complex community which cannot meaningfully be subdivided on a basis of size, although this has proved useful in practical terms.

Nearly all phyla are represented when both permanent inhabitants and temporary members of the meiofauna are taken into account.[553,792] Ciliata are the most important of the microfauna,[228] while nematodes dominate the meiofauna. Among the crustaceans, ostracods, copepods, and isopods are the most important group together with the Mystacocaridae, a group known only from the interstitial fauna. Other groups which are found are aberrant hydroids, gastotrichs, archiannelids, vermiform opisthobranch mollusks, occasional holothurian species, and a solitary bryozoan.

1. Population Density, Composition, and Distribution

On the average one can expect to find 10^6 m^{-2} meiofaunal organisms and a standing crop dry weight biomass of 1 to 2 g m^{-2}. Obviously these numbers will vary according to season, latitude, water depth, substrate, etc.[282] According to McLachlan et al.[557] average macrofauna:meiofaunal ratios are $1:10^5$ for numbers, 5:1 for biomass, and 1:1 for production. Meiofauna are thus quantitatively important, but still account for only approximately 23 μg organic carbon $m\ell^{-1}$ sediment compared with a value of approximately 90 μg organic carbon $m\ell^{-1}$ for the combined macrofauna and bacteria.[282] However there is evidence that they may be quantitatively more important in estuarine ecosystems where they play an important role in carbon flow.

Meiofaunal densities are often compared in terms of individuals per 10 cm². McIntyre,[553] Harris,[369,370] and Dye and Furstenberg[208] have provided summaries of available data on this basis. Analysis of this data leads to the conclusion that estuaries are considerably richer in meiofauna than are nonestuarine areas. Estuarine habitats may be divided into three areas on the basis of substrate and plant cover, i.e., sand flats, mud flats, and salt marshes, and these areas may be ranked in importance with the sand flats having the lowest meiofaunal populations and the salt marshes the highest.

The average meiofaunal density in sand flats appears to be about 1000/10 cm², but some areas may achieve this in nematodes alone.[228] Studies of South African sand flats have shown meiofaunal densities that vary from 187 to 1828/10 cm².[207] An increase in meiofaunal populations is usually found in mud flats and here the average is in the region of 3000/10 m², although densities as high as 6000/10 cm² have been recorded.[762] Salt marshes are characterized by high meiofaunal numbers. Teal and Wieser,[916] for instance, recorded 12,400/10 cm² in a Georgia salt marsh, while Wieser and Kaniwisher[1010] found 1785/10 cm² in a Massachusetts salt marsh. In Knysna Lagoon, South Africa, Dye[207] recorded densities between 493 and 19,682/10 cm² with an average density of 5232/10 cm². Possibly the highest recorded meiofaunal density of between 30,000 and 65,000/10 cm² has been found in the salt marshes of the Swartkops Estuary in South Africa.[208]

There are also differences in composition of the meiofauna according to sediment type. Fenchel[237] distinguishes three main faunal assemblages. In well-sorted sands with particles of diameter greater than 100 μm and having interstices that are filled with water rather than clay or silt, there is a rich interstitial fauna comprising ciliates, tardigrades, turbellarians, gastotrichs, oligochaetes, archiannelids, harpacticoids, ostracods, and others. In silty and clayey sediments the fauna changes completely and is dominated by nematodes which are capable of burrowing. However, there is a richer fauna inhabiting the upper 1 mm or so of these sediments, and harpacticoids, ostracods, foraminifera, and various annelids, along with nematodes, are found in abundance. Figure 14 schematically illustrates mud and sand cores and associated redox layers with the mean numbers of animals and biomass from 5 years of monthly subtidal data in North Inlet, South Carolina.[148] The marsh core has twice the meiofaunal biomass as the sand core and in the mud, all the fauna is concentrated in the upper 1 cm, whereas in the sand the fauna is distributed to a depth of 10 to 15 cm. The third meiofaunal community is that inhabiting the anoxic sediments below the redox potential discontinuity layer. Here occur quantities of anaerobic species of ciliates, together with a few species of flagellates, nematodes, turbellarians, rotifers, gnathostomulids, and gastrotrichs.

Nematodes are usually the dominant group accounting for more than 80% of the total meiofauna. In a study of an estuarine salt marsh in Barataria Bay, Louisiana, it was found that nematodes comprised 87.3% of the total meiofauna.[190] Total numbers of meiofauna were highest in early March and lowest in October with a mean of 2.9×10^6 m⁻². This is comparable to those of a South Carolina estuarine marsh where nematode numbers averaged 3×10^6 m⁻².[842] Platt and Warwick[726] have compared the relative importance of nematodes in the Lynher estuary, England, from the viewpoint of numbers, species, biomass, respiration, and production with other meiofaunal groups and the macrofauna (Figure 15). In terms of numbers, species, and production, they exceeded all other groups.

Sikora and Sikora[843] have examined the vertical distribution of nematodes in anaerobic salt marsh sediments in Barataria Bay. In these sediments the Eh of the uppermost sediments

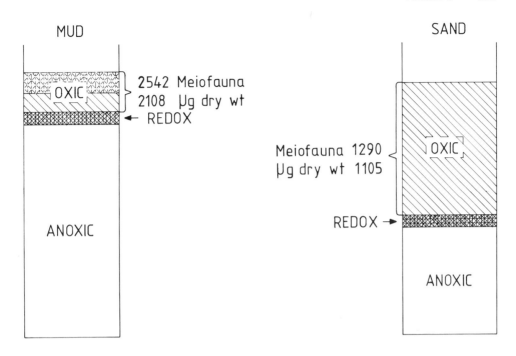

FIGURE 14.   A schematic diagram of the sediment at a muddy and sandy site in North Inlet, S.C., with numbers and biomass at the two sites. The oxidized layer in the mud is limited to 1 cm; in the sand it varies between 10 and 15 cm. (From Coull, B. C. and Bell, S. S., *Ecological Processes in Coastal and Marine Systems,* Livingston, R. J., Ed., Plenum Press, New York, 1979, 199. With permission.)

was 78.5 ± 46.9 mV (Figure 16) indicating that no true oxygen was present. Redox potential dropped rapidly to −110.7 ± 18.9 mV at 7 cm and to −146.7 ± 17.1 mV at 10 cm. Sikora and Sikora[843] found 45.6% of the total nematodes in the uppermost sections (0 to 1 cm). The next section (1 to 2 cm) contained 39.9%. Numbers dropped rapidly below this depth, with 12.3% found in the 2 to 3 cm sections and 1.0% in the 3 to 4 cm sections. According to Sikora and Sikora,[843] comprehensive reviews of redox equilibria in fresh water sediments,[303] marine waters and sediments,[610] and marsh sediments[1086] have demonstrated that no free oxygen occurs below +350 mV. They contend that the dark layer (redox discontinuity potential layer) shown in Figure 15 is the zone (= +25 mV) where ferric iron is reduced to ferrous iron,[1086] and is not the dividing line between "oxic" and "anoxic" sediments.

Sikora and Sikora[843] have compared the numbers of nematodes (Figure 17) in a subtidal salt marsh creek along a transect through low intertidal, midtidal, and high intertidal creek bank to the low marsh. The annual mean number of subtidal nematodes was 45% of the mean number of the high intertidal creek bank nematodes.[842] In the high marsh, or short *Spartina alterniflora* zone, the nematode numbers were lower than in the high intertidal tall *S. alterniflora* zone.[41] Total microbial biomass, as calculated from sediment ATP determinations, is higher in the tall *S. alterniflora* zone just above the high intertidal creek bank than in the short *S. alterniflora* zone.[116]

The subdominant meiofauna group is usually the harpacticoid copepods.[396] These are more strongly influenced by factors such as desiccation and oxygen availability. Thus, they occur near the surface of the sediment and are more abundant towards the lower tidal levels. Population density decreases considerably in muddy areas. In the sand flats of the Swartkops Estuary, South Africa, the meiofauna consists of nematodes (84%), harpacticoids (11.5%), and the remainder made up of oligochaetes, polychaetes, flatworms, and gastotrichs. In all

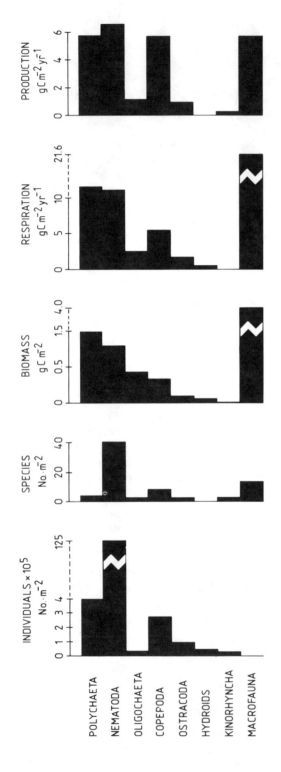

FIGURE 15.   Relative importance of nematodes compared with other meiofauna groups and macrofauna in the Lynher Estuary, England. (From Platt, H. M. and Warwick, R. M., *The Shore Environment*, Vol. 2, Price, J. H., Irvine, D. E. G., and Farnham, W. F., Eds., Academic Press, London, 1980, 746. With permission.)

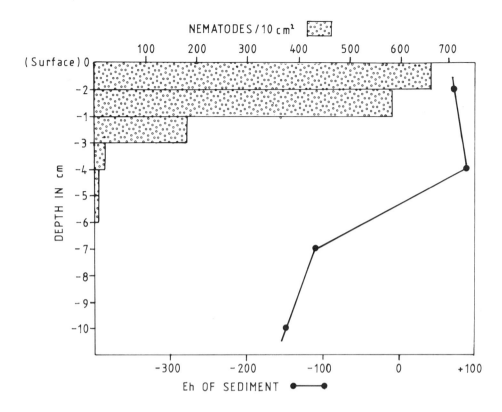

FIGURE 16. Distribution of nematodes with depth, and Eh of sediments in Barataria Bay. (From Sikora, W. B. and Sikora, J. P., *Estuarine Comparisons,* Kennedy, V. S., Ed., Academic Press, New York, 1982, 272. With permission.)

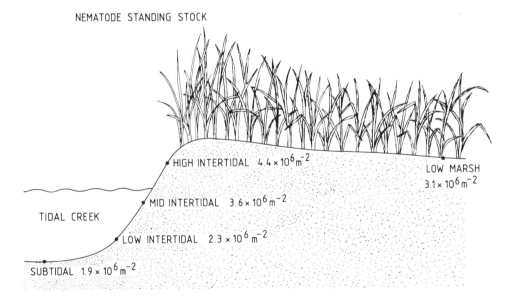

FIGURE 17. Distribution of nematodes along a transect from the subtidal creek bottom to the low marsh in North Inlet, S.C. (From Sikora, W. B. and Sikora, J. P., *Estuarine Comparisons,* Kennedy, V. S., Ed., Academic Press, New York, 1982, 274. With permission.)

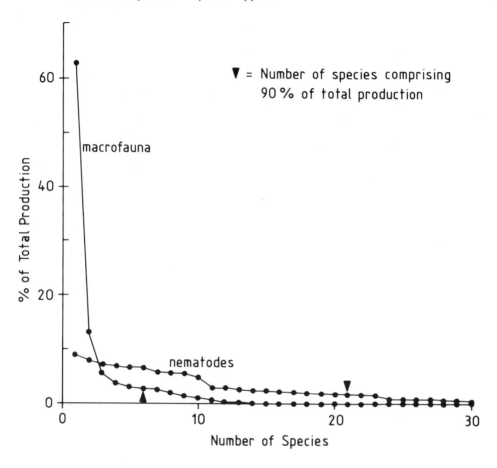

FIGURE 18. Partitioning of production among the 30 most important species in a macrofauna community and a community of meiofaunal nematodes with a similar number of species. Macrofauna from a *Venus* community based on data of Warwick et al.[977] Meiofaunal nematodes from an estuarine mud flat based on data of Warwick and Price[978] assuming that annual production is proportional to annual respiration. (From Warwick, R. M., *Marine Benthic Dynamics,* Tenore, K. R. and Coull, B. C., Eds., University of South Carolina Press, Columbia, 1980, 17. With permission.)

estuaries investigated, the tendency noted above for densities to reach a maximum at midtide levels and for the lowest densities to occur at or below low tide levels has been found.[916]

Where long-term studies of meiofaunal densities have been carried out they have revealed pronounced seasonal changes in composition and abundance. All studies show a pronounced meiofaunal peak in the summer.[370,426,708] Data from monthly samples of the meiofauna of South Inlet, South Carolina, show significant differences between the yearly densities. Over the period 1972 to 1974, peak summer densities ranged about 500 to 1000/10 cm² whereas over the period 1975 to 1977, peak numbers ranged from about 3500 to 6400/10 cm².

Warwick[976] has discussed the reasons for the high diversity of the meiofauna in comparison with the macrofauna at the same sites. Platt and Warwick[726] list nematode species diversities ranging from 40 for a Lynher Estuary mud flat to 145 for a fine sand beach in North Carolina.[674] For an intertidal mud flat in the Lynher Estuary, Warwick[976] has compared the diversities of meiofauna, nematodes, and macrofauna. Not only is the species diversity of the meiofauna much higher than that of the macrofauna, but it appears that the meiofauna are able to partition the total energy available to them rather evenly among the species in comparison with the macrofauna. Figure 18 shows the partitioning of production among the species of macrofauna[978] and among the nematode component of the meiofauna.[975] In com-

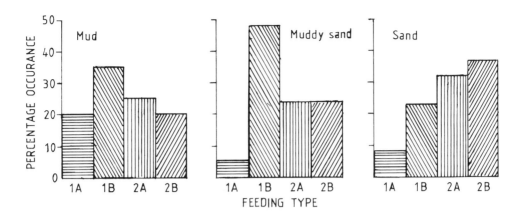

FIGURE 19.   Relative proportions of the four feeding groups of nematodes in mud, muddy sand, and clean sand from the Exe Estuary, Devon, England. 1A — selective deposit feeders; 1B — nonselective deposit feeders; 2A — epigrowth feeders, 2B — predators and omnivores. (From Platt, H. M. and Warwick, R. M., *The Shore Environment*, Vol. 2, Price, J. H., Irvine, D. E. G., and Farnham, W. F., Eds., Academic Press, London, 1980, 742. With permission.)

mon with other shallow water and estuarine communities studied, the macrofauna production is dominated by a few species, whereas that of the nematodes is partitioned remarkably evenly without dominance by a few species. The question then is how does the meiofauna maintain such high diversity, and how do they partition the available resources so equitably among the species and yet appear to avoid competitive interactions? The implications are, therefore, that in order for 40 species of nematodes to coexist in 1 ml of sediment (as they do in the Lynher Estuary), the sedimentary environment must be highly structured on a microscopic scale. Warwick[976] presents evidence supporting the view that some of the nematodes have intimate and specific relationships with the microbial populations.

Food partitioning may be the key to the high diversity, and most nematodes appear to be highly selective feeders.[937] On the basis of their feeding, Wieser[1009] classified nematodes into (1) selective deposit feeders, (2) nonselective deposit feeders, (3) epigrowth feeders, and (4) predators and omnivores. The relative proportions of these four feeding groups depend on the nature of available food which in turn is reflected by the nature of the sediment. For example, in the Exe Estuary, England (Figure 19), approximately similar proportions are found in mud, but in muddy sand nonselective deposit feeders equipped to ingest larger particles predominate. However, in clearer sandy areas, the numbers of all deposit feeders are reduced and forms dominate which are able to scrape food from surfaces, pierce other organisms to obtain their contents, or act as predators. Resource partitioning has led to behavioral and reproductive strategies enabling each species to make the most efficient use of the available food.[726]

2. Role of Meiofauna in Benthic Systems

The role of meiofauna in the energy flow in estuarine ecosystems will be discussed in Volume II, Chapter 2, Section IV. Here we will be concerned with the extent to which meiofauna form part of the diet of other animals. Initially, nematodes and other meiofauna in general were thought to be a trophic dead end, either acting in competition with macrofauna for primary food sources or functioning as nutrient recyclers.[388,553,580] More recently, alternative arguments have been put forward to suggest that nematodes, which were thought to have no nutritive value, may indeed be consumed directly as food and their production passed up the food chain.[1070] Possible nematode consumers are listed in Figure 20. Nonselective deposit feeders such as some polychaetes, echinoids, holothurians, and sipunculids

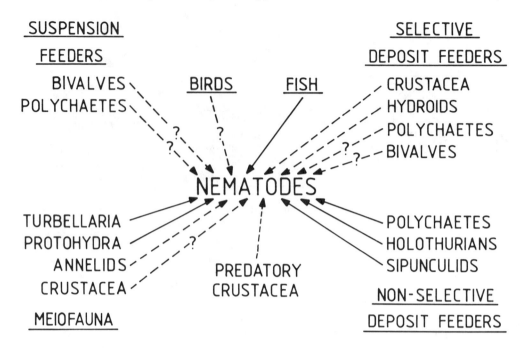

FIGURE 20.   Summary of known and potential nematophagous organisms. Continuous lines from known nematovores; broken lines from probable nematovores; and queried lines from possible nematovores. (From Platt, H. M. and Warwick, R. M., *The Shore Environment*, Vol. 2, Price, J. H., Irvine, D. E. G., and Farnham, W. F., Eds., Academic Press, London, 1980, 748. With permission.)

cannot avoid ingesting nematodes and other meiofauna animals. Gerlach[282] suggested that meiofauna (including Foraminifera) in subtidal sand contribute about 20% to the food of deposit-feeding macrofauna. As Gerlach[282] points out, "bacterial biomass is not very much higher than meiofaunal biomass; therefore meiofauna automatically has to be considered as an important source of food, if the concept of non-selective feeding is valid." Platt and Warwick[726] speculate that the reason why nematodes are not more frequently reported in the guts of consumers is that, once macerated, they are difficult to identify.

Fish, especially in their younger stages, are now known to take large quantities of meiofauna, especially harpacticoid copepods.[4,70,297,341,448] In tropical seagrass beds the emerald clingfish *Acyrtops beryllinus* and the goby *Gobionellus boleosoma* depend on meiofauna for food at all life stages.[94,318] Feller and Kaczynski[227] and Sibert et al.[1090] showed quite conclusively that juvenile salmon in British Columbian estuaries feed almost exclusively on harpacticoid copepods, and Odum and Heald[661] reported the meiobenthic copepods comprised 45% of North American grey mullet gut contents. Also Sikora[842] has reported that nematodes provide a significant proportion of the *in situ* food of the grazing glass shrimp *Palaemonetes pugio*.

Hicks[1078] investigated the biomass and production of the meiobenthic harpacticoid copepod *Parastenhelia megarostrum* in Pauatahanui Inlet, New Zealand, and its exploitation as a food resource by juvenile flatfish. In this inlet, *P. megarostrum* reaches a density of 263,000 m^{-2}, the highest recorded density for a meiobenthic harpacticoid copepod. Hicks calculated the mean annual density (March 1981 to March 1982) as $263/10 \ cm^{-2}$. A peak standing stock density of $1189/10 \ cm^{-2}$ excluding nauplii was recorded in September 1981. If nauplii are included the maximum population density is 1902. Therefore in January when 60% of the population was contributed to by nauplii, the mean sediment density was 442.0 ± 179.9 (Table 4). Biomass was estimated as 0.605 g ash free dry wt $m^{-2} \ year^{-1}$, or 0.242 g C $m^{-2} \ year^{-1}$. Annual production based on a P:B ratio of 15 and a 30% correction for nauplius

Table 4

INSTANTANEOUS ESTIMATE OF FEEDING PRESSURE BY FLATFISH ON THE HARPACTOCOID COPEPOD, *PARASTENHELIA MEGAROSTRUM*, JANUARY 1984, MANA BANK, PAUATAHANUI INLET, NEW ZEALAND

Mean no. fish m^{-2} = 2.10 ± 2.14 (n = 10)
Mean no. of *P. megarostrum* in guts = 264.8 ± 143.3 (n = 15)
Mean sediment density of *P. megarostrum*/0.10 cm^{-2} = 442.0 ± 179.9 (n = 5)
  Estimate = 2.1 × 264.8 = 556.1 : 442000 = 0.00126%
Daily gut turnover rate of fish = 3
  Daily removal rate = 0.00377%
  Annual removal rate = 1.38%

From Hicks, G. R. F., *N.Z. J. Ecol.*, 8, 125, 1985. With permission.

production gave an upper estimate of 9.074 g ash free dry wt m^{-2} $year^{-1}$, or 3.630 g C m^{-2} $year^{-1}$.

An assessment of the annual removal rate of *P. megarostrum* by flatfish predators is given in Table 4. O-group flatfish within a size range of about 8.0 to 35.0 mm standard length were found to feed predominantly on harpacticoid copepods with *P. megarostrum* comprising over 95% of the intake. The mean number of *P. megarostrum* found in the juvenile flatfish guts was 264.8 ± 143.3. Assuming a daily gut turnover rate of 3, it was estimated that the daily removal rate of *P. megarostrum* was 0.00377% and the annual removal rate 1.38%.

Sogard[862] has studied the feeding ecology of the spotted dragonet *Callionymus paucira-diatus* in seagrass beds in Biscayne Bay, Florida. He found that the fish fed almost exclusively on meiofauna, with harpacticoid copepods comprising an average 89% of the diet. He estimated that at an average fish density of 0.42 fish m^{-2},[861] an average harpacticoid density of 71,600 copepods m^{-2}, and a daily ration of 1000 copepods per fish, the fish would remove an estimated 0.59% of the harpacticoid standing crop each day. Bregnballe[70] found that juvenile plaice consumed 1.2% of the standing crop of harpacticoids per day, but concluded that the fish were too few in number to cause any major decrease in harpacticoid abundance. Alheit and Scheibel[4] estimated a maximum daily predation rate by juvenile tomtates *Haemulon auriolineatum* and concluded that it would have little effect on harpacticoid populations. There is thus general agreement that fish predation exerts little regulatory control on annual harpacticoid population, especially in view of their relatively rapid turnover rate. Such fish are therefore not food limited.

Coull and Bell[148] suggested that the importance of meiofauna as a food source for upper trophic levels is related to the grain size of the sediment in which they live. Studies which have found a significant consumption of meiofauna have concentrated on muddy or detrital sediments, while those concluding that meiofauna are not transferred to upper trophic levels have concentrated on sandy habitats. With the much deeper redox discontinuity layer in sands, meiofauna have more opportunity to burrow and escape predation.[148]

Meiofauna, and in particular the nematodes, play important roles in facilitating decomposition and influencing sediment stability.[726] The breakdown of dead organisms and detritus, and the consequent recycling of their constituent nutrients, is a vital role played by the bacterial flora. There is now evidence to show that this microbial activity is stimulated significantly by benthic organisms such as nematodes.[282,925] Not only may burrowing and feeding activities improve the exchange of metabolites and provide essential nutrients for bacteria, but by feeding on the bacteria the meiofauna thereby maintain the microorganisms in a "youthful" condition. In other words, they keep bacterial populations at the point of maximum growth (the log phase) or sustainable yield. Studies of the burrowing of marine nematodes and its effect on microenvironments within sediments has indicated that they play two opposing roles with regard to sediment stability. One is that they assist in bioturbation; the other is that by producing a network of mucus-lined burrows[153,790] the sediment is rendered more stable.

C. Macrofauna

There is enormous literature on the distribution and abundance of benthic macroinvertebrate animals. Traditional approaches to their study have involved the use of community concepts such as the classification of species associations, diversity, succession, and stability/time relationships, often involving sophisticated mathematical analyses. However, such approaches have not proved to be particularly useful in understanding the system functions of benthic communities.[309,421,528,1011] No attempt will be made here to discuss these ideas further; instead we shall explore the functional organization of macrobenthic communities with particular reference to faunal-sediment interactions, the capture and processing of food, and the role of disturbance in structuring benthic communities.

A convenient way of classifying macrobenthic animals is by their method of feeding. Pearson[701] described five trophic groups: motile predators, algal scrapers, surface deposit feeders, deposit swallowers, and suspension feeders. The first two groups are quite specialized and make up only a small proportion of the total. The bulk of the species are either suspension or deposit feeders. Suspension feeders filter particles that are in suspension in the water just above the sediment. Examples of suspension feeders are bivalve mollusks, sabellid polychaetes, sponges, and ascidians. For a review of the mechanisms of filter feeding see Jørgensen.[438] Surface-deposit feeders include those epifaunal animals (e.g., amphipods, isopods, and gastropods) that move freely over the sediment surface grazing on the organically rich surface sediments with their detrital particles and their associated microbial community, benthic microalgae, and meiofauna. Other surface-deposit feeders include species living in the sediment but feeding on the surface sediments such as some bivalve mollusks (e.g., tellinids), amphipods, crabs, and many polychaete species. Deposit swallowers are those species, principally polychaetes, that ingest sediment below the surface. These major feeding groups will be considered in turn.

1. Deposit Feeders

Deposit feeders are a dominant component of estuarine benthic communities where they are responsible for extensive sediment reworking and microbial grazing. There is a general tendency for deposit feeders to predominate in fine sediments with a high organic and clay content.[810] Filter feeders on the other hand reach their optimum abundance in coarser sediments with a mean grain size of 0.18 mm. In their study of the intertidal benthic fauna of a Florida estuary, Bloom et al.[54] identified species assemblages in which the numbers of deposit- and filter-feeding species were inversely related. Hughes and Thomas[415] demonstrated that 46% of the variance in occurrence of polychaetes and echinoderms in a coastal bay were associated with an area of soft mud. Analysis of distribution patterns by various mathematical procedures have identified species groups of macrofauna which correlate with differences in sediment texture.[98,430,809,1027]

Levinton[510] has recently reviewed the role of deposit-feeding invertebrates in soft-bottom benthic communities. He lists the impacts that dense populations of deposit feeders have on the benthic system as: (1) altering the texture and resuspension properties of the sediment;[780,781] (2) influencing the vertical distribution of chemical pore water properties;[440] (3) changing bacterial and microalgal standing stocks;[244,536] (4) influencing microbial population dynamics;[537] and (5) influencing the patterns of larval recruitment.[784,1036] The high densities of deposit feeders which may exceed 100,000 m^{-2} [475] suggest an important role in acceleration of detrital breakdown and the regulation of deposit-feeding populations by resource availability. Levinton[510] considers that three lines of evidence suggest that resources are limiting to deposit-feeding benthic populations: (1) correlations of population density with food-related parameters, such as organic content of the sediment, percent clay, and microbial abundance; (2) patterns of niche structure of coexisting deposit-feeding species which show little niche overlap; and (3) experimental demonstrations of intra- and interspecific interactions and resource depression by deposit feeders.

Coexistence of species requiring similar limiting resources implies a mechanism of niche partitioning. In recent years there have been a number of investigations of differences in resource exploitation among coexisting species.[233,234,509,535,995] The last of these has been selected for discussion. Whitlach[995] investigated the patterns of foraging and distribution for 19 species of infaunal polychaetes inhabiting sand and mud flats in Barnstable Harbor, Massachusetts. Two major feeding groups were recognized: surface-feeding forms — including all species which lived in vertically positioned tubes or burrows and collected food from the sediment-water interface with palps or tentacles (e.g., the spionids, *Polydora ligni*, *Scolecolepides viridis*, and *Prionospio heterobranchia*), and burrowing forms — constituting a rather loosely defined group which contained species feeding below the sediment surface (e.g., the capitellids *Capitella capitata* and *Heteromastus filiformis*, the orbiniids *Scoloplos acutus*, and the cirrutulid *Chaetozone* sp.). In both groups, between-habitat species richness was positively correlated with food-resource supplies in the sediment.

Food resources were allocated among the deposit feeders in two primary ways: with respect to particle size and among subsurface feeders according to vertical space utilization. The size distributions of particulate material ingested by the surface feeding and burrowing deposit feeders are summarized in Figures 21 and 22. The majority of species in both feeding modes ingested the smaller and more abundant size-fractions within the sediments. This is not surprising since as pointed out in Chapter 4, Section IV, the smaller particles have a greater microbial biomass associated with them.[155,533,623] While all of the surface-feeding polychaetes collected particulate materials from the sediment-water interface, burrowing species were found at different levels with the sediment. Estimates of overlap in food size and vertical position among species suggested that interspecific competition influenced community structure. When organic content of the sediments was high, species that were similar in resource use were able to coexist and species richness was high, while in areas with lower organic content, ecologically similar species were excluded and species richness was reduced. The excluded species generally exhibited the highest overlap in resource use when compared to other species in the community. Species having lesser amounts of resource overlap with other species in the system tended to be found ubiquitously throughout the study areas. This additive effect of competition, termed "diffuse competition",[545] inducing interspecific exclusion has been chronicled in both empirical and theoretical studies of competition.[404,545,722]

A convenient framework for the examination of niche partitioning is the organization of the distribution of a species into niche axes, or dimensions with each species occupying an ecological hypervolume.[953,1079] Levinton[510] envisages niche separation of benthic deposit feeders along the following resources axes:

1.   Sediment type.
2.   Living or feeding depth below the sediment-water interface — This has been demonstrated above for the polychaete deposit feeders in Barnstable Harbour. Levinton[509] examined niche structure in two deposit feeding communities in Quisset Harbor, Cape Cod, Massachusetts, and found a depth stratification with no two dominant deposit-feeding species coexisting at the same depth within a community type. Laboratory studies show direct interference interactions between dominants not typically co-occurring.[625,994]
3.   Particle size — as demonstrated in Figures 20 and 21, coexisting species may consume particles of differing diameters. A positive correlation has been demonstrated between

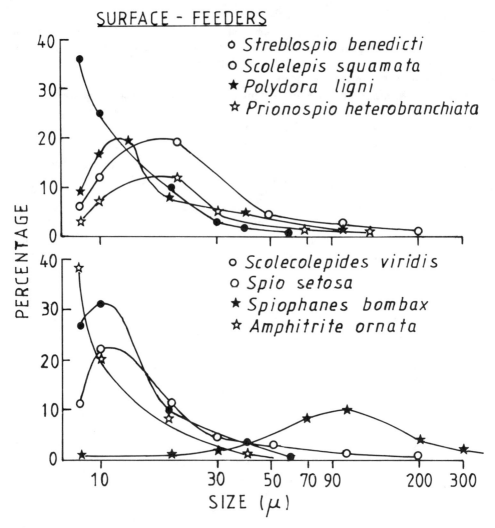

FIGURE 21.   Particle-size utilization curves for eight species of surface-feeding annelids found at Barnstable Harbor. Each curve has been quasinormalized through logarithmic transformation. Each curve represents the average size distribution of ingested particles in at least 20 individuals. Only a fraction of the data points have been included on each curve, and the curves have been arbitrarily divided into two sets to reduce visual confusion. Generally the data points did not deviate more than 3% from the eye-drawn lines. (From Whitlach, R. B., *J. Mar. Res.*, 38, 750, 1980. With permission.)

body length and ingested particle diameter for the tube-dwelling polychaete *Pectinaria gouldii*,[993] and for three species of the mud snail *Hydrobia*.[233,234]

4.   Food type — coexisting deposit feeders might conceivably feed on different types of food. Lopez and Kofoed[535] have studied the feeding behavior of four species of mud-flat snails *Hydrobia ulvae*, *H. ventrosa*, *H. neglecta*, and *Potamopyrgus jenkinsi*. The results of a species of feeding experiments are detailed in Table 5. *P. jenkensi* has a strikingly different feeding behavior to that of the *Hydrobia* spp. When the two similarly sized *H. ventrosa* and *H. neglecta* are compared it can be seen that the former is more specialized for feeding on fine sediment fractions. All species were found to be capable of feeding on material adhering to sediment particles by swallowing small particles and by browsing upon the particle surfaces, a feeding method which Lopez and Kofoed[535] call "epipsammic browsing". This is accomplished by taking particles into the buccal cavity, scraping off attached microorganisms, and then spitting out the particle. Of all the species, *H. ulvae* was the poorest browser on larger particles.

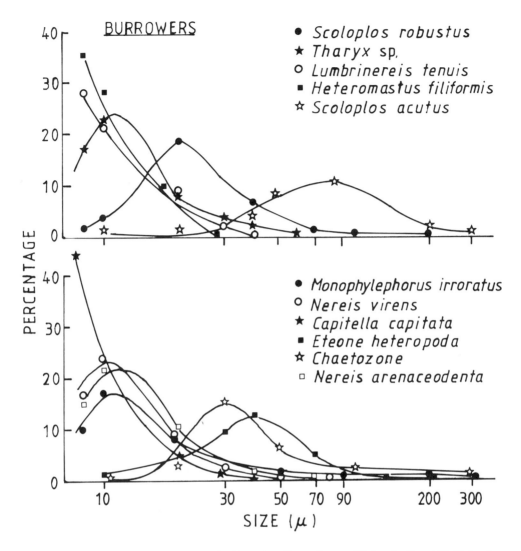

FIGURE 22.   Particle-size utilization curves for 11 species of burrowing annelids found at Barnstable Harbor (see Figure 21 caption for details). (From Whitlach, R. B., *J. Mar. Res.*, 38, 751, 1980. With permission.)

Table 5
FEEDING ACTIVITY OF FOUR SPECIES OF MUD-FLAT SNAILS

Species	Shell length (μm)	Particle size most rapidly fed upon (μm)	Maximum size of particles ingested (μm)
Hydrobia ulvae	4.0—4.5	<40, 40—80	~200
Hydrobia ventrosa	2.5—3.0	<40	~120
Hydrobia neglecta	2.5—3.0	40—80	~120
Potamopyrgus jenkinsi	3.0—3.5	80—160	~180

From Lopez, G. L. and Kofoed, L. H., *J. Mar. Res.*, 38, 585, 1980. With permission.

The differential trophic value of different kinds of detritus and different stages of degradation of detritus has been documented by several authors.[56,537,918,919] Tenore[922] has discussed the problem of estimating trophic transfer efficiencies (the amount of detritivore-production per amount of detritus supplied to the system) of detritus-based systems. Energy flow is complicated by the bacterial and micro- and meiofaunal components of what is in reality a detritus food web.[924] From feeding studies under laboratory conditions has arisen the concept of detritus "aging", i.e., the nutritional value of detritus is a function of time or the age of the detritus particles (see Chapter 4, Section V.B).

Tenore[922] has measured specific growth rate, population production, and trophic transfer efficiency for the deposit-feeding polychaetes *Capitella capitata* and *Nereis succinea* in laboratory systems receiving similar detritus food rations (1500 mg C m^{-2} day^{-1}). The opportunistic *Capitella* species more effectively exploited the available food because of their higher specific growth rate and a higher population density. The metabolic cost was high (production:respiration 0.66, vs. 2.50 for *Nereis*), but a greater proportion of the available detritus went into production in *Capitella*. Trophic transfer efficiency is the ratio between worms production and food supplied; 27% for *Capitella* and 3% for *Nereis*. Opportunistic species such as *Capitella* effectively exploit habitats with high organic conditions, not only by fast population increase, but also by achieving high population densities.

The more typical, larger, benthic deposit feeders, such as *Nereis succinea,* not only have a slower population response to increasing food, but behavioral interactions due to crowding (i.e., increased mortality) may limit population densities so that they do not maximally exploit available food.

Levinton[511] has analyzed the feeding and resources of the deposit-feeding snail *Hydrobia ventrosa*. For the details of the ecology of this species see Fenchel and Kofoed,[244] Levinton and Lopez,[515] Levinton et al.,[516] Lopez and Levinton,[536] and Levinton.[510] Laboratory microcosm studies show that the following parameters affect resource availability for the mobile deposit-feeding genus *Hydrobia*: (1) *fecal pellet breakdown* — snails have been shown to avoid ingestion of intact fecal pellets; (2) *renewal of microbial resources* such as diatoms and bacteria; (3) *space* — *Hydrobia* individuals feed more slowly under crowded conditions and emigration is increased; and (4) *particle size* — feeding rate decreases with increasing particle diameter. Experiments have shown that resource limitation affects Hydrobiid snails within the range of maximal field densities.

2. Suspension Feeders

Species feeding on organic particles suspended in the water are termed suspension feeders; their potential food sources are phytoplankton, zooplankton, suspended bacteria, organic aggregates (detritus), and dissolved organic matter. Zooplankton is relatively unimportant when compared to phytoplankton on a weight basis, but some suspension feeders such as the edible oyster *Ostrea edulis* may consume considerable amounts of zooplankton. Dissolved organic matter is quantitatively unimportant. Thus, phytoplankton and suspended organic matter are the prime available foods. In muddy substrates resuspended sediments, with their ample supplies of benthic microalgae, micro- and meiofauna, and organic matter, can provide a nutritious food source. However, as we shall see in the next section, the physical instability of muddy sediments which are reworked by deposit feeders tends to inhibit or exclude suspension feeders from the community.[781,1051]

Levinton[508] suggested that suspension feeders rely largely on sedimenting phytoplankton for food, since the gut contents of many species reflect what is available in the phytoplankton. However, phytoplankton is extremely variable, both in space and time, in terms of quantity, and the kinds of species present. Thus, as suspension feeders are fixed to the substrate and cannot move to more favorable feeding areas, they must be typical opportunists, rapidly exploiting favorable conditions by building up large populations. A consequence is that they are subject to equally dramatic population crashes,[507] as are well known in suspension-feeding bivalve populations. Levinton[507] argued that it is unlikely that two such species will compete for long enough to evolve niche specializations or competitively exclude each other, and he predicted that suspension feeders will have broad overlapping niches with few feeding specializations. However, there is evidence of interference interactions between bivalve species.

Jørgensen[438] observed that the response of suspension-feeding invertebrates to changes in the number and type of available food particles was variable, but that in general filtration declined at high-particle concentration and ceased entirely when the ciliary or feeding appendages became clogged. Winter[1096] proposed a stepped behavioral response in mollusks to changes in particle concentration. Food intake by mussels and oysters, for example, reached a plateau above certain concentrations while other suspension-feeding species demonstrated no threshold or decreased ingestion at high-particle density.

Many studies of suspension feeding have indicated that particle quality as well as its size and abundance influences the rate of ingestion. Conover[1068] cited numerous examples of adaptation of feeding rates which appeared to maximize useful ration. Thus, when suspension feeders consume inert or less nutritious particles, greater ingestion rates may ensue. One of the major impacts of suspension feeders on benthic communities is through biodeposition where large amounts of feces and pseudofeces are deposited in the surface sediments.

3. Biologic Modification of the Sediments

Rhoads[781] has reviewed the interactions of benthic invertebrates and sediments. The feeding activities of both deposit and suspension feeders are important in changing the properties of the sediments. Mobile deposit feeders move laterally and vertically within the sediments, causing mixing and transport of particles as well as interstitial water and dissolved gases. The fecal pellets they produce are deposited within the sediments. Surface-deposit feeders generally deposit their fecal pellets on the surface. Other sedentary-deposit feeders (mainly polychaetes) form dense populations of vertically orientated tubes. If these worms feed at the lower ends of their tubes, a massive transfer of sediment may take place to the surface. The tube mats may also physically bind and stabilize the sediment. Suspension feeders trap suspended seston, ingest it, and produce fecal pellets that become incorporated in the surface sediments. The shape and sculptured surface patterns of fecal pellets may be species specific. Fecal pellets produced by zooplankton also settle to the sediment surface.

The various methods whereby deposit feeders mix and recycle sediments is illustrated in Figure 23. The maldanid and the holothurian are termed conveyor-belt species by Rhoads and Young.[784] The feeding activities of these animals form a cone of fecal pellets around the anus, and a ring of unconsolidated sediments in a depression around the cone. Filter feeders are inhibited from colonizing the depressions, but colonize the sides of the cone in great numbers. These cones confer a spatial heterogeneity on an otherwise rather uniform environment. The rates of reworking of sediments by deposit feeders is considerable (Table 6). These rates become more impressive when multiplying the individual rates by the standing stocks found in estuaries which commonly are between $10^2 - 10^4$ individuals m^{-2}. Many estuaries have sedimentation rates between 1 and 2 mm year^{-1}.[802] Given this rate of accumulation and the rates of reworking shown in Table 6, it is not surprising to find that the surface sediments are passed through the benthos at least once, and in some cases, several times a year.[779,781] As a result, the surface layer of estuarine sediments is composed largely of fecal pellets (Figure 24). The pellet content of the sediment usually decreases with depth of burial.

Sediments are also profoundly modified by the activities of suspension feeders. Material filtered from the water is combined into aggregates and voided as feces or pseudofeces which settle to the bottom as biodeposits. The quantitative importance of biodeposition by

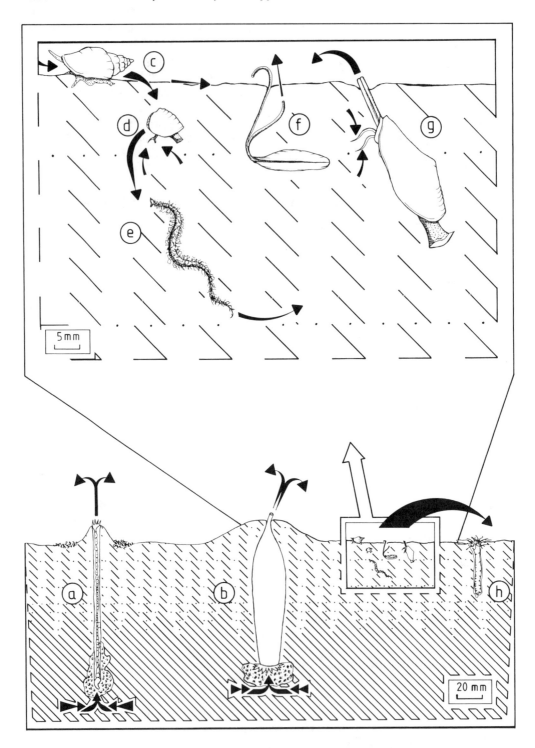

FIGURE 23.   Methods of mixing and recycling of sediments by deposit feeders: (a) maldanid polychaete; (b) holothurian; (c) gastropod *(Nassarius)*; (d) nuculid bivalve *(Nucula* sp.); (e) errant polychaete; (f) tellinid bivalve *(Macoma* sp.); (g) nuculanid bivalve *(Yoldia* sp.); and (h) anemone *(Cerianthis* sp.). Oxidized sediment lightly hatched; reduced sediment densely hatched. (From Rhoads, D. C., *Oceanogr. Mar. Biol. Annu. Rev.*, 12, 275, 1974. With permission.)

Table 6
RATES OF REWORKING OF MUD BY DEPOSIT FEEDERS

Organism	Location	Rate	Ref.
Polychaeta			
Clymenella torquata	Cape Cod, Mass.	274	778
Clymenella torquata	Beaufort, N.C.	246	573
Clymenella torquata	Passamaquoddy, New Brunswick	96	573
Pectinaria gouldii	Cape Cod, Mass.	400	311
Bivalvia			
Yoldia limatula (pseudo feces)	Cape Cod, Mass.	257	778
Nucula annulata (fecal pellets)	Cape Cod, Mass.	365	1051
Arthropoda			
Callianassa	Mugu Lagoon, Calif.	75-cm thick layer transport to surface by population/year	973

Note: (mℓ wet mud/individual/year).

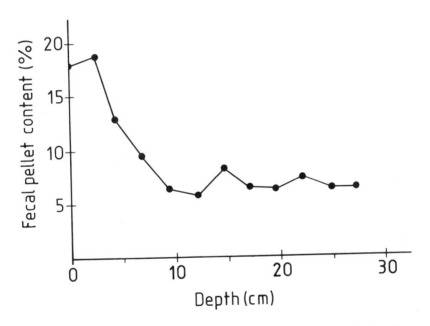

FIGURE 24.   Vertical gradient of fecal pellets in muds from Buzzards Bay, Mass. (From Rhoads, D. C., *Oceanogr. Mar. Biol. Annu. Rev.*, 12, 268, 1974. With permission.)

a variety of suspension-feeding invertebrates is given in Table 7. Haven and Morales-Alamo[376] estimated that a population of the estuarine oyster *Crassostrea virginica* on 0.405 ha may produce up to 981 kg (dry weight) of feces and pseudofeces weekly. They found that the material deposited contained 77 to 91% inorganic matter, 4 to 12% organic carbon, and 1.0 g kg[-1] of phosphorus.

The physical stability of the sediment is closely tied to the activities of the benthos. Mobile deposit feeders produce pelletal surfaces and sediments with an open fabric, high near-

Table 7
RATES OF BIODEPOSITION BY SUSPENSION FEEDING INVERTEBRATES

Organism	Location	Rate	Ref.
Bivalves			
Cardium edule	France	648 mg wet ind^{-1}day^{-1}	156
Cardium edule	Wadden Sea	1×10 m-tons day total pop.$^{-1}$year^{-1}	957
Mytilus edulis	Wadden Sea	25—175 $\times 10$ m-tons dry total pop.$^{-1}$year^{-1}	957
Crassostrea virginica	Japan	3 mg wet ind^{-1}day^{-1}	1080
Crassostrea virginica	Texas	0.7 ton dry acre^{-1}day^{-1}	544
Crassostrea virginica	Chesapeake Bay	18.4 mg dry ind^{-1}day^{-1}	376
Mya arenaria	Chesapeake Bay	25 mg dry ind^{-1}day^{-1}	376
Geukensia demissa	Chesapeake Bay	125 mg dry ind^{-1}day^{-1}	376
Barnacle			
Balanus eburneus	Chesapeake Bay	18.4 mg dry ind^{-1}day^{-1}	376
Turnicate			
Molgula manhattenensis	Chesapeake Bay	60 mg dry ind^{-1}day^{-1}	376

surface porosity, and low compaction and adhesion. As a consequence, such sediments are easily resuspended. Sedentary deposit and suspension feeders (especially tube dwellers), on the other hand, bind the bottom by the concentration of mud into tubes and by entrapping particles between the tubes, so that a 'tighter' fabric results with greater compaction and adhesion.[781] Bioturbation substantially increases sediment water content. Intensely bioturbated sediments are frequently resuspended by wind-driven waves and tidal streams. This resuspension cycle in Buzzards Bay, Massachusetts, is illustrated in Figure 25. Resuspension creates a turbidity maximum near the bottom which reaches a maximum during the ebb phase of the tidal cycle. These resuspended materials are rich in detritus and contain benthic microalgae. Tenore[920] considers that these resuspended materials might significantly influence the bioenergetics of shallow water benthos. Tenore found that a portion of the resuspended chlorophyll represented viable phytoplankton which resulted in a significant increase in primary production when the resuspended cells were exposed to light and nutrients in the water column. In addition, Tenore postulates that microbial heterotrophy associated with detrital particles might be enhanced, owing to their resuspension in oxygenated water. Thus, resuspension could result in a dual recycling of sedimented carbon and augmentation of the benthic food source (Figure 26).

4. Influence of Macrofaunal Activities on the Chemistry of the Sediments
Intensive biogenic mixing and irrigation of the 10 to 30 cm of the sediment takes place depending on the species present. Rhoads[781] lists the chemical processes and factors which may be influenced by macrofaunal activity as

1. Rate of exchange of dissolved or adsorbed ions; compounds and gases across the sediment-water interface
2. Magnitude and form of the vertical gradients of Eh, pH, and dissolved O_2
3. Transfer of reduced compounds from below the RPD layer to the aerated surface
4. Cycling of carbon, nitrogen, sulfur, and phosphorus
5. Concentration of elements in tissues and sediments which are dilute at ambient concentrations

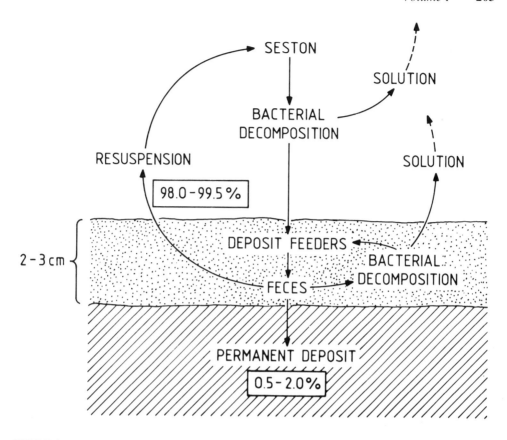

FIGURE 25.   Resuspension cycle in Buzzards Bay, Mass. Approximately 98.0 to 99.5% of annual sediment influx to the bay floor is resuspended and recycled within the upper 2-to 3-cm thick, bioturbated surface.[781,1049]

Space does not permit discussion of these points in detail and some illustrations only will be given. Molecular diffusion of dissolved O_2 into a fine-grained mud-bottom is limited to a distance of 1 or 2 mm in the absence of mixing. When bioturbating organisms are present (or dense populations of tubiculous organisms that pump water into the bottom), the rate of oxygen diffusion is greatly increased and oxygen penetration may reach depths of 20 or 30 mm.

Montague[601] has reviewed the influence of fiddler crab burrows on metablic processes in salt marsh sediments. Fiddler crab (*Uca* sp.) burrow densities range from 50 to 700 m^{-2} with density highest near creek banks.[451,490] They increase the surface area of salt marshes by 20 to 60% or more.[214,451,913] Water within burrows contains some oxygen and the walls of the burrows are aerobic, similar to the sediment-water interface at the marsh surface.[214] Burrows may improve water circulation through the subsurface sediment,[488,913] thus diluting salts that accumulate in sediments from cordgrass transpiration.[852] Lower soil salinity as we have seen can improve nutrient uptake and growth of *Spartina alterniflora*.[526,852] In excavating burrows, fiddler crabs also transport organic carbon from the subsurface anaerobic sediments to the surface where more rapid anaerobic decomposition occurs.[1072] Montague[602] found that fiddler crabs in the Sapelo Island salt marsh transported 26 g organic C m^{-2} from belowground in July 1979. Annual excavation was estimated at 157 g C m^{-2}, or 20% of the belowground production of *S. alterniflora*.

Montague[602] found that the mean burrow respiration was 2.1 mg O_2 hr^{-1}, accounting for 20 to 90% of the salt marsh sediment respiration depending on marsh wetness. Compared with unburrowed plots, burrows increased *S. alterniflora* standing stocks by 23% in the high

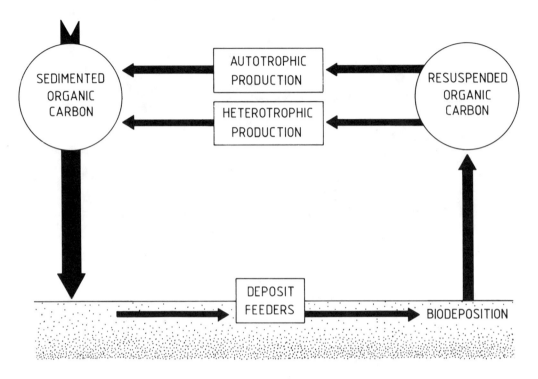

FIGURE 26.   Illustration of the possible role of resuspended organic carbon and augmentation of the food resource to the benthos by related autotrophic and heterotrophic production. (From Tenore, K. R., *Ecology of the Marine Benthos,* Coull, B. C., Ed., University of South Carolina Press, Columbia, 1977c, 48, With permission.)

marsh. The chemistry of the burrow water was different from that of the interstitial water. Salinity of the burrow water was 20 to 23‰, while that of the interstitial water was 37 to 45‰. Phosphate and ammonium concentrations (Figure 27) were significantly different from that of the interstitial water. In September, burrow water contained nearly three times the phosphate of the interstitial water. Ammonium concentrations in the tidal waters was usually 2 to 3 μmol N 1^{-1}.[347] Fiddler crabs excrete ammonia, and 10 g of fiddler crabs per m^{-2} [1025] probably regenerate much more than the 4.2 to 5.7 g N m^{-2} year^{-1} required to supply 19% of the production of *S. alterniflora.*[267,996]

Montague[602] considers that fiddler crab-burrowing activity provides probably five of six factors reviewed by Morris[612] that improve growth of short *S. alterniflora.* These factors are: (1) more oxygen, even though *S. alterniflora* supplies some oxygen internally to its roots;[588] (2) less hydrogen sulfide; (3) lower salinity;[621,851,852] (4) increased rates of nutrient diffusion through pore water;[622] (5) added nitrogen;[82,267,526,851,852] and (6) greater exchange capacity of sediments.

Three aspects of organism-sediment relations are important in recycling: (1) ingestion of pellets (coprophagy) which have been reconstituted by bacteria or microalgae; (2) vertical transport of organic detritus from below the RPD to the sediment-water interface; and (3) resuspension of nutrient-rich bottom sediments into the water column providing a potential food source for suspension feeders. Coprophagy will be considered in the next section. Suspension feeders play an important role in the cycling of nutrients in estuarine ecosystems. Kuenzler[494,495] measured both energy flow and the cycling of phosphorus in a population of the ribbed mussel *Geukensia (Modiolus) demissa* living in a salt marsh on the eastern coast of the U.S. (Figure 28). Numbers varied from 8 to 32 m^{-2}. Organic biomass (ash-free dry weight) averaged 12 g m^{-2} (equivalent to 60 kcal m^{-2} year^{-1}). However, the mussels filtered large quantities of water, removing considerable quantities of particulate matter and

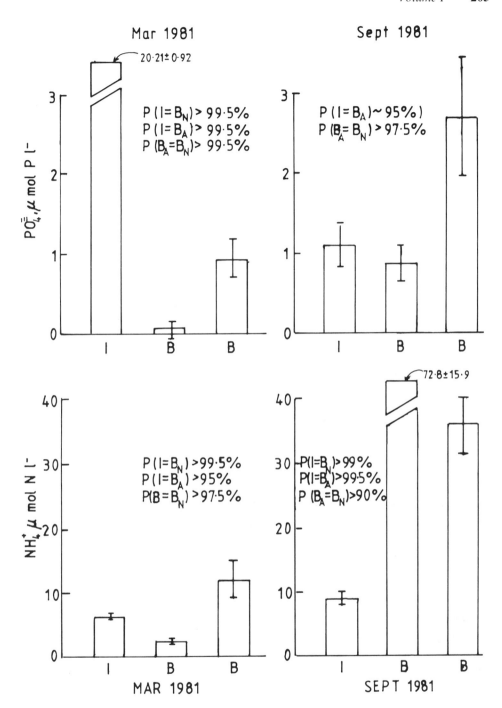

FIGURE 27. Phosphate and ammonium concentrations in interstitial water of a *Spartina alterniflora* marsh. I = water from natural fiddler crab burrows; B_N = water from artificial burrows; and B_A = interstitial water sampled in the vicinity of the artificial burrow test plots. Standard errors are indicated on each bar. (From Montague, C. L., *Estuarine Comparisons*, Kennedy, V. S., Ed., Academic Press, New York, 1982, 294. With permission.)

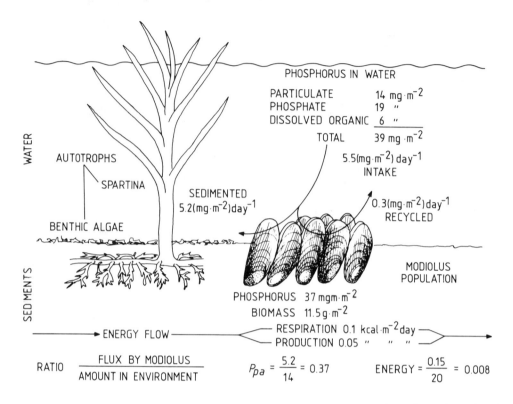

FIGURE 28.   The role of the salt-marsh mussel *(Geukensia (Modiolus) demissa)* in the cycling and retention of phosphorus in the estuarine ecosystem. (From Odum, E. P., *Ecology,* 2nd ed., Holt, Reinhart & Winston, New York, 1975, 143. With permission.)

sedimenting it onto the marsh surface. Of the total average phosphorus concentration in the water of 39 mg m^{-2}, the mussels removed 5.5 mg m^{-2} day^{-1}, and of this 5.2 mg m^{-2} day^{-1} were sedimented to the marsh surface via feces and pseudofeces. Kuenzler calculated that the turnover time of the 14 mg m^{-2} of particulate phosphorus in the water due to the action of the mussels was only 2.6 days. Other filter feeders in estuarine ecosystems are likely to have a similar impact.

V. ESTUARINE BIRDS

A. Introduction

An important consumer component is the estuarine birds. Recent studies indicate that the grazing pressure exerted by the birds may be considerable in northern temperate-zone estuaries.[213,681] They can be divided into two groups, the permanent residents such as gulls, cormorants, some ducks, and marsh rails, or migratory species such as wading birds, geese, or many ducks. Consequently the numbers and varieties of birds in any estuary varies throughout the year. Many estuaries are stop-over points for many species on their annual migration pathways. Other species overwinter in estuaries before migrating to breeding areas in the spring and summer.

The bird predators of estuaries are highly mobile and typically exhibit clear tidal rhythms of activity which are associated with water movements and the activity of the prey in relation to the tides. On the basis of their food and feeding methods, the birds can be divided into wading birds, water fowl, fishing birds, rails, and other marsh species. Wading birds can be divided into three groups: (1) the herons, egrets, and ibis which feed principally on small

Table 8
PRINCIPAL PREY SPECIES OF THE MAIN WADING BIRDS IN THE WASH (ENGLAND)

Bird species	Prey	
Oystercatcher *(Haematopus ostralegus)*	Bivalves	*Cardium edule*
		Mytilus edulis
Knot *(Calidris canutus)*	Bivalves	*Macoma balthica*
		Cardium edule
Dunlin *(Calidris alpina)*	Snail	*Hydrobia ulvae*
	Polychaetes	*Nereis divericolor*
Redshank *(Tringa totanus)*	Crustacea	*Carcinus maenas*
		Crangon sp.
	Snail	*Hydrobia ulvae*
	Polychaetes	*Nereis* spp.
Bar-tailed godwit *(Limosa lapponica)*	Polychaetes	*Lanice conchilega*
		Nereis spp.
	Bivalves	*Macoma balthica*
Turnstone *(Arenaris interpres)*	Bivalves	*Cardium edule*
		Among mussel beds
Grey plover *(Pluvialis squatarola)*	Polychaetes	*Lanice conchilega*
		Various spp.
Curlew *(Numenius arquata)*	Crustacea	*Carcinus maenas*
	Polychaetes	*Lanice conchilega*
		Arenicola marina

From Goss-Custard, J. D., Jones, R. E., and Newberry, P. E., *J. Appl. Ecol.,* 14, 681, 1977a. With permission.

fishes, crabs, and snails; (2) the shore birds, comprising resident stilts and migratory species such as plovers, curlews, turnstones, godwits, red shanks, dunlin, knot, sandpipers, whimbrels, stints, etc. which feed principally on small crustaceans and polychaete worms; and (3) the oystercatchers which feed principally on bivalves. The fishing birds such as cormorants and terns consume principally fishes but also take larger pelagic crustacea. Gulls take a wide variety of food items and are also scavengers on carrion. Ducks consume plant material (up to 75%) and take smaller amounts of benthic animals, mainly snails. Some swamp hens feed on the roots and shoots of marsh plants, while marsh rails feed on snails, crabs, other crustaceans, and insects. A number of the marsh birds such as swallows are insectivorous.

B. Feeding Ecology of Estuarine Birds

Goss-Custard et al.[314] have investigated the feeding of wading birds in the Wash, Great Britain. The principal prey species are listed in Table 8, from which it can be seen that the prey resources are divided between the species of waders with only limited overlap between the species. Within the Morecombe Bay estuarine area in Northern England there is a large (70,000+) overwintering population of the knot *(Calidris canutus)*. The birds feed principally on the bivalve *Macoma balthica*, supplemented by the mussel *Mytilus edulus* and the mud-flat snail *Hydrobia ulvae*. The choice of *Macoma* appears to be dictated by its availability. In the Morecombe Bay area the knot feed predominantly on the lower half of the intertidal zone and thereby feed mainly on *Macoma*, which are generally lower down the beach than *Hydrobia*. Prater[738] suggests that the knot have a clear preference for *Hydrobia*, but that at Morecombe Bay the greater accessibility of *Macoma* has made it the dominant food item. The preferred diet of the red shank *(Tringa totanus)* at 30 estuarine sites in the British Isles is the amphipod *Corophium volutator*, and the feeding rate of red shank depends mainly on the density of that prey in the muddy sediment.[315] Where the density of *Corophium* is low, red shank take polychaete worms *Nereis diversicolor* and *Nepthys hombergi* instead. The bivalves *Macoma balthica* and *Scrobicularia plana* are taken only sporadically.

Many shore birds have adapted their food-searching behavior to suit the tidal and seasonal rhythms of their preferred prey. It has been generally shown that the bill length of particular species of birds is adapted to particular species of prey organisms, but Evans[224] has suggested that more subtle behavioral adaptations also occur. Two basic strategies (with many gradations between these) are used by waders to detect and catch their prey.[724] One group, including many sandpipers and the oystercatchers, commonly use a tactile element in their foraging — the bill is placed in contact with the substratum at various locations as the bird walks over the mud flats. The birds may also use sight, sometimes chosing where to place the bill in response to visual cues. The other group of waders, typified by the plovers, use a totally visual method of foraging.[88,723] These normally stand still for a time while scanning the area for indications of available prey and may then run rapidly to peck at a prey item or, if none is seen, run to a new waiting position. Tactile foragers usually have long bills with high densities of pressure detectors along much of their length to aid probing into the substratum; plovers and other visual searchers have relatively short bills with tactile receptors mainly at the tip, this combination being compatible with rapid alignment of the head and bill towards the prey, hard impacting on the substratum surface, and precise gripping of the prey.[88,723]

One of the most intensively studied species of estuarine birds is the oystercatcher *(Haematopus ostralegus)* in northern European estuaries.[162,200,314,316,394,395,632,739,1062] In these estuaries oystercatchers feed on mussel beds when they are exposed at low tide by hammering on one of the bivalve shells until it splits open, or inserting the bill between the valves to cut the adductor muscle. Once opened the flesh is picked out and the shell discarded. In a study of oystercatcher feeding in The Netherlands, Zwarts and Drent[1062] found that when mussels taken by oystercatchers were compared with the size distribution of mussels on offer, there was a clear preference for the larger mussels; when compared to the "model mussel" of 44 mm in 1973, the relative risk of a mussel of 50, 54, or 58 mm being taken by oystercatchers was 3.6, 6.7, and 10.5 times as high, respectively. They also found that feeding rate decreased significantly as the density of the oystercatchers went up (Figure 29). Figure 29 presents an overview of the dynamics of oystercatcher feeding on a single study plot over the period 1971 to 1979. During this period the total number of oystercatchers in the area remained constant (Figure 29D). However, there was a marked decline in the size of the mussels over the period (Figure 29A) accompanied by a decline in the feeding density of the oystercatchers (Figure 29B). To compensate for the smaller-sized mussels the feeding rate (no. mussels per 10 min) increased but in spite of this increase the actual ash-free dry weight ingested declined (Figure 29C). Thus, the presence of conspecifics influences feeding rate, i.e., intake depends on the density of the oystercatchers as well as that of the prey. This coupled with the long-term decline of mussel stocks demonstrated during the study implies a continuous adjustment of feeding density to match the prevailing conditions.

A detailed study of the feeding activity of feeding ecology of the South Island Pied Oystercatcher *(Haematopus ostralegus finschii)* has been caried out on the Avon-Heathcote Estuary, Christchurch, New Zealand.[27-29] A maximum of some 4000 birds overwinter on this 8 km² estuary. The feeding behavior follows a regular tidal cycle. During high water the birds congregate on high water roosts while the periods of ebb- and flood-tides are spent feeding in a narrow band averaging about 3 m from the waters edge down to the maximum at which birds can stand (about 6 to 7 cm). While the range of prey taken includes cockles *(Chione stutchburyi)*, pipis *(Amphidesma australe)*, wedge-shells *(Macoma liliana)*, mud-flat snails *(Amphibola crenata)*, a range of polychaete worms, amphipods, shrimps, mud-flat crabs, and juvenile flounders, the predominant food item is the cockle. Baker found that when feeding, the birds were not uniformly distributed but were concentrated in areas

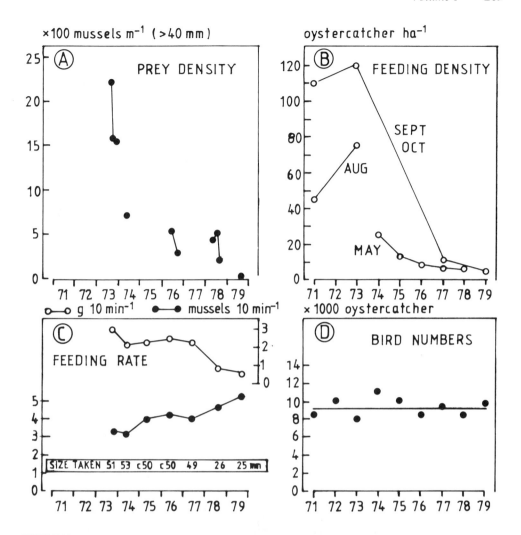

FIGURE 29.   Overview of events on the mussel bed plot 13 near Schiermonnikoog through the years. (A) number of mussels larger than 40 mm; dots within 1 year are connected; (B) number of oystercatchers in the plot; each point is the mean of several hundred counts; (C) intake rate (lower segment — mussels per 10-min feeding time) and in combination with data on average size taken. The feeding rate (upper segment — g AFDW ingested per 10 min) can be derived with the aid of the length/width relationships of the mussel populations concerned. Data for May (except 1973: August to September, and 1979: September), n = number of 10-min observation periods; note shift in prey size taken; and (D) number of oystercatchers present at the high water roosts for the SW Schiermonnikoog study area (August to November; number of counts indicated) corrected for seasonal trend within this period. (From Zwarts, L. and Drent, R. H., *Feeding and Survival Strategies of Estuarine Organisms,* Jones, N. V. and Wolff, W. F., Eds., Plenum Press, New York, 1981, 213. With permission.)

of high-prey density. The effect of prey concentration on distribution and feeding rate is shown in Table 9.

A detailed study of feeding rates and of the factors affecting the rates was carried out. Figure 30 depicts the feeding rate over a tidal cycle. Feeding rates were highest when the tide was retreating or advancing over the feeding grounds and lowest when the shellfish beds were completely exposed. Adverse weather can result in a substantial decline in the feeding rate. Feeding rates also vary with season and thus with ambient temperature. Mean monthly feeding rates were highest in the midwinter months of June, July, and August (40 + cockles hr^{-1} in July), and lowest in the midsummer months of December, January, and

Table 9
THE EFFECT OF PREY CONCENTRATION
(*CHIONE STUTCHBURYI*) IN THE
DISTRIBUTION AND FEEDING RATE OF THE
SOUTH ISLAND PIED OYSTERCATCHER
(*HAEMATOPUS OSTRALEGUS FINSCHI*)[28]

Mean number of birds/study site	Mean cockle concentration (no. m^{-2})	Mean feeding rate (cockles hr^{-1})	Number of observations
22	78	96	10
16	54	60	14
10	42	48	12
5	33	24	14
1	20	24	11

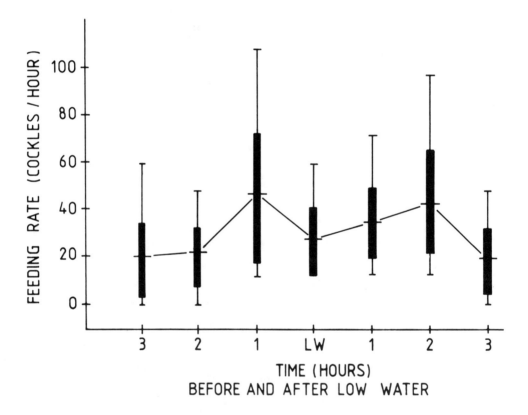

FIGURE 30.   Feeding rates of the South Island Pied Oystercatcher during a tidal cycle in the Avon-Heathcote Estuary, Christchurch, New Zealand. The thin vertical line indicates the range; the solid bars are standard deviation each side of the mean; and the horizontal line the mean.[28]

February (27 cockles hr^{-1} in February). Similar variations in feeding rates were noted for the European oystercatcher.[162,200,418] Energy requirements are greater in the colder months and this is met by the higher food intake. In addition to being affected by prey availability as mentioned above prey size also has a direct effect. Feeding rates of 108 cockles hr^{-1} were recorded when the model size of the cockles was 3.1 cm (1.9-mℓ flesh volume) as

Table 10
OYSTERCATCHER PREDATION ON COCKLES IN THE AVON-HEATHCOTE ESTUARY, NEW ZEALAND[28]

	Winter	Summer
No. of cockles per hour	40.9	29.0
No. of cockles per day	368	261
Grams of cockle flesh eaten per day	174.9	124.0
Percentage of body weight eaten per day	35.2	25.0
Mean daily winter cockle predation for 4000 oystercatchers	1,472,000	
Mean yearly food intake per oystercatcher	190,719	
Annual cockle predation by 4000 oystercatchers	438,876,000	

compared with a rate of 24 cockles hr^{-1} when the model size was 5.3 cm (6.0-mℓ flesh volume).

During the winter months the oystercatchers on the Avon-Heathcote Estuary ingested an average of 40.9 cockles hr^{-1}. Predated cockles had a mean valve length of 3.1 cm and a mean flesh volume of 1.9 mℓ. The average length of the daily feeding period was 12 hr but half of this occurred in darkness when the feeding rate was assumed to be half that during daylight hours.[395] Thus in winter the birds ingested an average 368 cockles day^{-1}, or an estimated intake of 174.9 g dry wt of cockle flesh, corresponding to 35.2% of their dry body weight (Table 10). In summer the daily cockle intake was reduced to 261 or 124.0 g dry wt of cockle flesh. The mean daily winter cockle predation by the oystercatcher population was 1,472,000 in this small estuary.

Workers in Great Britain[163,200,314,632] have investigated feeding rates of oystercatchers on cockles and their findings are similar to those detailed above. In Morecombe Bay it was found that the percentage annual cockle mortaility due to oystercatcher predation was 21.9% of the total population. In many estuaries in Great Britain, however, as mentioned above, oystercatchers feed on mussels *(Mytilus edulis)*.

The mussel bed community of the Ythan estuary in Scotland has been studied by Milne and Dunnet,[598] who recognized the mussel *(Mytilus edulis)*, the gammarid amphipod *(Marinogammarus marinus)*, and the shore crab *(Carcinus maenus)*, along with the tube-dwelling amphipod *Corophium volutator* as the dominant prey of birds (oystercatcher, eider duck, gulls, and turnstone) and fish (butterfish, blenny) (Figure 31). The partitioning of the *Mytilus* production among the various predators is shown in Figure 32. Annual gross production was 268 g dry wt m^{-2} or 1340.8 kcal m^{-2}. Of this, 600 kcal represented winter metabolism leaving a net production of 740 kcal m^{-2}. Of this, 480 kcal m^{-2}, or 65% of the net production was consumed by the bird predators. The feeding of these bird predators was separated in space and time and they fed on different sized mussels. Eider ducks feed when the mussels are submerged and take mussels averaging 18 mm long (2-year-old mussels). Oystercatchers and gulls feed when the mussel beds are exposed, the oystercatchers taking mussels averaging 33 mm (3- to 10-year-old mussels), while the gulls take mainly very small mussels 2 to 10 mm long (1-year-old mussels). Oystercatchers were estimated to consume 93 kcal m^{-2}, eider ducks 275 kcal m^{-2}, and gulls 112 kcal m^{-2}, representing 12.5, 37.2, and 14.5% of mussel net production, respectively.

C. Distribution and Ecology of South African Estuarine Birds

Siegfried[841] has recently reviewed the distribution and ecology of the estuarine avifauna of southern Africa. Tropical estuaries characteristically support resident populations of birds belonging mainly to the Ciconiiformes and Pelecaniformes. The herons, storks, ibises, cormorants, anhingas, and pelicans are predominantly at least tertiary consumers, taking

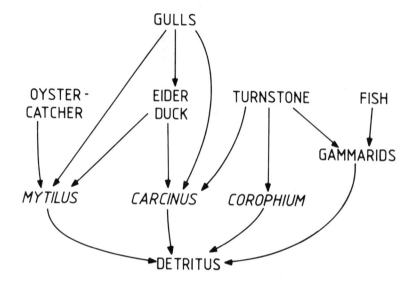

FIGURE 31.   The food web of the mussel bed *(Mytilus edulis)* community of the Ythan Estuary, Scotland. (From Milne, H. and Dunnet, G., *The Estuarine Environment,* Barnes, R. S. K. and Green, J., Eds., Elsevier Applied Science, London, 1972, 88. With permission.)

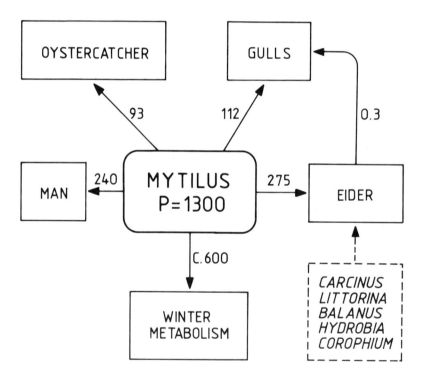

FIGURE 32.   The partitioning of the production of the mussel *(Mytilus edulis)* among the various predators of the mussel bed community in the Ythan Estuary, Scotland. (From Milne, H. and Dunnet, G., *The Estuarine Environment,* Barnes, R. S. K. and Green, J., Eds., Elsevier Applied Science, London, 1972, 99. With permission.)

Table 11
CLASSIFICATION OF BIRD SPECIES BY DIET IN SOUTHERN AFRICAN ESTUARIES DURING THE AUSTRAL SUMMER

Diet	Number of species		
	Migrants	**Residents**	**Totals**
Mainly fish and frogs	8	37	45
Mainly invertebrates	29	39	68
Mainly vegetable matter	0	14	14
Totals	37	90	127

From Siegfried, W. R., *Estuarine Ecology with Particular Reference to Southern Africa*, Day, J. H., Ed., A. A. Balkema, Rotterdam, 1981, 235. With permission.

mainly crabs and other large crustaceans, large gastropods, large insects, frogs, and fish. Siegfried[841] lists 37 migrant and 90 resident aquatic species making a total of 127 species. However, the total number of migratory individuals exceeded the total number of residents by a factor of 1.49 for those estuaries at which all birds were counted. Migratory sandpipers (Scolopacidae) are the most abundant birds in the summer (35.5%) followed by the Ciconiiformes (23.1%) (herons, egrets, storks, ibises, and flamingos). However, the resident pelicans, cormorants, and flamingos together make up 85% of the estuarine bird biomass whereas the migrants, in spite of their numerical abundance, account for only about 5%.

All the migrants which visit southern African estuaries are carnivores, primarily feeding on vertebrates, while resident invertebrate feeders are also common (Table 11). Taken together the migratory and resident carnivorous species make up 89% of the estuarine avifauna and 75% of its biomass. Within the resident community 40% of the species feed predominantly on fish and/or frogs, whereas 21% of the migrants do so. Among the species which feed on invertebrates, the greater flamingo alone accounts for 90% of the biomass.

At Langebaan Lagoon, the overall density of waders was 1564 birds km^{-2} yielding a biomass of 145 kg km^{-2}.[899] These figures are similar to those obtained for waders at the coastal wetland of Banc d'Arguin in Mauritania and at the Wash in England.[196] Summers[899] has calculated that the waders at Langebaan deposited 257.5 kg of feathers year^{-1}, or 0.06 kcal m^{-2} year^{-1} in the lagoon system. Additionally the birds returned annually 2.1 to 6.2 kcal m^{-2} as feces. The estimated energy removed from the system in the form of food by the wader community amounted to 20.8 kcal m^{-2} year^{-1}, or 4.2 g ash-free dry mass of animal food m^{-2} year^{-1}.[899] These figures are similar to those for the Wadden Zee where 18.5 kcal m^{-2} year^{-1} were removed by the carnivorous birds[906] and 17.1 kcal m^{-2} year^{-1} for the Grevelingen Estuary in the Netherlands.[1031]

The carnivorous birds studied by Swennen[906] made up 96% of the total avian biomass at Wadden Zee and their energy intake represented about 0.5% of the total amount of food (700 g dry wt m^{-2} year^{-1}) available for consumers. In a comparative study of the trophic role of the birds in the Grevelingen Estuary, Wolff et al.[1031] found that birds feeding on zoobenthos took about 6% of the total zoobenthos production of the estuary. At Langebaan Lagoon in southern Africa the curlew sandpiper population alone consumed 12.9% (91 kJ m^{-2} year^{-1}) of the estimated gross annual production of the intertidal zoobenthos living on the surface or in the upper 10 cm of the substrate.[745] There is evidence that avian predators in southern African estuaries can exert a considerable effect on the fish populations. Estimates for the fresh mass of fish taken by pelicans at Verlorenvlei (1070 ha), St. Lucia (30,050 ha), and Berg river estuaries (50 ha) are 6.8, 2.6, and 5.8 g m^{-2} year^{-1}, respectively. The annual average density and biomass of white pelicans at these three estuaries, taken together, are 5.2 birds km^{-2} and 55 kg m^{-2} $year^{-1}$, respectively. This biomass is at least four times greater than that given by Swennen[906] for all piscivorous birds in the Dutch Wadden Zee, suggesting that the ecological impact of fish-eating birds is greater in southern Africa than in north European estuaries.

VI. FACTORS CONTROLLING ESTUARINE BENTHIC COMMUNITY STRUCTURE

A. Introduction

In contrast to intertidal rocky shores where the nature and abundance of the biota are primarily influenced by tidal-dependent variations in environmental conditions, those living in sediments have to a large extent evaded the external environmental stresses of temperature, desiccation, and extremes of salinity. On the other hand the distribution and abundance of the sediment infauna is controlled by complex interactions between the physico-chemical and biological properties of the sediment in which they live.

The physico-chemical properties include sediment grain size, oxidation-reduction state, dissolved oxygen, organic content, and light. The biological properties include the following:

1. Food availability and feeding activity
2. Reproductive effects on dispersal and settlement
3. Behavioral effects which induce movement and aggregation
4. Interspecific competition — competitive exclusion effects
5. Predation effects and removal of certain species

As we have seen, estuarine soft-bottom communities have highly variable species composition. The local abundance of certain species may vary by orders of magnitude in time and space while at the same time the abundances of other species change little if at all. The underlying causes for these changes are one of the central questions in estuarine ecosystem dynamics. Year to year fluctuations in species composition may reflect catastrophic adult mortality brought about by climatic factors or competitive interactions among species. More often, however, the fluctuations are believed to be the result of factors affecting the success of spawning and dispersion patterns of larvae.[554,938]

Benthic samples taken in close proximity are often found to be less alike in species composition than widely separated samples.[211,1092] Such patchiness has been explained as the result of dispersal or concentration of planktonic larvae by currents,[226] small scale habitat heterogeneities, inter- and intraspecific competition, and amensalic interactions.[784] On the other hand Johnson[431] envisages the benthic community as a temporary mozaic with the different parts of a habitat disturbed at different times containing different faunas. Many authors have ascribed the spatially heterogeneous distribution patterns to the differential response of organisms to disturbance. This viewpoint considers that early and late colonizers of a new environment created by disturbance will have different adaptive strategies: r-strategies (opportunistic), or k-strategies (equilibrium).[325,546] The validity of these concepts will be discussed below.

In recent years experimental approaches have been used with varying degrees of success in efforts to understand the mechanisms involved in the determination of the structure of intertidal and shallow-water benthic communities. Three excellent recent reviews of this experimental approach are those of Dayton and Oliver[187] and Peterson.[713,714] Dayton and Oliver[187] point out that such controlled experiments offer a powerful means of testing hypotheses.

B. Interspecific Competition

As Dayton and Oliver[187] state, competition is a well known and often discussed process but it is often difficult to demonstrate in nature. A number of approaches have been used in the study of competition in estuarine and shallow water benthic communities. First, field studies have demonstrated intraspecific spacing and aggression among soft-bottom species.[136,752] A second approach is an experimental one in which laboratory and field experiments are used to test hypotheses concerning competition. In such experiments one looks to vary either the density of a presumed competitor or the abundance of a potentially limited resource. To evaluate the importance of competition, one examines the effect of the manipulation on (1) growth, (2), mortality, (3) recruitment, (4) migration, and (5) reproductive effort.[714] Most studies of interspecific competition have involved only two or perhaps three competing species. However, there is a need to have information on the impact of all the competing species in the system. "Diffuse competition" is the term used by MacArthur[545] to indicate the cumulative effects of all potential species in a given system.

Peterson[714] recognized four types of competitive interactions in soft-bottom sediments: (1) direct interference competition for space; (2) exploitative competition for food; (3) adult interference with larval settlement; and (4) indirect interference through alteration of the physical environment.

1. Space Competition

Manipulative and laboratory experiments have shown direct adult interference among several infaunal bivalves. Levinton[509] observed vertical and horizontal interspecific avoidance in laboratory aquaria between the bivalves *Yoldia limatula, Solemya velum,* and *Nucula proxima.* Direct interference competition for space has been invoked to explain complementary and largely nonoverlapping distributions of various pairs of infaunal macroinvertebrates, e.g., *Mya arenaria* and *Gemma gemma.*[66,812]

Peterson and Andre[715] performed an elegant species of experiments that demonstrated competitive interference between two bivalve species, *Sanquinolaria nuttallii* and *Tresus nuttallii.* These species have widely overlapping vertical stratification in the sediment, but disjunct horizontal distributions. Petersen and Andre transplanted known numbers of the two suspension feeding bivalves into the same sediment and observed a significantly higher emigration of *S. nuttallii* in the presence of *T. nuttallii* than in its absence. *Sanquinolaria* growth was also reduced by the presence of *Tresus.* In a second experiment, they substituted dead shells secured to wooden poles simulating clams with extended siphons for living *Tresus* and found similar results. *Sanquinolaria* had a lower growth rate in the presence of these artificial clams and siphons than they did in their absence. This artificial clam experiment suggests that the interaction is mediated by spatial interference and not by competition for food.

2. Exploitative Competition for Food

Levinton[508] has examined the abundance and composition of the food of soft-bottom benthic communities and its role as a limiting resource. He developed a hypothesis which states that suspension feeders live with unpredictable and fluctuating food supplies, whereas deposit feeders experience relatively constant food supplies. The food of suspension feeders can be plankton, dissolved organic matter, organic aggregates (detritus with its associated microbial community), and bacteria. The primary food source is sedimented plankton. Yet plankton is extremely variable, both in space and time, in terms of quantity and the species present. However, suspension feeders remain fixed to the substratum, apart from some bivalves that can move over limited distances in soft substrates and cannot move to better feeding areas. Thus they must typically be opportunists, rapidly exploiting favorable conditions by building up large populations. However, their populations fluctuate strongly over time and dramatic population crashes can occur.[507]

Deposit feeders, on the other hand, feed on plankton (live or dead) deposited on the bottom, benthic microalgae, dissolved organic matter, detritus with its attached microbial community, bacteria, and micro- and meiofauna. There is not only a greater range of potential food items but a major food source, organic detritus, is being constantly supplied to the sediments. There is a large "sink" of organic matter within the system and it is constantly being recycled. Deposit feeders produce large quantities of fecal pellets. Figure 33 depicts data on pellet formation in the intertidal mud-flat snail, *Hydrobia minuta*.[515] *H. minuta* will not reingest its own fecal pellets until they have been broken down (Figure 33B), thus rates of pellet formation and breakdown can be a very important population regulation factor for the species, since if the sediment becomes almost all pellets then the animals stop feeding. In some situations, between 70 and 90% of the surface sediments may be fecal pellets. Levinton argues from these data and the known behavior of *H. minuta* that, while there is a potentially large amount of organic matter present in the muddy sediments, most of it is not available because it is in the form of fecal pellets that deposit feeders will not reingest. The breakdown rates of pellets depend on bacteriaal action and the rates of bacterial activity are relatively constant. According to Levinton, bacteria provide the rate-controlling step for the generation of food, and the organic matter resides in the bottom as a "sink", buffering deposit feeders against fluctuations in food supply. Thus the food resource, although it may be limited, is predictable. Levinton suggests that deposit feeders have competed for this limiting resource over evolutionary time and that as a result feeding specializations have evolved. Also the presence of deposit-feeding food within the substratum allows different species to interact and outcompete other species, due to their greater efficiency at exploitation. The mobility of deposit feeders also permits competitive interference to result in exclusion. Levinton argues that species structures that are due to competitive interactions should be important in deposit-feeding communities and relatively unimportant in suspension-feeding communities.

3. Adult Interference with Larval Settlement

Woodin[1037] draws attention to the inadequacy of the traditional separation of benthic communities into those dominated by suspension feeders and those dominated by deposit feeders. There is, however, a third dominant infaunal group, the tube-building forms such as polychaetes, amphipods, tanaids, and phoronids. In contrast to dense assemblages of burrowing deposit feeders that destabilize the sediment, dense assemblages of tube builders stabilize the sediment regardless of the trophic type.[597,788,831,1051]

Thus three distinct types of dense infaunal assemblages have been described in soft substrata of marine sediments:

1. Infaunal deposit feeder assemblages of burrowing bivalves, decapods, holothurians, polychaetes, and irregular echinoids
2. Infaunal suspension-feeding bivalves
3. Infaunal tube-builder assemblages of varying trophic types comprised of tube-building and relatively sedentary amphipods, phoronids, polychaetes, and tanaids

All have sharp boundaries that are not associated with obvious physical ecotones such as abrupt changes in temperature, depth, or sediment grain size. All vary from assemblages that cover large areas to those that are in discrete patches within another assemblage type.

Woodin[1037] considers that the nonoverlapping distributions of the three assemblage types result from interactions between established individuals and settling, or newly settled larvae, rather than between two established individuals. Deposit feeders frequently feed on the surface sediments[508] and would engulf newly settled larvae in the process.[534,614] Thorson[935] and Hancock[354] have also suggested that much of the mortality of settling larvae is due to

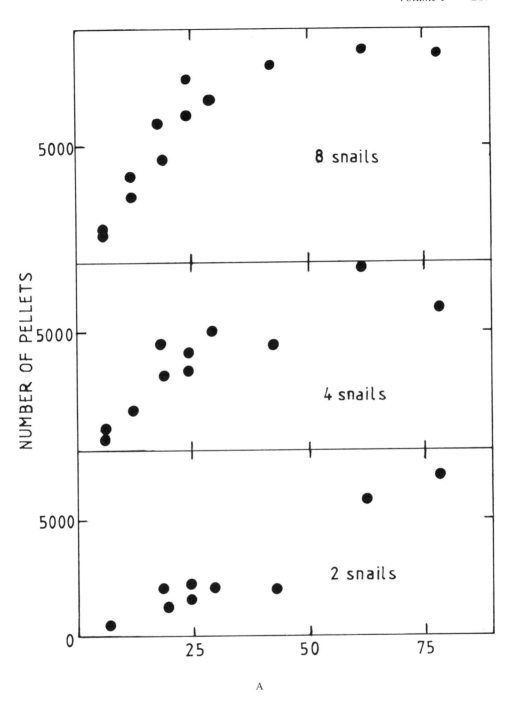

FIGURE 33.   (A) Fecal pellet production by the mud-flat snail *Hydrobia minuta* and (B) breakdown of fecal pellets in a laboratory experiment. (From Levington, J. S. and Lopez, G. R., *Oecologia*, 31, 180, 1977. With permission.)

filtration by suspension feeders. Fitch[251] described two populations of the Pismo clam, *Tivela stultonum*, in which settlement occurred in 1938, 1941, 1944, 1945, 1946, and 1947 in both populations. Since 1947 settlement has occurred in one population but not in the other. Fitch suggests that the density of the adults in the latter population was sufficient after 1947 to prevent settlement by adult straining of the spat out of the water column.

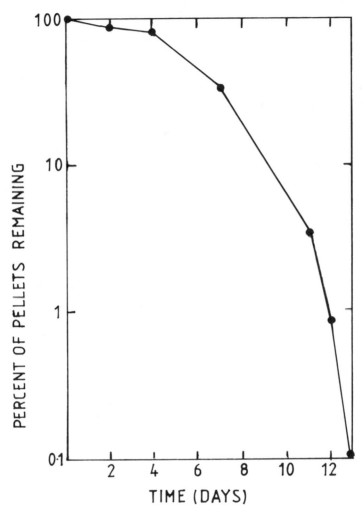

FIGURE 33B

Table 12 lists the three functional groups of benthic organisms distinguished above and the types of interactions that occur. Deposit feeders make the sediment more easily resuspended and ingest or disturb surface or near-surface larvae, such as those of suspension-feeding bivalves, as well as those of many deposit feeders. Suspension-feeding bivalves filter larvae as well as resuspended materials out of the water column. Tube-building forms, because of their tubes, reduce the ability of larvae to penetrate the sediment surface and, depending on the trophic type of the tube builder, ingest and/or defecate on the larvae. Woodin predicts on the basis of these adult-larval interactions that:

1.   Suspension-feeding epifaunal bivalves which brood their young and thus release them at sizes larger than those usually manageable by tube-building forms should reach their highest densities among the tube-builders.
2.   Small burrowing polychaetes should reach their highest densities among deposit feeders since they are susceptible to larval filtration by suspension feeders, and they are substrate limited because of tube density among tube-building forms. Woodin[1036] demonstrated spatial competition between the burrowing polychaete *Armandia brevis* and several tube-building polychaetes. The density of the burrowing polychaetes increased when the tube-building densities were reduced.

Table 12
FUNCTIONAL GROUPS OF INFAUNA AND TYPES OF INTERACTIONS THAT OCCUR

Functional group	Ingests or disturbs by its feeding activities surface or near surface larvae	Filters larvae out of water prior to settlement	Changes sediment	Larval type at settlement	Predicted dense co-occurring forms
Deposit-feeding bivalve	Yes	No	Makes more easily resuspended	May be surface or may burrow below surface	Burrowing polychaetes
Suspension-feeding bivalve	No	Yes	No, or much more slowly	Surface	None
Tube-building forms	Yes	Depends on trophic type	Stabilizes; reduces space below surface; increases epifaunal space on surface due to tubes	Surface	Epifaunal bivalves and tube epifauna

From Woodin, S. A., *J. Mar. Res.*, 34, 25, 1976. With permission.

3.   No infaunal forms should consistently attain high densities among densely packed suspension-feeding bivalves.
4.   As bivalves are perennials and can thus destroy their own larvae when filtering water or reworking sediments, their assemblages should be persistent but strongly age-class dominated.

There are two general components of the recruitment process: (1) habitat selection and (2) mortality during and after selection. Dayton and Oliver[187] point out that field experimental manipulations of sediment grade, water, and motion result in consistent species-specific recruitment patterns into artificial sedimentary habitats (plastic cups of defaunated sediment). These experimental recruitment patterns correspond remarkably well with the abundance patterns of established bottom patterns, and with larval recruitment into large (400 m²) defaunated bottom patches.[667] Generally, larvae settled into the particular artificial cup habitat which was the closest mimic to the natural sedimentary environment inhabited by the same species. Processes affecting larval recruitment are poorly understood, but may be some of the most important processes controlling the establishment and maintenance of soft-bottom communities.

4. Direct Interference through Alteration of the Physical Environment
Rhoads and Young[1087] noticed that high abundances of deposit feeders in Long Island Sound were associated both with fine sediments of high organic content and also with relatively low densities of suspension feeders. The boundaries between suspension- and deposit-feeding communities is often sharp. Rhoads and Young[1087] further noted that where deposit feeders were abundant the sediments have a surface layer of semifluid consistency made up chiefly of fecal pellets which is very readily resuspended by small water movements. The effect of sediment reworking by deposit feeders is shown in Figure 34. It can be seen that the water content of sediments dominated by suspension feeders is approximately 25% and shows very little variation with depth. On the other hand sediments dominated by deposit feeders have a high water content which reaches 70% in the semiliquid muds at the surface. From their observations Rhoads and Young developed the trophic group amensalism hypothesis which suggested that interactions between the two trophic groups (deposit feeders and suspension feeders) caused inhibition of one group, the suspension feeders, by resuspended sediments which clog the filtering mechanism and prevent larval settlement due to the absence of a suitable substrate for filter-feeding epifauna.
Building on the observations of Rhoads and Young,[1087] Wildish[1011] proposed the trophic group mutual exclusion hypothesis. ''It states (1) that sublittoral macrofaunal community, composition biomass and productivity are food limited; (2) the basic exclusion mechanism is a physical one: the current speed through its control on the supply of suspended food and its inhibitory effect on the development of the later stages of deposit feeding associations by control of oxygen exchange between sediments and bottom water and by removing biogenically resuspended sediments.'' According to this hypothesis the proportion of suspension to deposit feeders is decided primarily by the current speed spectra in the water immediately above the sediment, with predominantly deposit-feeding communities developing in low current speeds and predominantly filter-feeding communities developing in higher current speeds. Wildish and Kristmanson[1012] have presented this hypothesis in quantitative terms.
It is thus clear that a number of factors interact to control community composition, biomass, and productivity.

C. Role of Predators
Peterson[713] and Virnstein[960] have recently reviewed research relating to the role of predation

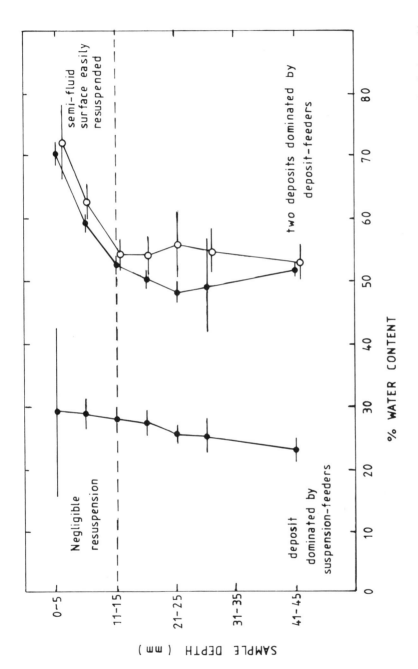

FIGURE 34.   Graphs showing the variations of water content with depth below the surface in a sediment dominated by suspension feeders and in two sediments dominated by deposit feeders. (From Newell, R. C., *The Biology of Intertidal Animals*, 3rd ed., Marine Ecological Surveys, Faversham, Kent, 1979, 42. With permission.)

Table 13
ABUNDANCES OF THREE TUBE-DWELLING AND ONE BURROWING SPECIES WITHIN 0.05 m^{-2} OF CAGED AND UNCAGED SEDIMENTS

	Tube-building species				Burrowing species
	Lumbrinerus inflata	*Axiothella rubrocinctata*	*Platynereis bicaniculata*	Total	*Armandia brevis*
No cage	168	123	358	649	47
	92	158	313	563	52
Cage	168	136	47	351	143
	132	153	25	310	160
	113	141	19	273	129
	64	104	54	222	139

Note: Samples taken from August to November-December.

From Woodin, S. A., *Ecol. Monogr.*, 44, 171, 1974. With permission.

and competitive exclusion in estuarine soft-bottom benthic communities. It is, however, only comparatively recently that experimental manipulation of natural benthic populations *in situ* has been used to advance our understanding of the role that predators play in the organization of these communities. This is in spite of the fact that as long ago as 1928 the Dane Blegrad used cages to exclude fish predators from the benthos in the Limfjord. Peterson[713] lists ten such studies carried out over the period 1964 to 1980.

Interpretation of the data obtained by the use of exclusion cages poses considerable problems due to the unnatural effects of the cages themselves.[959] For example, the cages shade the substrate, thus altering the light regime and possibly the growth of macro- and microalgae. Cages can also change the pattern of current flow and thus influence sedimentation patterns within the cages, increasing the rate of sedimentation, and changing the nature of the sediment by increasing the proportion of fine sediments. Fouling of the cage surface can increase local food concentrations. On the other hand food supplies for suspension feeders may be decreased. Furthermore, although the cages exclude large predators, they can serve as refuges into which predators can recruit as planktonic larvae or small individuals.[21] Thus, the results of predation exclusion experiments in the soft-sediment systems of estuaries must be interpreted with caution. Nevertheless such experiments have proved useful and have shown that the exclusion of predators from soft-bottom communities leads to a different set of changes to those which occur when predators are excluded from intertidal rocky shores.

The experiments which have been carried out can be grouped into two categories, those carried out on intertidal or subtidal soft-bottoms[133,713,771,817,958] and those carried out on seagrass beds.[672,770,959,1050,1052] As an example of the former, that of Woodin,[1036] in which cages were used to exclude predators on a muddy intertidal shore will be discussed. Table 13 shows the results of one such experiment over the course of which only *P. bicaniculata* and *A. brevis* reproduced. There is a clear change in the dominance pattern within the cages where the burrowing species increased in abundance. Woodin suggested that the tube-builders settle on the cage surface while the burrowers pass through the cage onto the sediment surface. Normally the tube-builders outcompete the burrowers for space by means of interference competition. In her species of experiments Woodin demonstrated two other important interactions: that when a predator (a crab *Cancer magister*) is placed inside a cage the abundance of the tube-builders decreases dramatically, whereas the burrowers maintain the same densities; and that variations in larval recruitment can strongly influence the community structure.

In 9 of the 11 studies on soft-bottoms cited by Peterson,[713] the density of macroinvertebrates was significantly higher inside the cages than in the control areas after some period of time. In most of the experiments the species richness also increased after predators were excluded, and remained higher than in the control areas after substantial periods of time at the elevated densities. Apparently adult-adult interactions, even at the relatively high densities achieved within the predator-exclusion cages, are incapable of simplifying the soft-sediment benthic community. This is in direct contrast to the currently accepted model of community organization developed from experimental work in marine intertidal communities[135,186,684] which would predict that significant simplification of the community should occur as a consequence of intense competition in such a system where density had increased substantially following the removal of predators.

Experiments in which large epibenthic predators have been excluded from estuarine or lagoonal seagrass beds tend to suggest that the removal of predators has very little effect on the macrobenthic community.[713] Peterson[713] advances the view that perhaps the infauna of densely vegetated areas represent a system that is naturally free of the effects of large epibenthic predators. Available data[672,928,974] comparing the infauna of seagreas beds with that of the surrounding unvegetated sediments suggests that in the seagrass beds the infauna have a much higher density and biomass (perhaps by an order of magnitude), yet far greater species richness with no obvious tendency toward monopolization by dominant species. This suggests that competitive exclusion is not operating in such systems.

Evans[225] has investigated a shallow soft-bottom in Gullmar Fjord, Sweden, in order to determine the impact of epibenthic predators and possible competition for food between co-occurring shrimp *(Crangon crangon)*, juvenile plaice *(Pleuronectes platessa)*, and sand goby *(Pomatoschistus minutus)*. Predation impact was calculated using estimates of gross production efficiencies and production rates. In 1976 and 1977 yearly predation amounted to 12 and 17%, respectively, of the total macrofaunal and meiofaunal production. This low exploitation rate contradicts the hypothesis that predation should be the major extrinsic determinant keeping population levels below the carrying capacity of the environment. Similar results have been obtained in other studies, e.g., Lockwood[530] calculated that the rate at which -0- group plaice cropped the available food in Filey Bay, Yorkshire (England) was considerably less than 1% of the mean standing stock day $^{-1}$ on any group of food items. Berge and Hesthagen[47] have shown experimentally that the common goby *Pomatoschistus microps* did not crop infauna to any extent sufficient to alter infaunal composition. Evans[225] concluded that shallow soft-bottoms provide an abundant food supply that will neither limit growth nor provide severe competition between epibenthic predators, although their diets may be almost identical. These conclusions appear to be at variance with the results of the predator exclusion experiments.

Peterson[713] points out that the reasons why competitive exclusion is not shown in caging experiments when predators are excluded might include the following: (1) the experiments have not been carried out for sufficient length of time for interference mechanisms to show up; (2) soft sediments provide much reduced opportunities for interference mechanisms to be employed (in contrast to the essentially two-dimensional rocky shore situation soft sediments are three-dimensional and there is greater opportunity for spatial separation); (3) the extreme importance of adult-larval interactions in soft sediments maintains densities below the carrying capacity even in the absence of predators; and (4) under food limitation marine invertebrates exhibit extreme plasticity in growth and reproduction and possess low-energy needs, both of which help to ensure high survivorship. Figure 35 illustrates the probable result of predation exclusion in two systems, one in which adult-larval interactions are important and the other in which they are not.

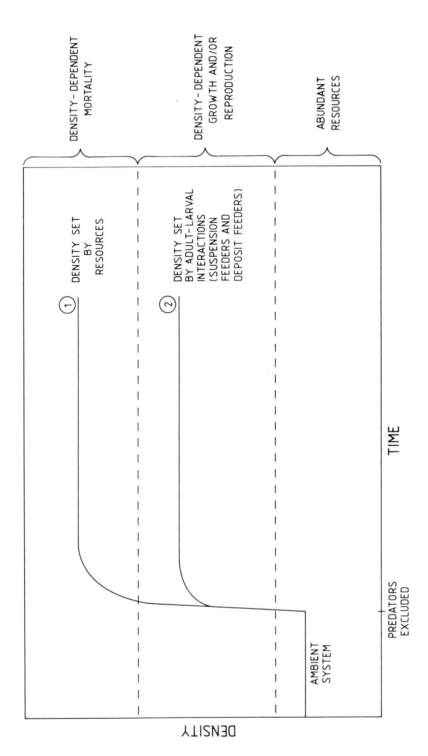

FIGURE 35.   The effects of removing predators from infaunal systems: (1) where adult-larval interactions are ineffective, and (2) where adult-larval interactions are important. In the ambient system, density is set by a combination of predation by mobile, epibenthic predators and adult-larval interactions. (From Peterson, C. H., *Ecological Processes in Coastal and Marine Systems*, Livingston, R. J., Ed., Plenum Press, New York, 1979, 249. With permission.)

In the caging experiments detailed above, fishes, large epibenthic crustacea, and birds are considered to be predators while the infaunal species are regarded as prey. However, as Ambrose[12] points out, many infaunal species (e.g., nemertine and polychaete worms) are themselves predators — and as prey for epibenthic predators and predators on other infauna — may function as intermediate predators. These predatory infaunal species also increase in the absence of epibenthic predators.[402,502,770,772] They are also capable of influencing the abundance of other infaunal species.[11,134,666,758,773,796] The polychaete *Nereis diversicolor* has been shown to reduce the abundance of juvenile cockles in the field[773] and *Macoma balthica* spat in the laboratory.[758] In experiments which measured colonization of sediment from which all organisms had been removed and infaunal predatory polychaetes selectively added, Commito[134] and Ambrose[11] found that *Nereis virens* reduced the abundance of a number of infaunal taxa. Oliver et al.[666] demonstrated that several species of phoxocephalid amphipods reduced infaunal densities by predation and/or disturbance.

Ambrose[12] has reviewed data from seven studies in which cages were used to exclude epibenthic predators.[133,402,502,618,770-772,900,1050] In these studies predatory infauna became proportionally more abundant after epibenthic predators were excluded from muddy sand and seagrass habitats. Ambrose[12] suggests that the increase could be a consequence of (1) preferential predation on predatory infauna by epibenthic predators; (2) preferential predation on predatory infauna and predation by predatory infauna on other infauna; (3) equal predation on predatory and nonpredatory infauna with additional predation by predatory infauna on other infaunal species; or (4) competition between predatory and nonpredatory infauna with predatory infauna out-competing nonpredatory infauna. Regardless of the mechanism the results suggested that predatory infauna should be considered separately from nonpredatory infauna when modeling interactions in soft-bottom communities. Ambrose[12] considers that if this is done then it may help to explain the paradox which exists between the results of experiments in rocky intertidal and soft-bottom communities.[713] As mentioned previously, while in the rocky intertidal, predation increases species diversity within communities by reducing competition among prey species and preventing dominant competitors from monopolizing a limiting resource and excluding other species from the community,[684] there is no evidence for competitive exclusion by dominant species following the exclusion of predators in soft-bottom communities. Ambrose[12] suggests that predation from predatory infauna within exclusion cages may help explain the failure of a competitive dominant to become established in response to the removal of epibenthic predators from soft-bottom communities.

D. Interactions of Deposit-Feeding Bivalves and Meiofauna

Interactions discussed above such as interspecific competition and predation result in population reduction or exclusion. There also exists the possibility of positive interactions. One excellent example of the latter is the study by Reise[775] of the biotic enrichment of intertidal sediments by experimental aggregates of the tellinid deposit-feeding bivalve *Macoma balthica*. Experimental aggregates of *Macoma* were established in a tidal flat on the coast of the Federal Republic of Germany. The tellinids stay buried at 3 to 6 cm while their long inhalent siphons pick up deposits at the surface. The short exhalent siphons terminate in the sediments. The experimental plots were core sampled after an interval of 30 days.

Figure 36 compares the vertical distribution of small zoobenthos (primarily meiofaunal species including Ciliates > 0.2 mm length) in sediment cores with and without *Macoma balthica*. It can be seen that there was a dramatic increase in abundance at the level of the *Macoma* individual. Reise[775] also found that interstitial Turbellarians showed a preference for the tellinid aggregates (Table 14). The difference between the control and aggregates was most pronounced in the subsurface component, but also at the surface the Turbellaria were also significantly more numerous in the experimental aggregates (Figure 37). The same

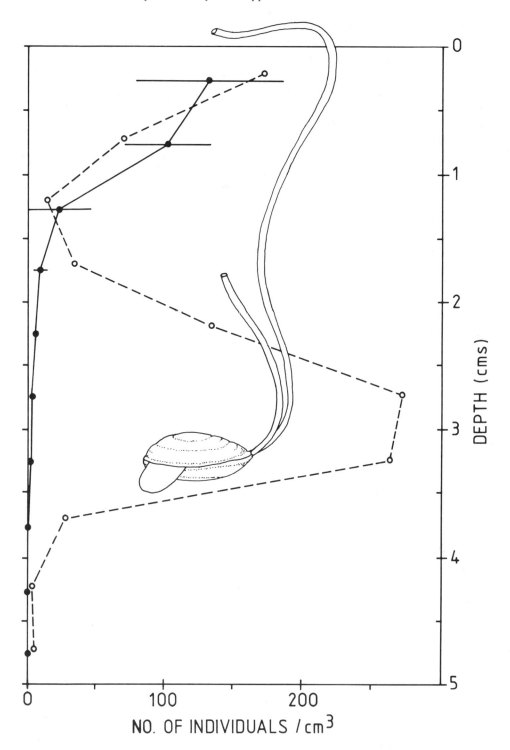

FIGURE 36.   Vertical distribution in 0.5 cm intervals of small zoobenthos abundance (including Ciliata >0.2 mm length) in the upper 5 cm of normal sediment (mean and standard deviation of 5 cores of 2 cm²) and in a single core (dashed line) which happened to include an individual of *Macoma balthica* positioned at 3.5-cm depth. (From Reise, K., *Mar. Ecol. Prog. Ser.*, 12, 231, 1983. With permission.)

Table 14
COMPONENTS OF THE
TURBELLARIAN ASSEMBLAGES IN
THE ABSENCE OF AND WITHIN
AGGREGATES OF *MACOMA BALTHICA*

Individuals 2 cm^{-2}(SD)	No *Macoma*	With *Macoma*
All turbellaria		
In 0 to 8 cm depth	17.3 (7.1)	41.7 (5.8)
In 0 to 2 cm depth	17.3 (7.1)	27.3 (7.4)
In 2 to 8 cm depth	0.0	14.4 (7.4)
Diatom feeders (8 spp.)	7.5 (3.6)	12.5 (3.7)
Bacteria feeders (5 spp.)	2.0 (1.1)	7.3 (2.7)
Predators (23 spp.)	7.7 (3.8)	21.0 (8.1)
Species richness (12 cm^{-2})		
In 0 to 2 cm	26	25
In 0 to 8 cm	0	11

Note: Means (SD = standard deviation) of 6 samples 2
cm^{-2}/0 to 8 cm.

From Reise, K., *Mar. Ecol. Prog. Ser.*, 12, 231, 1983. With
permission.

applied to species diversity and species richness. This strong dominance (equaling 54%) within the aggregates (Figure 37) appears at the sediment surface *(Macrostomum pusillum)*, over the entire vertical range *(Archilopsis unipunctata)*, and in the deep sediment *(Neoschizorhynchus parvorostro)*.

Small polychaetes also responded to the initiation of tellinid aggregates. The interstitial annelid *Macrophthalmus sczelkowskii* populated the transition zone between surface and subsurface when *Macoma balthica* is present (Figure 36). This hesionid polychaete feeds on bacteria and microalgae. The tube-dwelling *Pygiospio elegans* is significantly less abundant with the *Macoma* aggregates. Another tube-dwelling polychaete, *Spio filicornis*, was not affected.

The only case of competitive displacement is that of the polychaete *Pygiospio elegans*. Both *Macoma* and *Pygiospio* are surface deposit feeders. On the other hand the numbers and diversity of the micro- and meiofauna were enhanced by the presence of *Macoma balthica*. High abundance of meiofauna in micro-oxic zones generated by macrofauna in the subsurface sediments is a general phenomenon, explained by the provision of oxygen and by an increased microbial activity serving as food for the meiofauna.[10,15,774] In *Macoma* the exhalent siphon terminates below the surface and is responsible for the micro-oxic zones surrounding the shell. The exhalent current also probably produces localized concentrations of nutrients. These nutrients may stimulate diatom growth. Thus more food is provided for diatom- and bacteria-feeding meiofauna and also for *M. balthica* too. Both have diatoms as a common food source, and predatory Turbellaria prey on Ciliata, Nematoda, Copepoda, and others which in turn feed on diatoms and bacteria. Thus it may be concluded that *M. balthica* stimulates the growth of its own food, a process that has been termed "gardening".[282] Thus it is clear that structure and life habits of colonizing macrofauna are accompanied by diagenetic changes in sediment properties.[1098]

E. Disturbance and Natural Catastrophes

As we have seen, the distribution of benthic communities tends to be patchy. Johnson[431]

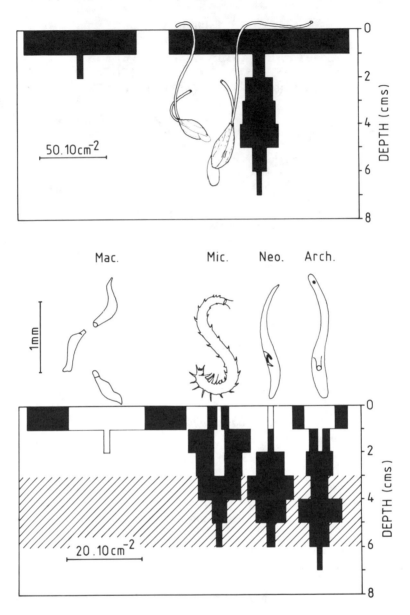

FIGURE 37. Vertical distribution of species abundance in normal sediment (white) and within aggregates of *Macoma balthica* (dark shaded with white areas inclusive). Hatched area: layer where tellinids are positioned. Mac = *Macrostomum pusillum* (diatom-feeder); Mic = *Micophthalmus* 2 spp. (diatom- and bacteria-feeder); Neo = *Neoschizorhynchus parvorostro* (bacteria-feeder); and Arch = *Archilopsis unipunctata* (predator). (From Reise, K., *Mar. Ecol. Prog. Ser.*, 12, 232, 1983. With permission.)

envisages the benthos as a temporal mozaic, with different parts of the habitat disturbed at different times containing different faunas. "The community", he says "is conceived of as continually varying in response to a history of disturbance. In his view, the community is a collection of relics (and recoveries) of former disasters." Sutherland[903] has likewise emphasized the role of history in determining the structure of a variety of natural communities. Gray[327,329] has discussed the question of stability in benthic communities. He recognizes two types of stability, neighborhood stability and global stability (Figure 38). The neighborhood model seems to represent adequately many cases of local temporal changes in

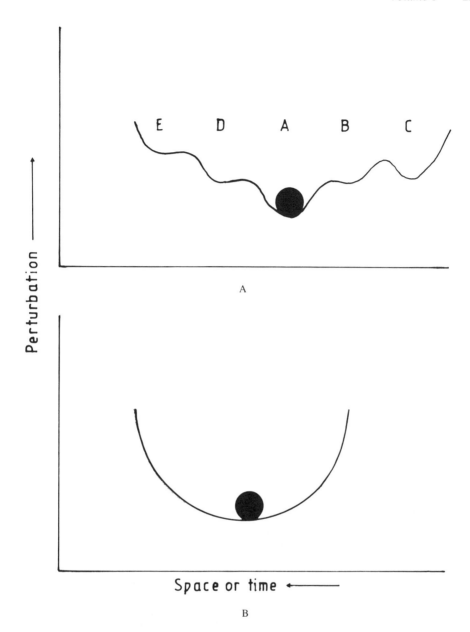

FIGURE 38. Models of stability in communities. The ball represents the community which can be perturbed from stable equilibrium; (A) in neighborhood stability there are many locally stable points (A to E); and (B) in global stability there is one unique stable point. (From Gray, J. S., *The Ecology of Marine Sediments*, Cambridge University Press, Cambridge, England, 1981, 80. With permission.)

benthic communities: at one point in time species A dominates, but is replaced by species B which may then go back to A or on to C, depending on which factors are operating. Scales of disturbance may be small, e.g., the temporary impact of a large predator, or large, e.g., defaunation following a red tide episode, an oil spill, intense bottom scouring, or sediment deposition as a consequence of a storm event.

There have been a number of studies of the colonization sequence in benthic communities.[547,703,782,783,845] There are progressive changes in trophic structure, and life habits of colonizing macrofauna are accompanied by diagenetic changes in sediment properties.[1098]

Pioneering species tend to be tubiculous or otherwise sedentary organisms that live near the sediment surface and feed on the surface or from the overlying water. Their influence on the sediments through biogenic reworking may be intense, but is restricted to the top centimeter or two.[615] In contrast, high order or equilibrium-successional stages tend to be dominated by errant infaunal deposit feeders that feed well below the surface and intensively rework the sediment to a depth of several centimeters[703,783] (Figure 22).

Much of the theoretical focus on the colonizing sequence has been on species-adaptive strategies, especially in relation to life history characteristics.[324,547,704,814,929] A central concept is that species with r-selected life history traits have adapted to respond in an opportunistic fashion following a disturbance and will dominate the initial stages of succession.[324,547] Grassle and Grassle[324] define the characteristics of opportunistic species as: initial response to disturbed conditions, ability to increase rapidly, huge population size, early maturation, and high mortality. The major opportunists are deposit-feeding polychaetes, and using the criteria outlined above, Grassle and Grassle[324] rank the opportunistic species that colonized the sediment of Falmouth Harbour following an oil spill (Figure 39) as follows: (1) *Capitella capitata*, (2) *Polydora ligni*, (3) *Syllides verrilli*, (4) *Microphthalmus abberans*, (5) *Streblospio benedicti*, and (6) *Mediomastus ambiseta*. Figure 38 shows a typical colonizing sequence.

McCall[547] has examined community patterns and adaptive strategies of the infaunal benthos of Long Island Sound. Samples of defaunated mud were placed on the bottom of the Sound to stimulate a local disaster and sampled after 10 days and periodically for up to 384 days. Some representative results are given in Tables 15 and 16. McCall recognizes three species groups: Group I, opportunistic species, Group III, equilibrium species, and Group II with characteristics intermediate between the other two groups. The attributes of Group I and Group III are summarized in Table 17. McCall found that the differences in the distribution and abundance of benthic communities in Long Island Sound could be explained in terms of these two different adaptive strategies, opportunistic or equilibrium.

Additional information on the responses of benthic communities to disturbance has come from a number of studies of the impact of organic pollution in the form of domestic sewage,[17,44,483,484] meatworks and cannery effluent,[474,481,869] and pulp mill effluent.[676,701-703,797,798] Pearson and Rosenberg[704] have recently reviewed the literature on macrobenthic succession in relation to organic enrichment. Figure 40 depicts the impact of domestic effluent on the numbers, biomass, and diversity of the benthic communities. It illustrates points common to all the above studies, i.e., a decrease in diversity, but a large increase in numbers and biomass in the vicinity of the outfall with a recovery to a more normal benthic fauna some distance from the outfall. The greater the load of organic matter the greater the distance before recovery takes place. Pearson and Rosenberg[704] present a generalized diagram of species number, abundance, and biomass (SAB) along an environmental gradient of organic enrichment (Figure 41).

Zajac and Whitlach[1053] have studied the responses of estuarine infauna to disturbance in a small estuary in Connecticut. In contrast to other studies, species responses to disturbances were not predictable and were quite variable. While some well-known opportunists such as *Polydora ligni* responded in typical fashion, others such as *Capitella capitata* and *Streblospio benedicti* were not significantly different in the experimental recolonization plots from ambient densities, and in some cases they were significantly below the ambient densities. Zajac and Whitlach[1053] consider that their results suggest that the extent of a species recolonization may be primarily dependent on ambient population densities.

Zajac and Whitlach[1054] found that infaunal successional patterns in the same estuary varied seasonally and along the estuarine environmental gradient. They found that the timing of disturbance greatly influenced succession. Succession after an early spring disturbance was characterized by peak species densities and numbers. Succession following a fall disturbance

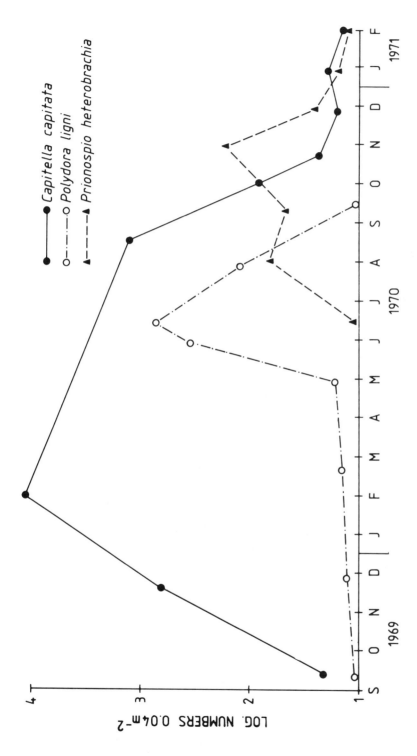

FIGURE 39.   Recolonization sequence following on oil spill in Massachusetts (Station B1 at 3-m depth). (From Gray, J. S., *The Ecology of Marine Sediments*, Cambridge University Press, Cambridge, England, 1981, 83. With permission.)

Table 15
A COMPARISON OF COLONIZATION OF EMPTY SEDIMENT AND
COLONIZATION OF SEDIMENT WITH ANIMALS PRESENT

	Colonization of empty sediment		Colonization of sediment with animals present	
	Colonization in August 1972	Colonization in August 1973 (partially empty)	Natural bottom in August 1973	1-year-old previously empty sediment in August 1973
Capitella capitata	36,120	170,585	5,868	8,940
Streblospio benedicti	418,315	12,575	13,310	4,015
Owenia fusiformis	0	86,650	240	0
Ampelisca abdita	5,130	45	15	472
Tellina agilis	275	60	206	55
Nucula proxima	205	35	205	75
Yoldia limatula	0	10	24	45
Nepthys incisa	0	95	266	15
Ensis directus	10	0	36	0
Totals	459,565	269,855	19,433	13,427

Note: Abundances are no m^{-2}.

From McCall, P. L., *J. Mar. Res.*, 35, 221, 1977. With permission.

Table 16
COLONIZATION OF DEFAUNATED SEDIMENTS
PLACED ON THE BOTTOM OF LONG ISLAND
SOUND[a]

Species	Peak abundance	Sampling interval (days)	Final abundance
Group I			
Streblospio benedicti	418,315	10	335
Capitella capitata	80,385	29—50	995
Ampelisca abdita	9,990	29—50	0
Group II			
Nucula proxima	3,735	50	50
Tellina agilis	1,400	86	0
Group III			
Nepthys incisa	220	175	120
Ensis directus	30	50—223	0

[a]   Number m^{-2}.

From McCall, P. L., *J. Mar. Res.*, 35, 221, 1977. With permission.

was abbreviated with few species at low densities, while after a summer disturbance intermediate trends were found. In the area they studied it was apparent that estuarine succession was quite variable and that re-establishment of community structure could occur over various time scales with no set serial stages. They concluded that the physical and biological processes appearing to be important determinants of estuarine succession may include: (1) timing of disturbance, (2) habitat in which the disturbance takes place, (3) reproductive periodicity of

Table 17
SUMMARY OF THE ATTRIBUTES OF OPPORTUNISTIC (GROUP I) AND
EQUILIBRIUM (GROUP III) ADAPTIVE MACROBENTHIC SPECIES TYPES

Group I	Group III
Opportunistic species	Equilibrium species
Many reproductions per year	Few reproductions per year
High recruitment	Low recruitment
Rapid development	Slow development
Early colonizers	Late colonizers
High death rate	Low death rate
Small	Large
Sedentary	Mobile
Deposit feeders (mostly surface feeders)	Deposit and suspension feeders
Brood protection; lecithophic larvae	No brood protection; planktotrophic larvae

From McCall, P. L., *J. Mar. Res.*, 35, 253, 1977. With permission.

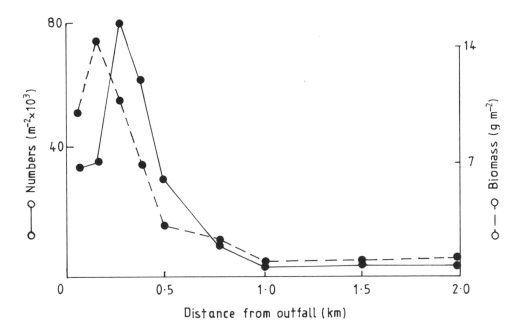

FIGURE 40. Illustrates the effects of sewage on the numbers and biomass of the benthic fauna of Kiel Bay. (From Gray, J. S., *The Ecology of Marine Sediments*, Cambridge University Press, Cambridge, England, 1981, 88. With permission.)

the infauna, (4) ambient population dynamics which generate the pool of recolonizers, and (5) abiotic and biotic factors (e.g., food and space resources affecting the preceding four factors).

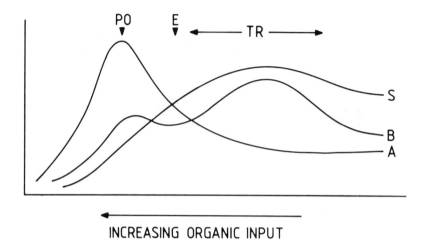

INCREASING ORGANIC INPUT

FIGURE 41.   Generalized SAB (species number, abundance, and biomass) of changes along a gradient of organic enrichment. S = species numbers; A = total abundance; B = total biomass; PO = peak of opportunists; E = ecotone point; and TR = transition zone. (From Pearson, T. H. and Rosenberg, R., *Oceanogr. Mar. Biol. Annu. Rev.*, 16, 234, 1978. With permission.)

REFERENCES

1. **Admiraal, W.,** Influence of light and temperature on growth rate of estuarine diatoms in culture, *Mar. Biol.,* 39, 1, 1977.
2. **Admiraal, W. and Peletier, H.,** Influence of organic compounds and light limitation on the growth rate of estuarine benthic diatoms, *Br. J. Phycol.,* 14, 197, 1977.
3. **Albright, L. J., Chocair, J., Masuda, K., and Valdes, D. M.,** *In situ* degradation of the kelp *Macrocystis integrifolia* and *Nereocystis leatheana* in British Columbia waters, *Nat. Can.,* 107, 3, 1980.
4. **Alheit, J. and Scheibel, W.,** Benthic harpacticoids as a food source for fish, *Mar. Biol.,* 70, 141, 1982.
5. **Alldredge, A. L.,** The chemical composition of macroscopic aggregatès in two neritic seas, *Limnol. Oceanogr.,* 24, 855, 1979.
6. **Allen, G. W.,** Estuarine destruction — a monument to progress, *Proc. 26th N. Am. Wildl. Conf.,* 29, 324, 1964.
7. **Allen, H. L.,** Primary productivity, chemo-organtrophy and nutritional interactions of epiphytic algae and bacteria on macrophytes in the littoral of a lake, *Ecol. Monogr.,* 41, 97, 1971.
8. **Aller, R. C.,** The Influence of Macrobenthos on Chemical Diagenesis of Marine Sediments, Ph.D. thesis, Yale University, New Haven, Conn., 1977, 600 pp.
9. **Aller, R. C.,** Relationships of tube-dwelling benthos with sediment and overlying water chemistry, in *Marine Benthic Dynamics,* Tenore, K. R. and Coull, B. C., Eds., University of South Carolina Press, Columbia, 1980, 385.
10. **Aller, R. C. and Yingst, J. Y.,** Biogeochemistry of tube-dwellings: a study of the sedentary polychaete *Amphitrite ornata* (Leidey), *J. Mar. Res.,* 35, 201, 1978.
11. **Ambrose, W. G., Jr.,** The Influence of the Predatory Polychaetes *Glycera dibranchiata* and *Nereis virens* on the Structure of a Soft-Bottom Community in a Maine Estuary, Ph.D. thesis, University of North Carolina, Chapel Hill, 1982.
12. **Ambrose, W. G., Jr.,** Role of predatory infauna in structuring marine soft-bottom communities, *Mar. Ecol. Prog. Ser.,* 17, 109, 1984.
13. **Anderson, F. Ø.,** Effects of nutrient level of the decomposition of *Phragmites communis* Trin, *Arch. Hydrobiol.,* 84, 42, 1978.
14. **Anderson, F. Ø. and Hargrave, B. T.,** Effects of *Spartina* detritus enrichment on aerobic/anaerobic benthic metabolism in an intertidal sediment, *Mar. Ecol. Prog. Ser.,* 16, 161, 1984.
15. **Anderson, J. G. and Meadows, P. S.,** Microenvironments in marine sediments, *Proc. R. Soc. Edinburgh,* 76B, 1, 1978.
16. **Andersson, M.,** Einige ernahrungsphysiologische Versuche mit *Ulva* und *Enteromorpha, Fysiogr. Sallsk. I Forhand. Lund, Bd.,* 12, 42, 1942.
17. **Anger, K.,** On the influence of sewage pollution on inshore benthic communities in the south of Kiel Bay. II. Quantitative studies on community structure, *Helgol. Wiss. Meeresunters.,* 24, 408, 1975.
18. **Antflinger, A. E. and Dunn, E. L.,** Seasonal patterns of CO_2 and water vapor exchange of three salt-marsh succulents, *Oecologia,* 43, 249, 1979.
19. **Arlt, G.,** Zur produktionsbiologischen Bedentung der Meiofauna in Küstengewasben. *Wiss. Z. Univ. Rostock.,* 22, 1141, 1973.
20. **Armstrong, W.,** Root aeration in the wetland condition, in *Plant Life in Anaerobic Environments,* Hook, D. D. and Crawford, R. M. M., Eds., Ann Arbor Science, Ann Arbor, Mich., 1978, 269.
21. **Arntz, W. E.,** Results and problems of an "unsuccessful" benthos cage experiment (Western Baltic), in *Biology of Benthic Organisms,* Keegan, B. F., O'Ceidigh, P., and Boaden, P. J. S., Eds., Pergamon Press, New York, 1977, 31.
22. **Aston, S. R. and Chester, R.,** The influence of suspended particles on the precipitation of iron in natural waters, *Estuarine Coastal Mar. Sci.,* 1, 225, 1973.
23. **Aston, S. R. and Chester, R.,** Estuarine sedimentary processes, in *Estuarine Chemistry,* Burton, J. D. and Liss, P. S., Eds., Academic Press, London, 1976, 37.
24. **Axelrod, D. M., Moore, K. A., and Bender, M. E.,** Nitrogen, Phosphorus, and Carbon Flux in Chesapeake Bay Marshes, Virginia Polytechnic Institute, Virginia Water Resour. Res. Cent. Bull., 79, 182, 1976.
25. **Azam, F., Fenchel, T., Field, J. G., Gray, J. S., Meyer-Reil, L. A., and Thingstad, F.,** The ecological role of water-column microbes in the sea, *Mar. Ecol. Prog. Ser.,* 10, 257, 1983.
26. **Bagge, P.,** Effects of pollution on estuarine ecosystems. I. Effects of effluents from wood processing industries on the hydrography, bottom and fauna of Saltkallefjord (W. Sweden). II. The succession of the bottom fauna communities in polluted estuarine habitats in the Baltic-Skagerrak region, *Merentutkimuslaitoksen Julk. Havsforskningsinst. Skr.,* 228, 1, 1969.
27. **Baker, A. J.,** Observations on the Winter Feeding of the South Island Pied Oystercatcher *(Haematopus ostralegus finschi),* B.Sc. Hons project in Zoology, University of Canterbury, Christchurch, New Zealand, 1966.

28. **Baker, A. J.,** The Comparative Biology of New Zealand Oystercatchers, M.Sc. thesis, University of Canterbury, Christchurch, New Zealand, 1969, 134 pp.

29. **Baker, A. J.,** Prey-specific feeding methods of New Zealand oystercatchers, *Notornis,* 21, 219, 1974.

30. **Barker, L. D.,** The Ecology of the Ctenophore *Mnemiopsis mccradyi* Mayer, in Biscayne Bay, Florida, University of Miami Tech. Rep., UM-RSMAS-73016, 1973, 131.

31. **Bakker, C., Phaff, W. J., Van Ewijk-Rosier, M., and De Pauw, N.,** Copepod biomass in an estuarine and a stagnant brackish environment in SW Netherlands, *Hydrobiologica,* 52, 3, 1977.

32. **Ball, M. C.,** Patterns of succession in a mangrove forest of Southern Florida, *Oecologia,* 44, 226, 1980.

33. **Barker, R. T., Kirby-Smith, W. W., and Parsley, P. E.,** Wetlands alterations for agriculture, in *Wetland Function and Values,* Greeson, P. E., Clark, J. R., and Clark, J. E., Eds., American Water Resources Association, Urbana, Illinois, 1980, 642.

34. **Barnes, R. S. K.,** *Coastal Lagoons,* Cambridge University Press, Cambridge, England, 1980, 106 pp.

35. **Bascand, L. D.,** The roles of *Spartina* species in New Zealand, *Proc. N. Z. Ecol. Soc.,* 17, 33, 1976.

36. **Batham, E. J.,** Rocky shore ecology of a New Zealand southern fjord, *Trans. R. Soc. N. Z. Zool.,* 6(20), 216, 1965.

37. **Bauersfield, P., Kifer, R. R., Purrant, N. W., and Sykes, J. E.,** Nutrient content of Turtle Grass *(Thalassia testudinum), Proc. Int. Seaweed Symp.,* 6, 637, 1969.

38. **Baylis, C. T. S.,** Some Observations on *Avicennia officinalis* Linn. in New Zealand, M.Sc. thesis, University of Auckland, Auckland, New Zealand, 1935.

39. **Beeftink, W. G.,** The coastal salt marshes of western and northern Europe: an ecological and phytoso-ciological approach, in *Wet Coastal Ecosystems,* Chapman, V. J., Ed., Elsevier, Amsterdam, 1977, 109.

40. **Beers, J. R. and Stewart, C. L.,** Microzooplankters in the plankton communities of the upper waters of the eastern tropical Pacific, *Deep-Sea Res.,* 18, 861, 1971.

41. **Bell, S. S.,** Meiofaunal-macrofaunal interactions in a high marsh habitat, *Ecol. Monogr.,* 50, 487, 1980.

42. **Bell, C. R. and Albright, L. J.,** Attached and free-floating bacteria in the Fraser River Estuary, British Columbia, Canada, *Mar. Ecol. Prog. Ser.,* 6, 317, 1981.

43. **Bell, S. S. and Coull, B. C.,** Experimental evidence for a model of juvenile macrofauna-meiofauna interactions, in *Marine Benthic Dynamics,* Tenore, K. R. and Coull, B. C., Eds., University of South Carolina Press, Columbia, 1979, 179.

44. **Bellan, G. and Bellan-Santini, D.,** Influence de la pollution sur les peuplements marins de la region de Marseille, in *Marine Pollution and Sea Life,* Ruvio, M., Ed., Fishing News (Books), London, 1972, 396.

45. **Bender, M. E. and Correll, D. L.,** Use of Wetlands as Nutrient Removal Systems, Publ. No. 29, Chesapeake Research Consortium, Baltimore, Md., 1974, 12 pp.

46. **Bent, E. J. and Goulder, R.,** Planktonic bacteria in the Humber Estuary, seasonal variation in population density and heterotrophic activity, *Mar. Biol.,* 62, 175, 1981.

47. **Berge, J. A. and Hesthagen, I. H.,** Effects of epibenthic macropredators on community structure in a eutrophicated shallow water area, with special reference to food consumption by the common goby *Pomatoschistus microps, Kiel. Meeresforsch.,* 5, 462, 1981.

48. **Berk, S. G., Brownlie, D. C., Colwell, R. R., Heinle, D. R., and Kling, H. J.,** Ciliates as a food source for marine planktonic copepods, *Microb. Ecol.,* 4, 27, 1977.

49. **Berman, T. and Holm-Hansen, O.,** Release of photoassimilated carbon as dissolved organic matter by marine phytoplankton, *Mar. Biol.,* 28, 305, 1974.

50. **Bittaker, H. F. and Iverson, R. L.,** *Thalassia testudinum* productivity: a field comparison of measurement methods, *Mar. Biol.,* 37, 39, 1976.

51. **Blaber, S. J. M.,** The food and feeding ecology of mullet in the St Lucia Lake system, in *St. Lucia Scientific Advisory Council Workshop,* Heydom, A. E. F., Ed., National Parks Board, Pietermaritzberg, 1976.

52. **Blackburn, D. J.,** The Feeding Biology of Tintinnid Protozoa and Some Other Inshore Microzooplankton, Ph.D. thesis, University of British Columbia, Vancouver, 1974, 224 pp.

53. **Blegrad, H.,** Quantitative investigations of bottom invertebrates in the Limfjord 1910 to 1927 with special reference to plaice food, *Rep. Dan. Biol. Stn.,* 34, 33, 1928.

54. **Bloom, S. A., Simon, J. L., and Hunter, V. D.,** Animal-sediment relations and community analysis of a Florida estuary, *Mar. Biol.,* 13, 43, 1972.

55. **Blum, U., Seneca, E. D., and Strand, L. M.,** Photosynthesis and respiration of *Spartina* and *Juncus* salt marshes in North Carolina: some models, *Estuaries,* 1, 228, 1978.

56. **Bobbie, R. J., Morrison, S. J., and White, D. C.,** Effects of biodegradability on the mass and activity of the associated estuarine microbiota, *Appl. Environ. Microbiol.,* 35, 179, 1978.

57. **Boesch, D. F.,** Species diversity of the macrobenthos in the Virginia area, *Chesapeake Sci.,* 13(3), 206, 1972.

58. **Boesch, D. F.,** Classification and community structure of macrobenthos in the Hampton Roads area, Virginia, *Mar. Biol.,* 21, 226, 1973.

59. **Boesch, D. F.,** Diversity, stability and response to human disturbance in estuarine ecosystems, in *Proc. 1st Int. Congr. Ecol.,* Pudoc, Wageningen, The Netherlands, 1974, 109.

60. **Boesch, D. F.,** A new look at the zonation of benthos along an estuarine gradient, in *Ecology of the Marine Benthos,* Coull, B. C., Ed., University of South Carolina Press, Columbia, 1977, 245.

61. **Boesch, D. F., Wass, M. L., and Virnstein, R. W.,** Dynamics of estuarine benthic communities, in *Estuarine Processes,* Vol. 1, Academic Press, New York, 1976, 177.

62. **Boonruang, P.,** The degradation rates of mangrove leaves of *Rhizophora apiculata* (Bl.) and *Avicennia marina* (Forsk.) Vierh. at Puhket Island, *Puhket Mar. Biol. Cent. Res. Bull.,* 26, 1, 1978.

63. **Boyle, C. D. and Patriquin, D. G.,** Carbon metabolism of *Spartina alterniflora* in relation to associated nitrogen fixing bacteria, *New Phytol.,* 89, 275, 1981.

64. **Boynton, W. R., Kemp, W. M., and Osborne, C. G.,** Nutrient fluxes across the sediment water interface of a coastal plain estuary, in *Estuarine Perspectives,* Kennedy, V. S., Ed., Academic Press, New York, 1980, 93.

65. **Boynton, W. R., Kemp, W. M., and Keefe, C. W.,** A comparative analysis of nutrients and other factors influencing estuarine phytoplankton production, in *Estuarine Comparisons,* Kennedy, V. S., Ed., Academic Press, New York, 1982, 69.

66. **Bradley, W. H. and Cooke, P.,** Living and ancient populations of the clam *Gemma gemma* in a Maine coastal tidal flat, *Fish. Bull.,* 58, 305, 1959.

67. **Bradshaw, J. S.,** Environmental parameters and marsh foraminifera, *Limnol. Oceanogr.,* 13, 26, 1968.

68. **Brafield, A. E.,** The oxygen content of interstitial water in sandy shores, *J. Anim. Ecol.,* 33, 97, 1964.

69. **Branch, G. M. and Branch, M. L.,** Competition in *Berabicium auratum* Gastropoda and its effects on microalgal standing stock in mangrove mudflats, *Oecologia,* 46, 106, 1980.

70. **Bregnballe, F.,** Plaice and flounders as consumers of the microscopic bottom fauna, *Medd. Dan. Fisk. Havunders.,* 3, 133, 1961.

71. **Briggs, I.,** The Feeding Ecology and Energetics of a Population of *Amphibola crenata,* M.Sc. thesis, University of Auckland, New Zealand, 1972.

72. **Briggs, I.,** Assessment of the Current Trophic Status of the Upper Waitemata Harbour. Upper Waitemata Harbour Catchment Study Working Rep. No. 33, Auckland Regional Authority, Auckland, 1982, 65 pp.

73. **Briggs, K. B., Tenore, K. R., and Hanson, R. B.,** The role of microfauna in detrital utilization by the polychaete, *Nereis succinea* (Frey and Leuckart), *J. Exp. Mar. Biol. Ecol.,* 36, 225, 1979.

74. **Briggs, S. V.,** Estimates of biomass in a temperature mangrove community, *Aust. J. Ecol.,* 2, 369, 1977.

75. **Brkich, S. W.,** Phytoplankton Productivity in the Barataria Bay Area of Louisiana, Ph.D. dissertation, Louisiana State University, Baton Rouge, 1972.

76. **Broome, S. W., Woodhouse, W. W., Jr., and Seneca, E. D.,** The relationship of mineral nutrients to growth of *Spartina alterniflora* in North Carolina. II. The effects of N, P, and Fe fertilizers, *Proc. Am. Soil Sci. Soc.,* 39, 301, 1975.

77. **Brown, D. H., Gibby, C. E., and Hickman, M.,** Photosynthetic rhythms in epipelic algal populations, *Br. J. Phycol.,* 7, 37, 1972.

78. **Brylinsky, M.,** Release of Dissolved Organic Matter by Marine Macrophytes, Ph.D. dissertation, University of Georgia, Athens, 1971.

79. **Bunt, J. S.,** Studies of mangrove litter fall in tropical Australia, in *Mangrove Ecosystems in Australia,* Clough, B. F., Ed., Australian National University Press, Canberra, 1982, 223.

80. **Bunt, J. S., Boto, K. G., and Boto, G.,** A survey method for estimating potential level of mangrove forest primary production, *Mar. Biol.,* 52, 123, 1979.

81. **Bunt, J. S. and Williams, W. T.,** Studies in the analysis of data from Australian tidal forests (mangroves). I. Vegetational sequences and their graphic representation, *Aust. J. Ecol.,* 5, 385, 1980.

82. **Buresh, R. J., DeLaune, R. D., and Patrick, W. H., Jr.,** Nitrogen and phosphorus distribution and utilization by *Spartina alterniflora* in a Louisiana Gulf coast marsh, *Estuaries,* 3, 111, 1980.

83. **Burkholder, P. R., Burkholder, L. M., and Rivero, J. A.,** Some chemical constituents of turtle grass *Thalassia tesudinum, Bull. Torrey Bot. Club,* 86, 88, 1959.

84. **Burkholder, P. R. and Doheny, T. E.,** The Biology of Eelgrass, with Special Reference to Hemstead and South Oyster Bays, Nassau County, Long Island, New York, Contrib. No. 3, Department of Conservation and Waterways, Town Hempstead, L.I., N.Y., 1968, 120 pp.

85. **Burrell, D. C. and Schubel, J. R.,** Seagrass ecosystem oceanography, in *Seagrass Ecosystems. A Scientific Perspective,* McRoy, C. P. and Helfferich, C., Eds., Marcel Dekker, New York, 1977, 195.

86. **Burrows, E. M.,** Assessment of pollution effects by the use of algae, *Proc. R. Soc. London, Ser. B:* 177, 295, 1971.

87. **Burton, J. D.,** Basic properties and processes in estuarine chemistry, in *Estuarine Chemistry,* Burton, J. D. and Liss, P. S., Eds., Academic Press, London, 1976, 1.

88. **Burton, P. J. K.,** *Feeding and Feeding Apparatus in Waders,* British Museum (Natural History), London, 1974.

89. **Butler, P. A.,** The problem of pesticides in estuaries, *Am. Fish. Soc. Spec. Publ.,* No. 3, 110, 1966.

90. **Bryne, P., Borengasser, M. J., Drew, G., Muller, R. A., Smith, B. L., Jr., and Wax, C. L.,** Barataria Basin: hydrologic and climatological processes. Louisiana State University, Center for Wetland Resources, Baton Rouge, Final Report to the State Planning Office, 1976.

91. **Cadée, G. C. and Hedgman, J.,** Primary production of the benthic microflora living on tidal flats in the Dutch Wadden Sea, *Neth. J. Sea Res.,* 8(2-3), 260, 1974.

92. **Cain, T. D.,** The Reproductive Cycle and Larval Tolerances of *Rangia cuneata* in the James River, Virginia, Ph.D. thesis, University of Virginia, Charlottesville, 1972.

93. **Cameron, W. M. and Pritchard, D. W.,** Estuaries, in *The Sea,* Vol. 2, Hill, N. M., Ed., John Wiley & Sons, New York, 1963, 306.

94. **Carle, K. J. and Hastings, P. A.,** Selection of meiofaunal prey by the darter goby, *Gobionellus boleosoma* (Gobiidae), *Estuaries,* 5, 316, 1983.

95. **Carpenter, E. J. and McCarthy, T. J.,** Benthic nutrient regeneration and high rate of primary production in continental shelf waters, *Nature,* 274, 188, 1978.

96. **Carricker, M. R.,** Ecology of estuarine benthic invertebrates: a perspective, in *Estuaries,* Publ. No. 83, Lauff, G. H., Ed., American Association for the Advancement of Science, Washington, D.C., 1967, 442.

97. **Carter, M. R., Burns, L. A., Cavinder, T. R., Dugger, K. R., Fore, P. L., Hicks, D. B., Revells, H. L., Schmidt, T. W., and Farley, R.,** Ecosystem Analysis of the Big Cypress Swamp and Estuaries, South Florida Ecological Study, U.S. Environmental Protection Agency, Atlanta, Ga., 1973.

98. **Cassie, R. M. and Michael, A. D.,** Fauna and sediments on an intertidal mudflat: a multivariate analysis, *J. Exp. Mar. Biol. Ecol.,* 2, 1, 1968.

99. **Cavalieri, A. J. and Huang, A. H. C.,** Accumulation of proline and glycine-betaine in *Spartina alterniflora* Losiel in response to NaCl and nitrogen in the marsh, *Oecologia,* 49, 224, 1981.

100. **Chalmers, A. G.,** Pools of Nitrogen in a Georgia Salt Marsh, Ph.D. dissertation, University of Georgia, Athens, 1977.

101. **Chalmers, A. G.,** Soil dynamics and the productivity of *Spartina alterniflora,* in *Estuarine Comparisons,* Kennedy, V. S., Ed., Academic Press, New York, 1982, 231.

102. **Chalmers, A. G., Haines, E. B., and Sherr, B. F.,** Capacity of a *Spartina* Marsh to Assimilate Nitrogen from Secondarily Treated Sewage, Tech. Completion Rep. USDI/OWRT Proj. No. A-057-GA, University of Georgia, Athens, 1976.

103. **Chambers, D. G.,** An Analysis of the Nekton Communities in the Upper Barataria Basin, Louisiana, M.S. thesis, Louisiana State University, Baton Rouge, 1980.

104. **Chapman, G.,** The thixotrophy and dilatancy of a marine soil, *J. Mar. Biol. Assoc. U.K.,* 28, 123, 1949.

105. **Chapman, G. and Rae, A. C.,** Excretion of photosynthate by a benthic diatom, *Mar. Biol.,* 3, 341, 1969.

106. **Chapman, V. J.,** *Coastal Vegetation,* Pergamon Press, Oxford, 1964.

107. **Chapman, V. J.,** Mangrove phytosociology, *Trop. Ecol.,* 11, 1, 1970.

108. **Chapman, V. J.,** *Salt Marshes and Salt Deserts of the World,* 2nd ed., J. Cramer, Leline, 1974, 392 pp.

109. **Chapman, V. J.,** *Coastal Vegetation,* 2nd ed., Pergamon Press, Oxford, 1976a, 292 pp.

110. **Chapman, V. J.,** *Mangrove Vegetation,* J. Cramer, Vaduz, 1976b, 447 pp.

111. **Chapman, V. J., Ed.,** *Wet Coastal Ecosystems,* Vol. 1, Elsevier, Amsterdam, 1977a, 428 pp.

112. **Chapman, V. J., Ed.,** Introduction, in *Wet Coastal Ecosystems,* Vol. 1, Elsevier, Amsterdam, 1977b, 1.

113. **Choi, C.,** Primary production and release of dissolved organic carbon from phytoplankton in the Western North Atlantic, *Deep-Sea Res.,* 19, 731, 1972.

114. **Christensen, B.,** Biomass and primary production of *Rhizophora apicutata* Bl. in a mangrove in Southern Thailand, *Aquat. Bot.,* 4, 43, 1978.

115. **Christian, R. R.,** Regulation of a Salt Marsh Soil Microbial Community: a Field Experimental Approach, Ph.D. thesis, University of Georgia, Athens, 1976, 142 pp.

116. **Christian, R. R., Bancroft, K., and Wiebe, W. J.,** Distribution of microbial adenosine triphosphate in salt marsh sediments at Sapelo Island, Georgia, *Soil Sci.,* 119(1), 89, 1975.

117. **Christian, R. R., Bancroft, K., and Wiebe, W. J.,** Resistance of the microbial community within salt marsh soils to selected perturbations, *Ecology,* 59, 1200, 1978.

118. **Christian, R. R. and Hall, J. R.,** Experimental trends in sediment microbial heterotrophy: radioisotope techniques and analysis, in *Ecology of Marine Benthos,* Coull, B. C., Ed., University of South Carolina Press, Columbia, 1977, 67.

119. **Christian, R. R., Hanson, R. B., Hall, J. R., and Wiebe, W. J.,** Aerobic microbes and meiofauna, in *The Ecology of a Salt Marsh,* Pomeroy, L. R. and Wiegert, R. G., Eds., Springer-Verlag, New York, 1981, 113.

120. **Christian, R. R. and Wiebe, W. J.,** Aerobic microbial community metabolism in *Spartina alterniflora* soils, *Limnol. Oceanogr.,* 23, 238, 1978.

121. **Christian, R. R. and Wiebe, W. J.,** Three Experimental Regimes in the Study of Sediment Microbial Ecology, Litchfield, C. D. and Seyfried, P. L., Eds., American Society for Testing Materials STP 673, 1979.

122. **Christie, N. D.,** Primary production in Langebaan Lagoon, in *Estuarine Ecology with Particular Reference to South Africa,* Day, J. H., Ed., A.A. Balkema, Rotterdam, 1981, 101.

123. **Chrzanowski, T. H., Stevenson, L. H., and Spurner, J. D.,** Transport of dissolved organic carbon through a major creek of the North Inlet ecosystem, *Mar. Ecol. Prog. Ser.,* 13, 167, 1983.

124. **Clarke, L. D. and Hannon, N. J.,** The mangrove swamp and salt marsh community of the Sydney district. II. The holocoenotic complex with particular reference to physiography, *J. Ecol.,* 57, 213, 1969.

125. **Clifford, H. T. and Specht, R. L.,** *The Vegetation of North Stradbroke Island, Queensland,* University of Queensland Press, St. Lucia, 1979.

126. **Costerton, J. W., Geesey, G. C., and Cheng, K.-J.,** How bacteria stick, *Sci. Am.,* 238, 86, 1978.

127. **Clough, B. F., Andrews, T. J., and Cowan, I. R.,** Physiological processes in mangroves, in *Mangrove Ecosystems in Australia,* Clough, B. F., Ed., Australian National University Press, Canberra, 1982, 193.

128. **Clough, B. F. and Attiwell, P. M.,** Nutrient cycling in a community of *Avicennia marina* in a temperate region of Australia, in *Proc. Int. Symp. on Biology and Management of Mangroves,* Walsh, G. E., Snedaker, S. C., and Teas, H. J., Eds., Institute of Food and Agricultural Sciences, University of Florida, Gainesville, 1975, 137.

129. **Clough, B. F. and Attiwell, P. M.,** Primary production of mangroves, in *Mangrove Ecosystems in Australia,* Clough, B. F., Ed., Australian National University Press, Canberra, 1982, 211.

130. **Coles, S. M.,** Benthic microalgal populations on interstitial sediments and their role as precursors to salt marsh development, in *Ecological Processes in Coastal Environments,* Jeffries, R. L. and Davy, A. J., Eds., Blackwell Scientific, Oxford, 1979, 25.

131. **Colijn, F. and van Buurt, G.,** Influence of light and temperature on the photosynthesis rate of marine benthic diatoms, *Mar. Biol.,* 31, 209, 1975.

132. **Colijn, F. and Dijkema, K. A.,** Species composition of benthic diatoms and distribution of chlorophyll *a* on an intertidal flat in the Dutch Wadden Sea, *Mar. Ecol. Prog. Ser.,* 4, 9, 1981.

133. **Commito, J. A.,** Predation, Competition, Life-History Strategies, and the Regulation of Estuarine Soft-Bottom Community Structure, Ph.D. thesis, Duke University, Durham, N.C., 1976.

134. **Commito, J. A.,** The importance of predation by infaunal polychaetes in controlling the structure of a soft-bottom community in Maine, U.S.A., *Mar. Biol.,* 68, 77, 1982.

135. **Connell, J. H.,** Effects of competition, predation by *Thais lapillus* and other factors on natural populations of the barnacle *Balanus balanoides, Ecol. Monogr.,* 31, 61, 1961.

136. **Connell, J. H.,** Territorial behaviour and dispersion in some marine invertebrates, *Res. Popul. Ecol.,* 5, 87, 1963.

137. **Connor, M. S. and Edger, R. K.,** Selective grazing by the mud snail *Illyanassa obsoleta, Oecologia,* 53, 271, 1982.

138. **Conover, R. J.,** Assimilation of organic matter by zooplankton, *Limnol. Oceanogr.,* 11, 338, 1966.

139. **Corbett, C. J. and McLaren, A. J.,** The biology of *Pseudocalanus, Adv. Mar. Biol.,* 15, 1, 1978.

140. **Corner, E. D. S. and Cowey, C. B.,** Biochemical studies on the production of marine zooplankton, *Biol. Rev.,* 43, 393, 1968.

141. **Correll, D. L.,** Estuarine productivity, *Bioscience,* 28(10), 646, 1978.

142. **Correll, D. L., Faust, M. A., and Devera, D. J.,** Phosphorus flux and cycling in estuaries, in *Estuarine Research,* Vol. 1, Cronin, L. E., Ed., Academic Press, New York, 1975, 108.

143. **Correll, D. L., Pierce, J. W., and Faust, M. A.,** A quantitative study of the nutrient, sediment, and coliform bacterial constituents of water runoff from the Rhode River watershed, in *Non Point Sources of Water Pollution,* Water Resources Research Center, Virginia Polytechnic Institute and State University, Blacksberg, 1976, 131.

144. **Costerton, J. W. and Colwell, R. R.,** Native Aquatic Bacteria: Enumeration, Activity and Ecology, American Society for Testing and Materials, Philadelphia, 1979.

145. **Costerton, J. W., Geesey, G. C., and Cheng, K.-J.,** How bacteria stick, *Sci. Am.,* 238, 86, 1978.

146. **Cotton, A. D.,** On the growth of *Ulva latissima* in excessive quantity with special reference to Belfast Loch, Bot. Rep. R. Comm. on Sewage Disposal, 7th Rep., Appendix IV, 1911.

147. **Coull, B. C.,** Estuarine meiofauna, a review: trophic relationships and microbial interactions, in *Estuarine Microbial Ecology,* Stevenson, L. H. and Colwell, R. R., Eds., Belle Baruch Coastal Research Institute, University of South Carolina Press, Columbia, 1973, 499.

148. **Coull, B. C. and Bell, S. S.,** Perspectives in meiofaunal ecology, in *Ecological Processes in Coastal and Marine Systems,* Livingston, R. J., Ed., Plenum Press, New York, 1979, 189.

149. **Coupland, R. T.,** Productivity of grassland ecosystems, in Productivity of World Ecosystems, National Academy of Sciences, Washington, D.C., 1975, 44.

150. **Craig, N. J. and Day, J. W., Jr.,** Cumulative impact studies in the Louisiana coastal zone: eutrophication and land loss, Center for Wetland Resources, Louisiana State University, Baton Rouge, Final Rep. to Louisiana State Planning Office, Baton Rouge, 1977.

151. **Crawford, C. C., Hobbie, J. E., and Webb, K. L.,** Utilization of dissolved organic compounds by micro-organisms in an estuary, in *Estuarine Microbial Ecology,* Stevenson, L. H. and Colwell, R. R., Eds., Belle Baruch Coastal Research Institute, University of South Carolina, 1973, 169.

152. **Crawford, C. C., Hobbie, J. E., and Webb, K. L.,** The utilization of dissolved free amino acids by estuarine microorganisms, *Ecology,* 55, 551, 1974.

153. **Cullen, D. J.,** Bioturbation of superficial marine sediments by interstitial marine animals, *Nature,* 242, 323, 1973.

154. **Cuzon du Rest, R. P.,** Distribution of zooplankton in the salt marshes of southeastern Louisiana, *Publ. Inst. Mar. Sci. Univ. Tex.,* 9, 132, 1963.

155. **Dale, N. C.,** Bacteria in intertidal sediments: factors related to their distribution, *Limnol. Oceanogr.,* 19, 509, 1974.

156. **Damas, D.,** Le rôle des organismes dans la formation des vases marineus, *Ann. Soc. Geol. Belgique,* 58, 143, 1935.

157. **Darley, W. M., Dunn, E. L., Holmes, K. S., and Laren, H. G., III,** A ^{14}C method for measuring primary microalgal productivity in air, *J. Exp. Mar. Biol. Ecol.,* 25, 207, 1976.

158. **Darley, W. M., Montagne, C. L., Plumley, F. G., Sage, W. W., and Psalidas, A. T.,** Factors limiting edaphic algal biomass and productivity in a Georgia salt marsh, *J. Phycol.,* 17, 122, 1981.

159. **Darnell, R. M.,** Trophic spectrum of an estuarine community based on studies of Lake Pontchartrain, La., *Ecology,* 42, 553, 1961.

160. **Darnell, R. M.,** The organic detritus problem, in *Estuaries,* Publ. No. 83, Lauff, G. H., Ed., Am. Assoc, Ad. Sci., Washington, D.C., 1967a, 374.

161. **Darnell, R. M.,** Organic detritus in relation to the estuarine system, in *Estuaries,* Publ. No. 83, Lauff, G. H., American Association for the Advancement of Science, American Association for the Advancement of Science, Washington, D.C., 1967b, 376.

162. **Davidson, P. E.,** A study of the oystercatcher *(H. ostrelagus)* in relation to the fishery for cockles *(Cardium edule)* in the Burry Inlet, South Wales, *Minist. Agric. Fish. Food, Fish. Invest.,* Ser. II, 25(7), 1, 1967.

163. **Davidson, P. E.,** The oystercatcher — a pest of shell fisheries, in *The Problems of Birds as Pests,* Symp. Inst. Biol., 17, 1968, 141.

164. **Davies, J. L.,** A morphogenic approach to world shorelines, *Z. Geomorphol.,* 8, 127, 1964.

165. **Davies, J. L.,** *Geographical Variation in Coastal Development,* Oliver and Boyd, Edinburgh, 1972.

166. **Davies, J. M.,** Energy flow through the benthos in a Scottish sea lock, *Mar. Biol.,* 31, 353, 1975.

167. **Davis, M. W. and Lee, H.,** II, Recolonization of sediment-associated microalgae and effects of estuarine infauna on microalgal production, *Mar. Ecol. Prog. Ser.,* 11, 227, 1983.

168. **Davis, M. W. and McIntire, C. D.,** Effects of physical gradients on the production dynamics of sediment-associated algae, *Mar. Ecol. Prog. Ser.,* 13, 103, 1983.

169. **Dawes, C. J., Moon, R. E., and La Claire, J. W.,** II, Photosynthetic responses of the red alga, *Hypnea musciformis* (Wulfen) Lamaraux (Gigartinales), *Bull. Mar. Sci.,* 26, 467, 1976.

170. **Dawes, C. J., Moon, R. E., and Davis, M. A.,** The photosynthetic and respiratory rates and tolerances of benthic algae from a mangrove and salt marsh estuary: a comparative study, *Estuarine Coastal Mar. Sci.,* 6, 175, 1978.

171. **Dawes, C. J., Bird, K., Durako, M., Goddard, R., Hoffman, W., and McIntosh, R.,** Chemical fluctuations due to seasonal and cropping effects on an algal-seagrass community, *Aquat. Bot.,* 6, 79, 1979.

172. **Day, J. H.,** The ecology of South African estuaries. I. General considerations, *Trans. R. Soc. S. Afr.,* 35, 475, 1951.

173. **Day, J. H.,** The origin and distribution of estuarine animals in South Africa, *Monogr. Biol.,* 14, 159, 1964.

174. **Day, J. H.,** The biology of the Knysna estuary, South Africa, in *Estuaries,* Publ. No. 85, Lauff, G., Ed., American Association for the Advancement of Science, Washington, D.C., 1967, 397.

175. **Day, J. H.,** The ecology of the Morrumbene estuary, Mocambique, *Trans. R. Soc. S. Afr.,* 41, 43, 1974.

176. **Day, J. H.,** What is an estuary?, *S. Afr. J. Sci.,* 76, 198, 1980.

177. **Day, J. H., Ed.,** *Estuarine Ecology with Particular Reference to Southern Africa,* A.A. Balkema, Rotterdam, 1981a, 411 pp.

178. **Day, J. H., Ed.,** The nature, origin and classification of estuaries, in *Estuarine Ecology with Particular Reference to Southern Africa,* A.A. Balkema, Rotterdam, 1981b, 1.

179. **Day, J. H., Ed.,** Estuarine currents, salinities and temperatures, in *Estuarine Ecology with Particular Reference to Southern Africa,* A.A. Balkema, Rotterdam, 1981c, 27.

180. **Day, J. H., Ed.,** Estuarine sediments, turbidity and the penetration of light, in *Estuarine Ecology with Particular Reference to Southern Africa,* A.A. Balkema, Rotterdam, 1981d, 45.

181. **Day, J. H., Ed.,** The chemistry and fertility of estuaries, in *Estuarine Ecology with Particular Reference to Southern Africa,* A.A. Balkema, Rotterdam, 1981e, 57.

182. **Day, J. H., Ed.,** The estuarine flora, in *Estuarine Ecology with Particular Reference to Southern Africa,* A.A. Balkema, Rotterdam, 1981f, 77.

183. **Day, J. H., Blaber, S. J. M., and Wallace, J. H.,** Estuarine fishes, in *Estuarine Ecology with Particular Reference to Southern Africa,* Day, J. H., Ed., A.A. Balkema, Rotterdam, 1981g, 197.

184. **Day, J. W., Jr., Smith, P., Wagner, P., and Starve, W.,** Community Structure and Carbon Budget of a Salt Marsh and Shallow Bay Estuarine System in Louisiana, Sea Grant Publ. No. LSU-SG-72-04, Center for Wetland Resources, Louisiana State University, New Orleans, 1973, 79 pp.

185. **Day, J. W., Jr., Hopkinson, C. S., and Conner, W. H.,** An analysis of environmental factors regulating community metabolism and fisheries production in a Louisiana estuary, in *Estuarine Comparisons,* Kennedy, V. S., Ed., Academic Press, New York, 1982, 121.

186. **Dayton, P. K.,** Competition, disturbance and community organization: the provision and subsequent utilization of space in a rocky intertidal community, *Ecol. Monogr.,* 41, 351, 1971.

187. **Dayton, P. K. and Oliver, J. S.,** An evaluation of experimental analyses of population and community patterns in benthic marine environments, in *Marine Benthic Dynamics,* Tenore, K. R. and Coull, B. C., Eds., University of South Carolina Press, Columbia, 1979, 93.

188. **de la Cruz, A. A. and Hackney, C. T.,** Energy value, elemental composition, and productivity of belowground biomass of a *Juncus* tidal marsh, *Ecology,* 58, 1165, 1977.

189. **DeLaune, R. D. and Patrick, W. H., Jr.,** Nitrogen and phosphorus cycling in a Gulf Coast salt marsh, in *Estuarine Perspectives,* Kennedy, V. S., Ed., Academic Press, New York, 1980, 143.

190. **DeLaune, R. D., Smith, C. J., Patrick, W. H., Jr., Sikora, W. B., Sikora, J. P., and Humbrick, G. A., III,** The use of chemical dispersants as a method of marsh restoration following an oil spill, Rep. on Research and Engineering, Center for Wetland Resources, Louisiana State University, Baton Rouge, 1981.

191. **den Hartog, C.,** The structural aspect in the ecology of seagrass communities, *Helgol. Wiss. Meeresunters.,* 15, 648, 1967.

192. **den Hartog, C.,** Structure, function and classification in seagrass communities, in *Seagrass Ecosystems. A Scientific Perspective,* McRoy, C. P. and Helfferich, C., Eds., Marcel Dekker, New York, 1977, 89.

193. **Department of Conservation and Environment,** Management of Peel Inlet and Harvey Estuary, Bull. 170, Dept. of Conserv. and Environ., Perth, Western Australia, 1984.

194. **De Sylva, D.,** Ecology and Distribution of Postlarval Fishes in Southern Biscayne Bay, Florida, Prog. Rep. Environmental Protection Agency, Div. of Water Research, 1970.

195. **De Sylva, D.,** Nektonic food webs in estuaries, in *Estuarine Research,* Vol. 1, Cronin, L. E., Ed., Academic Press, New York, 1975, 420.

196. **Dick, W. J. A.,** Oxford and Cambridge Mauritanian Expedition, 1973 Rep., Cambridge, England, 1975.

197. **Dillon, R. C.,** A Comparative Study of the Primary Productivity of Estuarine Phytoplankton and Macrobenthic Plants, Ph.D. dissertation, University of North Carolina, Chapel Hill, 1971, 112 pp.

198. **Drake, C. B. and Read, M.,** Carbon dioxide assimilation, photosynthesis efficiency, and respiration in a Chesapeake Bay salt marsh, *J. Ecol.,* 69, 405, 1981.

199. **Drew, E. A.,** Botany, in *Underwater Science,* Woods, J. D. and Lythgoe, J. N., Eds., Oxford University Press, London, 1971.

200. **Drinnan, R. E.,** The winter-feeding of the oystercatcher *(Haematopus ostralegus)* on the edible cockle *(Cardium edule), J. Anim. Ecol.,* 26, 441, 1957.

201. **Duff, S. and Teal, J. M.,** Temperature changes and gas exchanges in Nova Scotia and Georgia salt marsh muds, *Limnol. Oceanogr.,* 10, 67, 1965.

202. **Duke, N. C., Bunt, J. W., and Williams, W. T.,** Mangrove litter fall in North-eastern Australia. I. Annual totals by component in selected species, *Austr. J. Bot.,* 29, 547, 1981.

203. **Duke, T. W. and Rice, T. R.,** Cycling of elements in estuaries, *Proc. Gulf Caribb. Fish. Inst.,* 19, 59, 1967.

204. **Dunn, R.,** The Effects of Temperature on the Photosynthesis, Growth and Productivity of *Spartina townsendii* (sensu lato) in Controlled and Natural Environments, Ph.D. thesis, University of Essex, Colchester, England, 1981.

205. **Durbin, A. G. and Durbin, E. G.,** Standing stock and estimated production rates of phytoplankton and zooplankton in Narragansett Bay, Rhode Island, *Estuaries,* 4, 24, 1981.

206. **Durbin, E. G., Durbin, A. G., and Smayda, T. J.,** Seasonal studies on the relative importance of different size fractions of phytoplankton in Narragansett Bay, *Mar. Biol.,* 32, 271, 1975.

207. **Dye, A. H.,** Ecophysiological Study of the Benthic Meiofauna of the Swartkops Estuary, M.Sc. thesis, University of Port Elizabeth, South Africa, 1977.

208. **Dye, A. H. and Furstenberg, J. P.,** Estuarine meiofauna, in *Estuarine Ecology with Particular Reference to Southern Africa,* Day, J. H., Ed., A.A. Balkema, Rotterdam, 1981.

209. **Dyer, K. R.,** Sedimentation in estuaries, in *The Estuarine Environment,* Barnes, R. S. K. and Green, J., Eds., Applied Science, London, 1972, 10.

210. **Dyer, K. R.,** *Estuaries: a Physical Introduction,* John Wiley & Sons, New York, 1973.

211. **Eagle, R. A.,** Benthic studies in Liverpool Bay, *Estuarine Coastal Mar. Sci.,* 1, 285, 1973.
212. **Eagle, R. A.,** Natural fluctuations in a soft bottom benthic community, *J. Mar. Biol. Assoc. U.K.,* 55, 865, 1975.
213. **Ebbinge, C. C. H., Canters, K., and Drent, R.,** Foraging routines and estimated daily food intake in Barnacle Geese wintering in the northern Netherlands, *Wildfowl,* 26, 5, 1975.
214. **Edwards, J. M. and Frey, R. W.,** Substrate characteristics within a holocene salt marsh, Sapelo Island, Georgia, *Senckenbergiana Marit.,* 9, 215, 1977.
215. **Edwards, S. F. and Welsh, B. L.,** Trophic dynamics of a mud snail [*Ilyanassa obsoleta* (Say)] population on an intertidal mudflat, *Estuarine Coastal Mar. Sci.,* 14, 663, 1982.
216. **Edzwald, J. K., Upchurch, J. B., and O'Melia, C. R.,** Coagulation in estuaries, *Environ. Sci. Technol.,* 8, 58, 1974.
217. **Eilers, H. P.,** Production ecology in an Oregon coastal salt marsh, *Estuarine Coastal Mar. Sci.,* 8, 399, 1979.
218. **Eppley, R. W.,** Temperature and phytoplankton growth in the sea, *Fish. Bull.,* 70, 1063, 1972.
219. **Eriksson, S., Sellei, C., and Wallström, K.,** Structure of the plankton community of the Öregrundsgrepen (South-west Bosnian Sea), *Helgol. Wiss. Meeresunters.,* 30, 582, 1977.
220. **Erkenbrecher, C. W. and Stevenson, H.,** The influence of tidal flux on microbial biomass in salt marsh creeks, *Limnol. Oceanogr.,* 20(4), 618, 1975.
221. **Ernst, W.,** ATP als indikator für die Biomass mariner Sedimente, *Oecologia,* 5, 56, 1970.
222. **Evans, G. C.,** *The Quantitative Analysis of Plant Growth,* Blackwell Scientific, Oxford, 1972, 734 pp.
223. **Evans, M. S. and Grainger, E. H.,** Zooplankton in a Canadian Arctic estuary, in *Estuarine Perspectives,* Kennedy, V. S., Ed., Academic Press, New York, 1980, 199.
224. **Evans, P. R.,** Energy balance and optimal foraging strategies in shore birds: some implications for their distribution and movements in the non-breeding season, *Ardea,* 64, 117, 1976.
225. **Evans, S.,** Production, predation and food niche segregation in a marine shallow soft-bottom community, *Mar. Ecol. Prog. Ser.,* 10, 147, 1983.
226. **Fager, E. W.,** Marine sediments: effects of a tube-building polychaete, *Science,* 143, 356, 1964.
227. **Feller, R. J. and Kaczynski, V. W.,** Size selective predation by juvenile chum salmon *(Onchorynchus chaeta)* on epibenthic prey in Puget Sound, *J. Fish. Res. Board Can.,* 32, 1419, 1975.
228. **Fenchel, T.,** The ecology of the marine microbenthos. IV. Structure and function of the benthic ecosystem, its chemical and physical factors and the microfauna communities with special reference to the ciliated protozoa, *Ophelia,* 6, 1, 1969.
229. **Fenchel, T.,** Studies on the decomposition of organic detritus derived from the turtle grass *Thalassia testudinum, Limnol. Oceanogr.,* 15, 14, 1970.
230. **Fenchel, T.,** The reduction-oxidation properties of marine sediments and the vertical distribution of the microfauna, *Vie Milieu,* 22, 509, 1971.
231. **Fenchel, T.,** Aspects of decomposer food chains in marine benthos, *Sonderh. Verh. Dtsch. Zool. Ges.,* 65, 14, 1972.
232. **Fenchel, T.,** Aspects of the decomposition of seagrasses, Int. Seagrass Workshop, Leiden, Netherlands, 1973.
233. **Fenchel, T.,** Factors determining the distribution patterns of mud snails (Hydrobiidae), *Oecologia,* 20, 1, 1975a.
234. **Fenchel, T.,** Character displacement and coexistence in mud snails (Hydrobiidae), *Oecologia,* 20, 19, 1975b.
235. **Fenchel, T.,** The quantitative importance of benthic microflora of an Arctic tundra pond, *Hydrobiologica,* 46, 445, 1975c.
236. **Fenchel, T.,** Aspects of decomposition in seagrasses, in *Seagrass Ecosystems: A Scientific Perspective,* McRoy, C. P. and Helfferich, C., Eds., Marcel Decker, New York, 1977, 123.
237. **Fenchel, T.,** The ecology of the micro- and meiobenthos, *Annu. Rev. Ecol. Syst.,* 9, 99, 1978.
238. **Fenchel, T.,** Ecology of heterotrophic microflagellates. I. Some important forms and their functional morphology, *Mar. Ecol. Prog. Ser.,* 8(3), 211, 1982a.
239. **Fenchel, T.,** Ecology of heterotrophic microflagellates. II. Bioenergetics and growth, *Mar. Ecol. Prog. Ser.,* 8(3), 225, 1982b.
240. **Fenchel, T.,** Ecology of heterotrophic microflagellates. III. Adaptations to heterogeneous environments, *Mar. Ecol. Prog. Ser.,* 9(1), 25, 1982c.
241. **Fenchel, T.,** Ecology of heterotrophic microflagellates. IV. Quantitative occurrence and importance as bacterial consumers, *Mar. Ecol. Prog. Ser.,* 9(1), 35, 1982d.
242. **Fenchel, T.,** Suspended marine bacteria as food source, in *Flow of Energy and Materials in Marine Ecosystems,* Fasham, M. J., Ed., Plenum Press, New York, 1982e.
243. **Fenchel, T. and Blackburn, T. H.,** *Bacteria and Mineral Cycling,* Academic Press, London, 1979.
244. **Fenchel, T. and Kofoed, L. H.,** Evidence for exploitative interspecific competition in mud snails (Hydrobiidae), *Oikos,* 27, 367, 1976.

245. **Fenchel, T. M. and Riedl, R. J.,** The sulfide system: a new biotic community underneath the oxidized layer of marine sand bottoms, *Mar. Biol.,* 7, 255, 1970.

246. **Ferguson, R. L. and Murdock, M. B.,** Microbial ATP and organic carbon in the sediments of the Newport River estuary mudflats, in *Estuarine Research,* Vol. 1, Cronin, L. E., Ed., Academic Press, New York, 1975, 229.

247. **Festa, J. F. and Hansen, P. D.,** Turbidity maxima in partially mixed estuaries: a two-dimensional model, *Estuarine Coastal Mar. Sci.,* 7, 347, 1978.

248. **Finn, J. T.,** Measures of ecosystem structure and function derived from analysis of flows, *J. Theor. Biol.,* 56, 563, 1976.

249. **Finn, J. T.,** Flow Analysis: a Method for Tracing Flows Through Ecosystem Models, Ph.D. dissertation, University of Georgia, Athens, 1977.

250. **Finn, J. T. and Leschine, T. M.,** Does salt marsh fertilization enhance shellfish production? An application of flow analysis, *Environ. Manage.,* 4, 193, 1980.

251. **Fitch, J. E.,** A relatively unexploited population of Pismo clams, *Tivela stultorum* (Moore, 1823) (Veneridae), *Proc. Malacol. Soc. London,* 36, 309, 1965.

252. **Flemer, D. A.,** Primary production in the Chesapeake Bay, *Chesapeake Sci.,* 11, 117, 1970.

253. **Fogg, G. E.,** The extracellular products of algae, *Oceanogr. Mar. Biol. Annu. Rev.,* 4, 195, 1966.

254. **Fogg, G. E., Nalewajko, C., and Watt, W. D.,** Extracellular products of phytoplankton photosynthesis, *Proc. R. Soc. London, Ser. B.,* 162, 517, 1965.

255. **Foster, W. A.,** Studies on the Distribution and Growth of *Juncus roemarianus* in South Eastern Brunswick County, North Carolina, M.S. thesis, North Carolina State University, Raleigh, 1968.

256. **Forster, G. L.,** Indications regarding the source of combined nitrogen for *Ulva lactuca, Ann. Mo. Bot. Gard.,* 1, 229, 1914.

257. **Fralick, R. A. and Mathieson, A. C.,** Physiological ecology of four *Polysiphonia* species (Rhodophyta, Ceramiales), *Mar. Biol.,* 29, 29, 1975.

258. **Frankenberg, D.,** Oxygen in a tidal river: low tide concentration correlates with location, *Estuarine Coastal Mar. Sci.,* 4, 455, 1915.

259. **Frankenberg, D., Coles, S. L., and Johannes, R. E.,** The potential significance of *Callianassa major* faecal pellets, *Limnol. Oceanogr.,* 12, 113, 1967.

260. **Frankenberg, D. and Smith, K. L.,** Coprophagy in marine animals, *Limnol. Oceanogr.,* 12, 443, 1976.

261. **Fredericks, A. D. and Sackett, W. M.,** Organic carbon in the Gulf of Mexico, *J. Geophys. Res.,* 75, 2199, 1970.

262. **Fuhrman, J. A., Ammerman, J. W., and Azam, F.,** Bacterioplankton of the coastal zone: distribution activity and possible relationships with phytoplankton, *Mar. Biol.,* 60, 201, 1980.

263. **Fuhrman, J. A. and Azam, F.,** Bacterioplankton secondary production estimates for coastal waters of British Columbia, Antarctica, and California, *Appl. Environ. Microbiol.,* 39, 1085, 1980.

264. **Furnas, M. J., Hitchcock, C. L., and Smayda, T. J.,** Nutrient-phytoplankton relationships in Narragansett Bay during the 1974 summer bloom, in *Estuarine Processes,* Vol. 1, Wiley, M. L., Ed., Academic Press, New York, 1976, 118.

265. **Gak, D. S., Romanova, E. P., Romaneko, V. I., and Sorokin, Yu. I.,** Estimation of changes in the number of bacteria in isolated water samples, in *Microbial Production and Decomposition in Fresh Waters,* Sorokin, Yu. I. and Kudota, H., Eds., Blackwell Scientific, Oxford, 1972, 77.

266. **Gallagher, J. L.,** Sampling macro-organic matter profiles in salt marsh plant root zones, *Proc. Am. Soil Sci. Soc.,* 38, 154, 1974.

267. **Gallagher, J. L.,** Effects of an ammonium nitrate pulse on the growth and elemental composition of natural stands of *Spartina alterniflora* and *Juncus roemerianus, Am. J. Bot.,* 62, 644, 1975.

268. **Gallagher, J. L.,** Zonation of wetlands vegetation, in *Coastal Ecosystem Management,* Clark, J. R., Ed., John Wiley & Sons, New York, 1977, 752.

269. **Gallagher, J. L.,** Growth and element compositional responses of *Sporobolus virginicus* (L.) Kunth. to substrate salinity and nitrogen, *Am. Midl. Nat.,* 102, 68, 1979.

270. **Gallagher, J. L., Pfeiffer, W. J., and Pomeroy, L. R.,** Leaching and microbial utilization of dissolved organic matter from leaves of *Spartina alterniflora, Estuarine Coastal Mar. Sci.,* 4, 467, 1976.

271. **Gallagher, J. L. and Plumley, F. G.,** Underground biomass profiles and productivity in Atlantic coastal marshes, *Am. J. Bot.,* 66, 156, 1979.

272. **Gallagher, J. L., Plumley, F. G., and Wolf, P. L.,** Underground biomass dynamics and substrate selective properties of Atlantic coastal salt marsh plants, MS. Report: TR-D-77-28, U.S. Army Engineer Waterways Experimental Station, Vicksberg, 1978.

273. **Gallagher, J. L., Reimold, R. J., Linthurst, R. A., and Pfeiffer, W. J.,** Aerial production, mortality and mineral accumulation-export dynamics on *Spartina alterniflora* and *Juncus roemerianus* plant stands, *Ecology,* 61, 303, 1980.

274. **Gallagher, J. L., Reimold, R. J., and Thompson, D. E.,** Remote sensing and salt marsh productivity, *Proc. 38th Annu. Meet. Am. Soc. Photogrammetry,* p.338, 1972.

275. **Gallagher, J. L., Reimbold, R. J., and Thompson, D. E.,** A comparison of four remote sensing media for assessing salt marsh productivity, in *Proc. 8th Int. Symp. on Remote Sensing of the Environ.,* Cook, J., Ed., Environmental Research Inst., University of Michigan, Ann Arbor, 1973, 1287.

276. **Galloway, B. J. and Strawn, K.,** Seasonal abundance and distribution of marine fishes at a hot-water discharge in Galveston Bay, Texas, *Contrib. Mar. Sci. Univ. Tex.,* 18, 71, 1974.

277. **Galloway, W. E.,** Process framework for describing the morphologic and stratigraphic evaluation of deltaic depositional systems, in *Deltas, Models for Exploration,* Broussard, M. L., Ed., Houston Geological Society, Houston, 1975, 87.

278. **Garber, J. H.,** ^{15}N tracer study of the short-term fate of particulate organic nitrogen at the surface of coastal marine sediments, *Mar. Ecol. Prog. Ser.,* 16, 89, 1984.

279. **Gardner, L. R.,** Runoff from an intertidal marsh during tidal exposure-recession curves and chemical characteristics, *Limnol. Oceanogr.,* 20, 81, 1975.

280. **Geesey, G. G., Mutch, R., Costerton, J. W., and Green, R. B.,** Sessile bacteria: an important component of the microbial population in small mountain streams, *Limnol. Oceanogr.,* 23, 1214, 1978.

281. **Gerlach, S. A.,** On the importance of marine meiofauna for benthos communities, *Oecologia,* 6, 151, 1971.

282. **Gerlach, S. A.,** Food chain relationships in subtidal silty sand marine sediments and the role of meiofauna in stimulating bacterial productivity, *Oecologia,* 33, 55, 1978.

283. **Gessner, F.,** Hydrobotanik, *In Die physiologischen Grundlagen der Pflanzenverbreitung im Wasser,* Vol. 2, VEB Deutscher Verlag der Wissenschaften, Berlin, 1959.

284. **Gessner, F. and Hammer, L.,** Die Primärproduction in Mediterranen *Caulerpa-Cymodoce* -Wiesen, *Bot. Mar.,* 2, 157, 1960.

285. **Gessner, F. and Hammer, L.,** Investigaciones sobre el clima de la luz en las regiones marinas de las costa Venezolar, *Bol. Inst. Oceanogr.,* 1, 263, 1980.

286. **Gieskes, W. W. C., Kmay, G. W., and Baars, M. A.,** Current ^{14}C methods for measuring primary production: gross underestimates in oceanic waters, *Neth. J. Sea Res.,* 13(1), 58, 1979.

287. **Gill, A. M.,** Australia's mangrove enclaves: a coastal resource, *Proc. Ecol. Soc. Austr.,* 8, 126, 1975.

288. **Gill, A. M. and Tomlinson, P. B.,** Studies on the growth of red mangroves (*Rhizophora mangle* L.). I. Habit and general morphology, *Biotropica,* 1(1), 1, 1969.

289. **Gill, A. M. and Tomlinson, P. B.,** Studies on the growth of red mangrove (*Rhizophora mangle* L.). II. Growth and differentiation of aerial roots, *Biotropica,* 3, 63, 1971.

290. **Gill, A. M. and Tomlinson, P. B.,** Aerial roots: an array of forms and functions, in *The Development and Function of Roots,* Torrey, J. G. and Clarkson, D. T., Eds., Academic Press, London, 1975, 237.

291. **Gill, A. M. and Tomlinson, P. B.,** Studies on the growth of red mangrove (*Rhizophora mangle* L.). IV. The adult root system, *Biotropica,* 9, 145, 1977.

292. **Gillespie, M. C.,** Analysis and treatment of zooplankton of estuarine waters of Louisiana, in *Cooperative Gulf of Mexico Estuarine Inventory and Study Louisiana. Phase IV. Biology,* Louisiana Wildlife and Fish. Comm., 1971, 108.

293. **Gillespie, P. A. and MacKenzie, A. L.,** Autotrophic and heterotrophic processes on an intertidal mud-sandflat, Delaware Inlet, Nelson, New Zealand, *Bull. Mar. Sci.,* 31(3), 648, 1981.

294. **Giurgevich, J. R. and Dunn, E. L.,** Seasonal patterns of CO_2 and water vapor exchange of *Juncus roemerianus* Scheele in a Georgia salt marsh, *Am. J. Bot.,* 65, 502, 1978.

295. **Giurgevich, J. R. and Dunn, E. L.,** Seasonal patterns of CO_2 and water vapor exchange of the tall and short height forms of *Spartina alterniflora* in a Georgia salt marsh, *Oecologia,* 43, 139, 1979.

296. **Gleason, M. L. and Zieman, J. C.,** Influence of tidal inundation on internal oxygen supply of *Spartina alterniflora and Spartina patens, Estuarine Coastal Shelf Sci.,* 13, 47, 1981.

297. **Godin, J.-G. J.,** Daily patterns of feeding behaviour, daily rations, and diets of juvenile pink salmon, (*Oncorhynchus gorbuscha*) in two marine bays of British Columbia, *Can. J. Fish. Aquat. Sci.,* 38, 10, 1981.

298. **Goering, J. J.,** Denitrification in the oxygen minimum layer of the eastern tropical Pacific Ocean, *Deep-Sea Res.,* 15, 157, 1968.

299. **Goering, J. J. and Parker, P. L.,** Nitrogen fixation by epiphytes on seagrasses, *Limnol. Oceanogr.,* 17, 320, 1972.

300. **Golley, F. B.,** Energy flux in ecosystems, in *Ecosystem Structure and Function,* Wiens, J. A., Ed., Oregon State University Press, Corvallis, 1972, 69.

301. **Golley, F. B., McGinnis, J. T., Clements, R. G., Childs, G. I., and Duever, M. J.,** *Mineral Cycling in a Tropical Moist Forest Ecosystem,* University of Georgia Press, Athens, 1975.

302. **Golley, F. B., Odum, H. T., and Wilson, R. F.,** The structure and metabolism of a Puerto Rican red mangrove forest in May, *Ecology,* 43, 9, 1962.

303. **Golterman, H. L.,** *Methods of Chemical Analysis of Freshwaters,* IBP Handbook No. 8, Blackwell Scientific, Oxford, 1969, 172 pp.

304. **Good, R. E.,** Salt marsh vegetation, Cape May, New Jersey, *Bull. N. J. Acad. Sci.,* 10, 1, 1965.

305. **Good, R. E.,** An Environmental Assessment of the Proposed Reconstruction of State Route 152 (Somers Point-Long Point), Atlantic County, New Jersey, Rep. to E. Lionel Pavlo Engineering, New York, 1977.

306. **Good, R. E. and Frasco, B. R.,** Estuarine Evaluation Study: a Four Year Report on Production and Decomposition Dynamics of Salt Marsh Communities: Manahawkin Marshes, Ocean County, New Jersey, Rep. N. J. Depart. Environmental Protection, Division of Fisheries, Game and Shell Fisheries, Trenton, 1979.

307. **Good, R. E., Good, N. F., and Frasco, B. R.,** A review of the primary production and decomposition dynamics of the belowground marsh component, in *Estuarine Comparions,* Kennedy, V. S., Ed., Academic Press, New York, 1982, 139.

308. **Good, R. E. and Walker, R.,** Relative Contribution of Saltwater and Reshwater Tidal Marsh Communities to Estuarine Productivity, Fin. Rep. (071171) submitted to Rutgers Univ. Center for Coastal and Environ. Stud., New Brunswick, N.J., Unpublished manuscript, 1977.

309. **Goodman, D.,** The validity of the diversity-stability hypothesis, *Proc. 1st Int. Congr. Ecol.,* p.75, 1974.

310. **Gordon, A. R., Jr.,** Dispersion of contaminants in an estuarine port area, James River, *Proc. Int. At. Energy Conf. Vienna,* 2, 152, 1960.

311. **Gordon, D. C.,** The effects of the deposit feeding polychaete *Pectinaria gouldii* on the intertidal sediments of Barnstable Harbour, *Limnol. Oceanogr.,* 11, 327, 1966.

312. **Gordon, D. C.,** Some studies on the distribution and composition of particulate organic carbon in the North Atlantic Ocean, *Deep-Sea Res.,* 17, 233, 1970.

313. **Gorsline, D. S.,** Contrasts on coastal bay sediments on the Gulf and Pacific coasts, *Estuaries,* Publ. No. 83, Lauff, G., Ed., Am. Assoc. Adv. Sci., Washington, D.C., 1967, 211.

314. **Goss-Custard, J. D., Jones, R. E., and Newberry, P. E.,** The ecology of the Wash. I. Distribution and diet of wading birds (Charadii), *J. Appl. Ecol.,* 14, 681, 1977.

315. **Goss-Custard, J. D., Kay, D. G., and Blindell, R. M.,** The density of migratory and overwintering Redshank, *Tringa totanus,* and Curlew, *Numenius arquata,* in relation to the density of their prey in S.E. England, *Estuarine Coastal Mar. Sci.,* 5, 497, 1977.

316. **Goss-Custard, J. D., Le V dit Durell, S. E. A., McGrorty, S., Reading, C. J., and Clarke, R. T.,** Factors affecting the occupation of mussel *(Mytilus edulis)* beds by oystercatchers *(Haematopus ostralegus)* on the Exe River, Devon, in *Feeding and Survival Strategies of Estuarine Organisms,* Jones, N. V. and Wolff, W. J., Eds., Plenum Press, New York, 1981, 217.

317. **Gosselink, J. G.,** Growth of *Spartina patens* and *S. alterniflora* as influenced by salinity and source of nitrogen, *Coast. Stud. Bull.,* 5, 97, 1970.

318. **Gould, W. R.,** The biology and morphology of *Acyrtops berrylinus,* the emerald clingfish, *Bull. Mar. Sci.,* 15, 165, 1965.

319. **Goulder, R.,** Relationships between suspended solids and standing crops and activities of bacteria during neap-spring-neap tidal cycle, *Oecologia,* 24, 83, 1976.

320. **Goulder, R., Blanchard, A. S., Sanderson, P. L., and Wright, B.,** Relationships between heterotrophic bacteria and pollution in an industrial estuary, *Water Res.,* 14, 591, 1980.

321. **Goulter, P. F. E. and Alloway, W. G.,** Litterfall and decomposition in a mangrove stand, *Avicennia marina* (Forsk.) Vierh., in Middle Harbour, Sydney, *Aust. J. Mar. Freshwater Res.,* 30, 541, 1979.

322. **Grainger, E. H.,** Zooplankton from the Arctic Ocean and adjacent waters, *J. Fish. Res. Board Can.,* 22, 543, 1965.

323. **Grant, R. R. and Patrick, R.,** Tinacum marsh as a water purifier, in *Two Studies of Tinicum Marsh,* McCormick, J., Ed., The Conservation Foundation, Washington, D.C., 1970, 105.

324. **Grassle, J. F. and Grassle, J. P.,** Opportunistic life histories and genetic systems in marine benthic polychaetes, *J. Mar. Res.,* 32(2), 253, 1974.

325. **Grassle, J. F. and Sanders, H. L.,** Life histories and the role of distrubance, *Deep-Sea Res.,* 20, 643, 1973.

326. **Gray, J. S.,** Animal-sediment relationships, *Oceanogr. Mar. Biol. Annu. Rev.,* 12, 223, 1974.

327. **Gray, J. S.,** The stability of benthic ecosystems, *Helgol. Wiss. Meeresunters.,* 30, 427, 1977.

328. **Gray, J. S.,** Pollution-induced changes in populations, *Trans. R. Soc. London, Ser. B,* 286, 545, 1979.

329. **Gray, J. S.,** *The Ecology of Marine Sediments,* Cambridge University Press, Cambridge, England, 1981, 185 pp.

330. **Green, J.,** *The Biology of Estuarine Animals,* Sidgwick and Jackson, London, 1968, 401 pp.

331. **Greenway, M.,** The grazing of *Thalassia testudinum* (König) in Kingston Harbour, Jamaica, *Aquat. Bot.,* 2, 117, 1976.

332. **Greve, W. and Reiners, F.,** The impact of prey-predator waves from estuaries on the planktonic marine ecosystem, in *Estuarine Perspectives,* Kennedy, V. S., Ed., Academic Press, New York, 1980, 405.

333. **Grindley, J. R.,** Effect of low-salinity water on the vertical migration of estuarine plankton, *Nature,* 203, 781, 1964.

334. **Grindley, J. R.,** The vertical migration behaviour of estaurine plankton, *Zool. Afr.,* 7, 13, 1972.
335. **Grindley, J. R.,** The plankton of Swartkops estuary, in S. Afr. Natl. Conf. on Mar. and Fresh Water Res., Port Elizabeth, CSIR, Port Elizabeth, 1976.
336. **Grindley, J. R.,** Residence time tests, in *Knysma Lagoon Model Investigation 1,* NRIO, Stellenbosch, South Africa, 1977, 28.
337. **Grindley, J. R.,** Environmental Effects of the Discharge of Sewage into Knysma Estuary, School of Environmental Studies, University of Capetown, South Africa, 1978.
338. **Grindley, J. R.,** Estuarine plankton, in *Estuarine Ecology with Particular Reference to Southern Africa,* Day, J. H., Ed., A. A. Balkema, Rotterdam, 1981, 117.
339. **Grindley, J. R. and Woolridge, T.,** The plankton of Richards Bay, *Hydrobiol. Bull.,* 8, 201, 1974.
340. **Grontved, J.,** Underwater macrovegetation in shallow coastal waters, *Rapp. P. V. Reun. Cons. Int. Explor. Mer,* 24, 32, 1958.
341. **Grossman, G. D., Coffin, R., and Moyle, P. B.,** Feeding ecology of the bay goby (Pices: Gibiidae): effects of behavioural, ontogenetic, and temporal variation in diet, *J. Exp. Mar. Biol. Ecol.,* 44, 47, 1980.
342. **Guide, V. and Villa, O.,** Chesapeake Bay Nutrient Impact Study, Tech. Rep. No. 47, Annapolis Field Office, U.S. Environmental Protection Agency, 1972, 80 pp.
343. **Gunter, G.,** Some relations of estuaries to the fisheries of the Gulf of Mexico, in *Estuaries,* Publ. No. 83, Lauff, G., Ed., Am. Assoc. Adv. Sci., Washington, D.C., 1967, 621.
344. **Haas, L. W. and Webb, K. L.,** Nutritional mode of several nonpigmented microflagellates from the York River estuary, Virginia, *J. Exp. Mar. Biol. Ecol.,* 39, 125, 1979.
345. **Haines, E. B.,** Relation between the stable carbon isotope composition of fiddler crabs, plants and soils in a salt marsh, *Limnol. Oceanogr.,* 21, 880, 1976.
346. **Haines, E. B.,** Growth dynamics of cordgrass *Spartina alterniflora* Losiel on control and sewage sludge fertilized plots in a Georgia salt marsh, *Estuaries,* 2, 50, 1979b.
347. **Haines, E. B.,** Nitrogen pools in Georgia coastal waters, *Estuaries,* 2, 34, 1979c.
348. **Haines, E. B. and Dunn, E. L.,** Growth and resource allocation response of *Spartina alterniflora* Loisel to three levels of $NH_4 - N$, Fe and NaCl in solution culture, *Bot. Gaz.,* 137, 224, 1976.
349. **Hair, M. E. and Basset, C. R.,** Dissolved and particulate humic acids in an east coast estuary, *Estuarine Coastal Mar. Sci.,* 1, 107, 1973.
350. **Hall, C. A. S. and Day, J. W., Jr.,** Systems and models: terms and basic princples, in *Ecosystem Modelling in Theory and Practice,* Hall, C. A. S. and Day, J. W., Jr., Eds., John Wiley & Sons, New York, 1977, 6.
351. **Hammer, L.,** Die primarproduktion in Golf von Cariaco (Ost-Venezuela), *Int. Rev. Gesamten Hydrobiol.,* 52, 757, 1967.
352. **Hammer, L.,** Salzgehalt und Photosynthese bei marinen Pflanzen, *Mar. Biol.,* 1, 185, 1968.
353. **Hammer, L.,** Anaerobiosis in marine algae and some marine phaerograms, *Proc. 6th Int. Seaweed Symp., 1968,* 7, 414, 1969.
354. **Hancock, D. A.,** The role of predators and parasites in a fishery for the mollusc, *Cordium edule* L, *Proc. Adv. Study Inst. on the Dynamics of Populations,* den Boer, P. J. and Gradwell, G. R., Eds., Osterbeck, 1970, 298.
355. **Hannon, B.,** The structure of ecosystems, *J. Theor. Biol.,* 41, 535, 1973.
356. **Hansen, J. A.,** Effects of Physical Factors on Fermentation in Salt Marsh Soils, M.S. thesis, University of Georgia, Athens, 1979.
357. **Hanson, R. B.,** Comparison of nitrogen fixation activity in tall and short *Spartina alterniflora, Appl. Environ. Microbiol.,* 33, 846, 1977a.
358. **Hanson, R. B.,** Organic nitrogen and caloric content of detritus. II. Microbial biomass and activity, *Estuarine Coastal Shelf Sci.,* 14, 325, 1982.
359. **Hanson, R. B. and Snyder, J.,** Microbial heterotrophic activity in a salt-marsh estuary: biological activity and chemical measurements, *Limnol. Oceanogr.,* 25, 633, 1979.
360. **Hanson, R. B. and Wiebe, W. J.,** Heterotrophic activity associated with particulate size fractions in a *Spartina alterniflora* salt-marsh estuary, Sapelo Island, Georgia, U.S.A. and the continental shelf waters, *Mar. Biol.,* 43, 321, 1977.
361. **Happ, G., Gosselink, J. C., and Day, J. W., Jr.,** The seasonal distribution of carbon in a Louisiana estuary, *Estuarine Coastal Mar. Sci.,* 5, 695, 1977.
362. **Hardisky, M. A.,** A comparison of *Spartina alterniflora* primary production estimated by destructive and non-destructive techniques, in *Estuarine Perspectives,* Kennedy, V. S., Ed., Academic Press, New York, 1980, 223.
363. **Hargrave, B. T.,** Epibenthic algal production and community respiration in the sediments of Marion Lake, *J. Fish. Res. Board Can.,* 26, 2003, 1969.
364. **Hargrave, B. T.,** The effect of a deposit-feeding amphipod on the metabolism of benthic microflora, *Limnol. Oceanogr.,* 15, 21, 1970.

365. **Hargrave, B. T.,** An energy budget for a deposit-feeding amphipod, *Limnol. Oceanogr.,* 16, 99, 1971.

366. **Harlin, M. M.,** Transfer of products between epiphytic marine algae and host plants, *J. Phycol.,* 9, 243, 1973.

367. **Harlin, M. M.,** Seagrass epiphytes, in *Handbook of Seagrass Biology: an Ecosystem Perspective,* Phillips, R. C. and McRoy, C. P., Eds., Garland STPM Press, New York, 1980, 117.

368. **Harper, R. M.,** Some dynamic studies of Long Island vegetation, *Plant World,* 21, 38, 1978.

369. **Harris, P. R.,** The distribution and ecology of the interstitial meiofauna on a sandy beach at Whitsand Bay, Cornwall, *J. Mar. Biol. Assoc. U.K.,* 52, 375, 1972a.

370. **Harris, P. R.,** Seasonal changes in the meiofaunal population of an intertidal sand beach, *J. Mar. Biol. Assoc. U.K.,* 52, 389, 1972b.

371. **Harrison, P. G. and Mann, K. H.,** Chemical changes during the seasonal cycle of growth and decay in eelgrass (*Zostera marina*) on the Atlantic coast of Canada, *J. Fish. Res. Board Can.,* 32, 615, 1975a.

372. **Harrison, P. G. and Mann, K. H.,** Detritus formation from eelgrass (*Zostera marina* L.): the relative effects of fragmentation, leaching and decay, *Limnol. Oceanogr.,* 20, 924, 1975b.

373. **Hartman, R. T. and Brown, D. L.,** Changes in internal atmosphere of submersed vascular holophytes in relation to photosynthesis, *Ecology,* 48, 252, 1967.

374. **Hass, L. W. and Webb, K. L.,** Nutritional mode of several nonpigmented flagellates microflagellates from the York River Estuary, Virginia, *J. Exp. Mar. Biol. Ecol.,* 39, 125, 1979.

375. **Hatcher, B. G. and Mann, K. H.,** Above-ground production of the marsh cordgrass *(Spartina alterniflora)* near the northern end of its range, *J. Fish. Res. Board Can.,* 32(1), 83, 1975.

376. **Haven, D. S. and Morales-Alamo, R.,** Aspects of biodeposition by oysters and other invertebrate filter feeders, *Limnol. Oceanogr.,* 11(4), 487, 1966.

377. **Hayes, M. O.,** Morphology of sand accumulation in estuaries: an introduction to the symposium, in *Estuarine Research,* Vol. 2, Cronin, L. E., Ed., Academic Press, New York, 1975, 3.

378. **Head, G. C.,** Shedding of roots, in *The Shedding of Roots,* Kozlowski, T. T., Ed., Academic Press, New York, 1973, 237.

379. **Head, P. C.,** Organic processes in estuaries, in *Estuarine Chemistry,* Burton, J. D. and Liss, P. S., Eds., Academic Press, New York, 1976, 53.

380. **Heald, E. J.,** The Production of Organic Detritus in a South Florida Estuary, Ph.D. thesis, University of Miami, Coral Gables, 1969.

381. **Heald, E. J.,** The production of organic detritus in a South Florida estuary, University of Miami Sea Grant Tech. Bull. 6, Coral Gables, 1971.

382. **Hedgpeth, J. W.,** Ecological aspects of the Laguna Madre, a hypersaline estuary, in *Estuaries,* Publ. No. 83, Lauff, G., Ed., Am. Assoc. Adv. Sci., Washington, D.C., 1967, 408.

383. **Heinbokel, J. F.,** Studies on the functional role of tintinnids in the Southern California Bight. I. Grazing and growth rates in laboratory cultures, *Mar. Biol.,* 47, 177, 1978a.

384. **Heinbokel, J. F.,** Studies on the functional role of tintinnids in the Southern California Bight. II. Grazing and growth rates in field populations, *Mar. Biol.,* 47, 191, 1978b.

385. **Heinle, D. R. and Flemer, D. A.,** Carbon requirements of a population of the estuarine copepod, *Eurytemora affinis, Mar. Biol.,* 31, 235, 1975.

386. **Heinle, D. R., Flemer, D. A., and Ustach, J. F.,** Contribution of tidal marshlands to mid-Atlantic estuarine food chains, in *Estuarine Processes,* Vol. 2, Wiley, M., Ed., Academic Press, New York, 1976, 309.

387. **Heinle, D. R., Harris, R. P., Ustach, J. F., and Flemer, D. A.,** Detritus as food for estuarine copepods, *Mar. Biol.,* 40(4), 341, 1977.

388. **Heip, C. and Smol, N.,** On the importance of *Protohydra leuckarti* as a predator of meiobenthic populations, *Proc. 10th Eur. Mar. Biol. Symp. (Ostend),* 2, 285, 1976.

389. **Hellebust, J. A.,** Excretion of organic compounds by cultured and natural populations of marine phytoplankton, in *Estuaries,* Lauff, G., Ed., Am. Assoc. Adv. Sci., Washington, D.C., 1967, 361.

390. **Hellebust, J. A.,** Extracellular products, in *Algal Physiology and Biochemistry,* Stewart, W. D. P., Ed., University of California Press, Berkeley, 1974, 838.

391. **Hellebust, J. A.,** Osmoregulation, *Annu. Rev. Plant Physiol.,* 27, 485, 1976.

392. **Henriques, P. R.,** Selected Ecological Aspects of the Manukau Harbour, Ph.D. thesis, University of Auckland, New Zealand, 1977.

393. **Henriques, P. R.,** The Vegetated Tidelands of the Manukau Harbour, Rep. Manakau Harbour Study, May, 1978.

394. **Heppleston, P. B.,** Anatomical observations on the bill of the oystercatcher *(Haematopus ostralegus occidentalis)* in relation to feeding behaviour, *Condor,* 161, 519, 1970.

395. **Heppleston, P. B.,** The feeding ecology of oystercatchers *(Haematopus ostralegus* L.) in winter in northern Scotland, *J. Anim. Ecol.,* 40, 561, 1971.

396. **Hicks, G. R. F. and Coull, B. C.,** The ecology of marine meiobenthos, *Oceanogr. Mar. Biol. Annu. Rev.,* 21, 67, 1983.

397. **Hirota, J.,** Quantitative natural history of *Pleurobrachia brachei* A. Agassiz in La Jolla Bight, *Fish. Bull.,* 72, 295, 1974.

398. **Hobbie, J. E., Darley, R. J., and Jasper, S.,** Use of nuclepore filters for counting bacteria by fluorescence microscopy, *Appl. Environ. Microbiol.,* 33, 1225, 1977.

399. **Hobbie, J. E. and Lee, C.,** Microbial production of extracellular material: importance in benthic ecology, in *Marine Benthic Dynamics,* Tenore, K. R. and Coull, B. C., Eds., University of South Carolina Press, Columbia, 1980, 341.

400. **Hodgkin, E. P., Birch, P. B., Black, R. E., and Humphries, R. B.,** The Peel-Harvey Estuarine System Study (1976—1980), Rep. No. 9, Depart. of Conser. and Environ., Perth, Western Australia, 1980.

401. **Hodgkin, E. P. and Rippingdale, R. J.,** Interspecies conflicts in estuarine copepods, *Limnol. Oceanogr.,* 16, 573, 1971.

402. **Holland, A. F., Mountford, N. K., Hiegel, M. H., Kaumeyer, K. R., and Mihursky, J. A.,** Influence of predation on infaunal abundance in upper Chesapeake Bay, U.S.A., *Mar. Biol.,* 57, 221, 1980.

403. **Holm-Hansen, O. and Booth, R. C.,** The measurement of adenosine triphosphate in the ocean and its ecological significance, *Limn. Oceanogr.,* 11, 510, 1966.

404. **Holt, R. T.,** Predation apparent competition and the structure of prey communities, *Theor. Popul. Biol.,* 12, 197, 1970.

405. **Hopkins, J. T.,** The diatom trail, *J. Quekett Microsc. Club,* 30, 209, 1967.

406. **Hopkins, T. L.,** The plankton of the St. Andrews Bay System, Florida, *Publ. Inst. Mar. Sci. Univ. Tex.,* 11, 12, 1966.

407. **Hopkinson, C. S., Day, J. W., and Gael, B. T.,** Respiration studies in a Louisiana salt marsh, *Ann. Centro. Ciencias Mar Limno, Univ. Nac. Anto Mexico,* 5, 225, 1978.

408. **Hough, R. A. and Wetzel, R. G.,** The release of dissolved organic carbon from submerged aquatic macrophytes. Diel, seasonal and community relationships. *Int. Vereinigung Theoret. Angewardte Lumnol.,* 19, 939, 1975.

409. **Houghton, R. A. and Woodwell, G. M.,** The Flax Pond ecosystem study: exchanges of carbon dioxide between a salt marsh and the atmosphere, *Ecology,* 61, 1434, 1980.

410. **Howarth, R. W. and Giblin, A.,** Sulphate reduction in the marshes of Sapelo Island, *Limnol. Oceanogr.,* 28, 70, 1983.

411. **Howarth, R. W. and Hobbie, J. E.,** The regulation of decomposition and heterotrophic microbial activity in salt marsh soils: a review, in *Estuarine Comparisons,* Kennedy, V. S., Ed., Academic Press, New York, 1982, 183.

412. **Howarth, R. W. and Teal, J. M.,** Sulphate reduction in a New England salt marsh, *Limnol. Oceanogr.,* 24, 999, 1979.

413. **Howarth, R. W. and Teal, J. M.,** Energy flow in a salt marsh ecosystem: the role of reduced inorganic sulphur compounds, *Am. Nat.,* 116, 862, 1980.

414. **Howes, B. L., Howarth, R. W., Teal, J. M., and Valiela, I.,** Oxidation-reduction potentials in a salt-marsh: spatial patterns and interactions with primary production, *Limnol. Oceanogr.,* 26, 350, 1981.

415. **Hughes, R. G. and Thomas, M. L.,** The classification and ordination of shallow-water benthic samples from Prince Edward Island, Canada, *J. Exp. Mar. Biol. Ecol.,* 7, 1, 1971.

416. **Hughes, R. N., Pear, D. L., and Mann, K. H.,** Use of multivariate analysis to identify functional components of the benthos in St. Margarets Bay, Nova Scotia, *Limnol. Oceanogr.,* 17, 111, 1972.

417. **Hulbert, E. M.,** Competition for nutrients by marine phytoplankton in oceanic, coastal and estuarine regions, *Ecology,* 51, 475, 1970.

418. **Hulschur, J. B.,** Scholeksters en Lamellibranchiatea in de Waddenzee, *De Levende Nat.,* 67, 86, 1964.

419. **Huntsman, S. A. and Barker, R. T.,** Primary production off N.W. Africa: the relationship to wind and nutrient conditions, *Deep-Sea Res.,* 24, 25, 1977.

420. **Hussey, A.,** The Net Primary Production of an Essex Salt Marsh with Particular Reference to *Pucinella maritima,* Ph. D. thesis, University of Essex, Colchester, England, 1980.

421. **Jacobs, J.,** Diversity, stability and maturity in ecosystems influenced by human activities, *Proc. 1st Int. Cong. Ecol.,* 1, 94, 1974.

422. **Jannasch, W. J.,** Current concepts in aquatic microbiology, *Int. Vereinigung Theoret. Angewardte Limnol.,* 17, 25, 1969.

423. **Jannasch, H. W. and Pritchard, P. H.,** The role of inert particulate matter in the activities of aquatic microorganisms, in *Detritus and its Role in Aquatic Ecosystems,* Melchorri-Santalini, U. and Hopton, J., Eds., Memoir Inst. Italiano Idrobiol., (Suppl. 29), 1972, 289.

424. **Jansson, A.-M. and Kautsky, N.,** Quantitative survey of hard bottom communities in a Baltic archipelago, in *Biology of Benthic Organisms,* Keegan, B. F., O'Ceidigh, P., and Boaden, P. J. S., Eds., Pergamon Press, London, 1977, 359.

425. **Jansson, B.-O., Wilmot, W., and Wulff, F.,** Coupling of the subsystems — the Baltic Sea a case study, in *Flows of Energy and Materials in Marine Ecosystems,* Fasham, M. J. R., Ed., Plenum Press, New York, 1984, 549.

426. **Jansson, B. P.,** Quantitative and experimental studies of the interstitial fauna of four Swedish beaches, *Ophelia,* 5, 1, 1968.
427. **Jeffers, J. N. R.,** *An Introduction to Systems Analysis: with Ecological Implications,* Edward Arnold, London, 1978, 198 pp.
428. **Jeffries, R. L., Davy, A. J., and Rudmik, T.,** The growth strategies of coastal halophytes, in *Ecological Processes in Coastal Environments,* Jeffries, R. L. and Davy, A. J., Eds., Blackwell Scientific, Oxford, 1977, 243.
429. **Jennings, J. N. and Bird, E. C.,** Regional geomorphological charactertistics of some Australian estuaries, *Estuaries,* Publ. No. 85, Lauff, G., Ed., Am. Assoc. Adv. Sci., Washington, D.C., 1967, 121.
430. **Johnson, R. G.,** Animal-sediment relations in shallow water benthic communities, *Mar. Geol.,* 11, 93, 1971.
431. **Johnson, R. G.,** Conceptual models in benthic communities in *Models in Paleobiology,* Schopf, T. J. M., Ed., Freeman Cooper, San Franscisco, 1973, 148.
432. **Joint, I. R.,** Microbial production of an estuarine mudflat, *Estuarine Coastal Mar. Sci.,* 7, 185, 1978.
433. **Joint, I. R.,** Growth and survival of estuarine microalgae, in *Feeding and Survival Strategies of Estuarine Organisms,* Jones, N. V. and Wolff, W. J., Eds., Plenum Press, New York, 1981, 17.
434. **Joint, I. R. and Morris, R. J.,** The role of bacteria in the turnover of organic matter in the sea, *Oceanogr. Mar. Biol. Annu. Rev.,* 20, 65, 1982.
435. **Joris, C.,** On the role of heterotrophic bacteria in marine ecosystems: some problems, *Helgol. Wiss. Meeresunters.,* 30, 611, 1977.
436. **Jones, J. A.,** Primary Productivity of the Tropical Marine Turtle Grass, *Thalassia testudinum* Konig, and its Epiphytes, Doctoral dissertation, University of Miami, Fla., 1968.
437. **Jones, J. C.,** A Guide to Methods for Estimating Microbial Numbers and Biomass in Fresh Water, No. 39, Freshwater Biol. Assoc. Sci. Publ., The Ferry House, Ambleside, Cumbria, U.K., 1979.
438. **Jørgensen, B. B.,** *Biology of Suspension Feeding,* Pergamon Press, Oxford, 1966, 357 pp.
439. **Jørgensen, B. B.,** The sulfur cycle of a coastal marine sediment (Limfjorden, Denmark), *Limnol. Oceanogr.,* 22, 814, 1977a.
440. **Jørgensen, B. B.,** Bacterial sulphate reduction within reduced microniches of oxidized marine sediments, *Mar. Biol.,* 41, 7, 1977b.
441. **Jørgensen, B. B. and Fenchel, T.,** The sulfur cycle in a marine sediment model system, *Mar. Biol.,* 24, 189, 1974.
442. **Jørgensen, N. L. G.,** Uptake of L-valine and other amino acids by the polychaete *Nereis virens, Mar. Biol.,* 52, 42, 1979.
443. **Joseph, E. B.,** Analysis of a nursery ground, in *Proc. Worshop on Egg, Larval and Juvenile Stages of Fish in Atlantic Coastal Estuaries,* Tech. Publ. No. 1, Pachero, A. L., Ed., NMFS, Mid-Atlantic Coast Fisheries Center, Highlands, N.J., 1973, 118.
444. **Joshi, A. C.,** A suggested explanation of the presence of viviparity on the seashore, *J. Ecol.,* 21, 209, 1933.
445. **Joshi, G. V., Jamale, B. B., and Bhoslae, L.,** Ion regulation in mangroves, in *Proc. Int. Symp. Biol. Manage. Mangroves,* Vol. 2, Walsh, G. E., Snedaker, S. C., and Teas, H. J., Eds., Institute of Food and Agricultural Sciences, University of Florida, Gainesville, 1975, 595.
446. **Juniper, S. K.,** Stimulation of sediment bacterial grazing in two New Zealand intertidal inlets, *Bull. Mar. Sci.,* 31, 691, 1981.
447. **Juniper, S. K.,** Regulation of Microbial Production in Intertidal Mudflats — the Role of *Amphibola crenata,* a Deposit-Feeding Gastropod, Ph.D. thesis, University of Canterbury, Christchurch, New Zealand, 1982, 152 pp.
448. **Kaczynski, V. W., Feller, R. J., Clayton, J., and Gerke, R. J.,** Trophic analysis of juvenile pink and chum salmon (*Oncorhynchus gorbuscha* and *O. keta*) in Puget Sound, *J. Fish. Res. Board Can.,* 30, 1003, 1973.
449. **Kaplan, W. I., Valiela, I., and Teal, J. M.,** Denitrification in a salt marsh ecosystem, *Limnol. Oceanogr.,* 24, 726, 1979.
450. **Karl, D. M.,** Cellular nucleotide measurements and applications in microbial ecology, *Microbial. Rev.,* 44, 739, 1980.
451. **Katz, L. C.,** Effects of burrowing by the fiddler crab, *Uca pugnax* (Smith), *Estuarine Coastal Mar. Sci.,* 11, 233, 1980.
452. **Keefe, C. W.,** Marsh production; a summary of the literature, *Contrib. Mar. Sci. Univ. Tex.,* 16, 163, 1972.
453. **Keefe, C. W. and Boynton, W. R.,** Standing crop of salt marshes surrounding Chincoteague Bay, Maryland-Virginia, *Chesapeake Sci.,* 14, 117, 1973.
454. **Keefe, C. W., Boynton, W. R., and Kemp, W. M.,** A review of phytoplankton processes in estuarine environments, Unpublished rep., UMCEES Ref. No. 81-193 CBL, Chesapeake Biological Laboratory, Solomon, Md., 1981.

455. **Keller, M.,** Growth and Distribution of Eelgrass (*Zostera marina* L.) in Humboldt Bay, California, M.A. thesis, Humboldt State College, Arcata, Calif., 1963, 53 pp.

456. **Kelly, J. R. and Nixon, S. W.,** Experimental studies of the effect of organic deposition on the metabolism of a coastal marine bottom community, *Mar. Ecol. Prog. Ser.,* 17, 157, 1984.

457. **Kemp, W. M. and Boynton, W. R.,** External and internal factors regulating metabolic rates in an estuarine benthic community, *Oecologia,* 51, 19, 1981.

458. **Kemp, W. M., Wetzel, R. L., Boynton, W. R., D'Elia, C. F., and Stephenson, J. C.,** Nitrogen cycling and estuarine interfaces: some current concepts and research directions, in *Estuarine Comparisons,* Kennedy, V. S., Ed., Academic Press, New York, 1982, 209.

459. **Kepkay, P. E., Cooke, R. C., and Novitsky, J. A.,** Microbial autotrophy: a primary source of organic carbon in marine sediments, *Science,* 204, 68, 1979.

460. **Kepkay, P. E. and Novitsky, J. A.,** Microbial control of organic carbon in marine sediments: coupled chemoautotrophy and heterotrophy, *Mar. Biol.,* 55, 261, 1980.

461. **Ketchum, G. H.,** Phytoplankton nutrients in estuaries, in *Estuaries,* Publ. No. 83, Lauff, G. H., Ed., Am. Assoc. Adv. Sci., Washington, D.C., 1967, 329.

462. **Khafji, A. K. and Norton, T. A.,** The effects of salinity on the distribution of *Fucus ceranoides, Estuarine Coastal Mar. Sci.,* 8, 433, 1979.

463. **Khailov, K. M. and Burlakova, Z. P.,** Release of dissolved organic matter by marine seaweeds and distribution of their total organic production to inshore communities, *Limnol. Oceanogr.,* 14, 521, 1969.

464. **King, C. M.,** *Introduction to Marine Geology and Geomorphology,* Edward Arnold, London, 1975, 370 pp.

465. **King, G. M. and Klug, M. J.,** The relationship of soil water movement, sulphide concentration and *Spartina alterniflora* production in a Georgia salt marsh, Unpublished manuscript, Kellogg Biol. Stn., Michigan State University, Hickory Corners, Minn., 1982.

466. **King, G. M. and Wiebe, W. J.,** Methane release from soils in a Georgia salt marsh, *Geochim. Cosmochim. Acta,* 42, 343, 1978.

467. **Kirby, C. J.,** The Annual Net Primary Production and Decomposition of the Salt Marsh Grass *Spartina alterniflora* Loisel in the Barataria Bay Estuary of Louisiana, Ph.D. dissertation, Louisiana State University, Baton Rouge, 1972, 87 pp.

468. **Kirby, C. J. and Gosselink, J. C.,** Primary production in a Louisiana Gulf Coast *Spartina alterniflora* marsh, *Ecology,* 47, 1052, 1976.

469. **Kinne, O.,** Salinity. Invertebrates, in *Marine Ecology,* Vol. 1 (Part 2), Kinne, O., Ed., Wiley-Interscience, New York, 1971, 800.

470. **Kita, T. and Harada, E.,** Studies on the epiphytic communities. I. Abundance and distribution of microalgae and small animals on *Zostera* blades, *Publ. Seto Mar. Biol. Lab.,* 10, 245, 1962.

471. **Kjeldsen, C. K. and Pinney, H. K.,** Effects of variations in salinity and temperature on some estuarine macroalgae, in *Proc. 7th Int. Seaweed Symp.,* 1972, 301.

472. **Klug, M. J.,** Detritus-decomposition relationships, in *Handbook of Seagrass Biology: an Ecosystem Perspective,* Phillips, R. C. and McRoy, C. P., Eds., Garland STPM Press, New York, 1980, 225.

473. **Knauer, G. A. and Ayers, A. V.,** Changes in carbon, nitrogen and adenosine triphosphate, and chorophyll *a* in decomposing *Thalassia testudinum, Limnol. Oceanogr.,* 23, 408, 1977.

474. **Knox, G. A.,** Effects of Freezing Works Effluent on Shakespeare Bay, Queen Charlotte Sound, Rep. No. 2, Estuarine Research Unit, University of Canterbury, Christchurch, New Zealand, 1973, 33 pp.

475. **Knox, G. A.,** The role of polychaetes in benthic soft-bottom communities, in *Essays on Polychaetous Annelids in Memory of Dr. Olga Hartman,* Reisch, D. J. and Fauchild, K., Eds., Allan Hancock Foundation, University of Los Angeles, 1977, 547.

476. **Knox, G. A.,** The estuarine zone: an overview, *Soil Water,* April, p.13, 1980.

477. **Knox, G. A.,** *Estuarine Ecology. Upper Waitemata Harbour Catchment Study Specialist Report,* Auckland Regional Authority, Auckland, 1983a, 151 pp.

478. **Knox, G. A.,** *Energy Analysis: Upper Waitemata Harbour Catchment Study Specialist Report,* Auckland Regional Authority, Auckland, 1983b, 266.

479. **Knox, G. A.,** The ecology of the Kaituna marsh, Pelorus Sound, with special reference to the introduced cord grass, *Spartina,* in *Nutrient Processing and Biomass Production in New Zealand Estuaries,* Gillespie, P. A., Ed., Water and Soil Misc. Publ. No. 60, Water and Soil Conservation Organization, Wellington, 1983c, 32.

480. **Knox, G. A.,** Primary productivity of three emergent macrophyte species, *Spartina townsendii, Juncus maritimus* and *Leptocarpus simplex* in Pelorus Sound, New Zealand, Unpublished manuscript, University of Canterbury, Christchurch, New Zealand, 1984.

481. **Knox, G. A. and Bolton, L. A.,** The Ecology of Shakespeare Bay, Queen Charlotte Sound, New Zealand, Rep. No. 8, Estuarine Research Unit, University of Canterbury, Christchurch, New Zealand, 1979, 141 pp.

482. **Knox, G. A., Bolton, L. A., and Fenwick, G. D.,** The Ecology of Brooklands Lagoon and the Waimakariri River Estuary, Rep. No. 16, Estuarine Research Unit, University of Canterbury, Christchurch, New Zealand, 1978, 125 pp.

483. **Knox, G. A. and Fenwick, G. D.,** A Quantitative Study of the Benthic Fauna off Clive, Hawke Bay, Rep. No. 14, Estuarine Research Unit, University of Canterbury, Christchurch, New Zealand, 1978, 91 pp.

484. **Knox, G. A. and Fenwick, G. D.,** Zonation of inshore benthos off a sewage outfall in Hawke Bay, New Zealand, *N. Z. J. Mar. Freshwater Res.,* 15, 417, 1982.

485. **Knox, G. A. and Kilner, A. R.,** The Ecology of the Avon-Heathcote Estuary, Rep. No. 1, Estuarine Research Unit, University of Canterbury, Christchurch, New Zealand, 1973, 358 pp.

486. **Korringa, P.,** Estuarine fishes of Europe as affected by man's multiple activities, in *Estuaries,* Publ. No. 83, Lauff, G., Ed., Am. Assoc. Adv. Sci., Washington, D.C., 1967, 658.

487. **Korringa, P.,** The edge of the North Sea as a nursery ground and shellfish area, in *North Sea, Science,* Goldberg, E. D., Ed., MIT Press, Cambridge, Mass., 1973, 361.

488. **Kraeuter, J. N. and Wolf, P. L.,** The relationship of marine macroinvertebrates to salt marsh plants, in *Ecology of Halophytes,* Reimold, R. J. and Queen, W. H., Eds., Academic Press, New York, 1974, 449.

489. **Krank, K.,** Sediment deposition from flocculated suspensions, *Sedimentology,* 22, 111, 1975.

490. **Krebs, C. T. and Valiela, I.,** Effect of experimentally applied chlorinated hydrocarbons on the biomass of the fiddler crab, *Uca pugnax* (Smith), *Estuarine Coastal Mar. Sci.,* 6, 375, 1978.

491. **Kremer, J. N. and Nixon, S. W.,** *A Coastal Marine Ecosystem: Simulation and Analysis,* Springer-Verlag, Berlin, 1978, 217 pp.

492. **Kremer, P.,** Population dynamics and ecological energetics of a pulsed zooplankton predator, the ctenophore *Mnemiopsis leidyi,* in *Estuarine Processes,* Vol. 1, Wiley, M., Ed., Academic Press, New York, 1976, 197.

493. **Kruczynski, W. L., Subrahmanyam, C. B., and Drake, S. H.,** Studies on the plant community of a north Florida salt marsh, *Bull. Mar. Sci.,* 28, 316, 1978.

494. **Kuenzler, E. J.,** Structure and energy flow of a mussel population in a Georgia salt marsh, *Limnol. Oceanogr.,* 6, 191, 1961a.

495. **Kuenzler, E. J.,** Phosphorus budget of a mussel population, *Limnol. Oceanogr.,* 6, 400, 1961b.

496. **Kuenzler, E. J.,** Mangrove swamp systems, in *Coastal Ecological Systems of the United States,* Odum, H. T., Copeland, B. J., and McMahon, E. A., Eds., Institute of Marine Sciences, University of North Carolina, Chapel Hill, 1968, 353.

497. **Kutner, M. B.,** Seasonal variation and phytoplankton distribution in Cananéia region, Brazil, Publ. 361, Inst. Oceanogr., University of Sao Paulo, Brazil, 1976.

498. **Lasker, R.,** Feeding, growth, respiration, and carbon utilization of a euphausiid crustacean, *J. Fish. Res. Board Can.,* 23, 1291, 1966.

499. **Laszlo, E. and Morgenau, H.,** The emergence of integrating concepts in contemporary science, *Philos. Sci.,* 39, 252, 1972.

500. **Lauff, G.,** *Estuaries,* Publ. No. 83, Am. Assoc. Adv. Sci., Washington, D.C., 1967.

501. **Leach, J. H.,** Epibenthic algal production in an intertidal mudlfat, *Limnol. Oceanogr.,* 15, 514, 1970.

502. **Lee, H.,** Predation and Opportunism in Tropical Soft-Bottom Communities, Ph.D. thesis, University of North Carolina, Chapel Hill, 1978.

503. **Lee, W. Y. and McAlice, B. J.,** Sampling variability of marine zooplankton in a tidal estuary, *Estuarine Coastal Mar. Sci.,* 8, 565, 1979.

504. **Leontif, W. W.,** *Input-Output Economics,* Oxford University Press, London, 1966, 257 pp.

505. **Letts, E. A. and Richards, E. H.,** Report on green seaweeds (especially *Ulva latissima*) in relation to the pollution of waters in which they occur, R. Comm. Sewage Disposal, 5th Rep., Appendix 3, Section 2, 1911.

506. **Levins, R.,** *Evolution in Changing Environments,* Princeton University Press, Princeton, N.J., 1969, 120 pp.

507. **Levinton, J. S.,** The paleoecological significance of opportunistic species, *Lethaia,* 3, 69, 1970.

508. **Levinton, J. S.,** Stability and trophic structure in deposit-feeding and suspension-feeding communities, *Am. Nat.,* 106, 472, 1972.

509. **Levinton, J. S.,** The ecology of deposit-feeding communities: Quisset Harbor, Massachusetts, in *Ecology of Marine Benthos,* Coull, B. C., Ed., University of South Carolina Press, Columbia, 1977, 191.

510. **Levinton, J. S.,** Deposit-feeders, their responses, and the study of resource limitation, in *Ecological Processes in Coastal and Marine Systems,* Livingston, R. J., Ed., Plenum Press, New York, 1979a, 117.

511. **Levinton, J. S.,** The effect of density upon deposit-feeding populations, movement, feeding and floating of *Hydrophobia ventrosa* Montagu (Gastropoda: Prosobranchia), *Oecologia,* 43, 27, 1979b.

512. **Levinton, J. S.,** Particle feeding by deposit feeders: models, data and a prospectus, in *Marine Benthic Dynamics,* Tenore, K. R. and Coull, B. C., Eds., University of South Carolina Press, Columbia, 1980, 423.

513. **Levinton, J. S.,** The body size-prey size hypothesis: the adequacy of body size as a vehicle for character displacement, *Ecology,* 63, 869, 1982.

514. **Levinton, J. S. and Bianchi, T. S.,** Nutrition and food limitation of deposit-feeders. I. The role of microbes in the growth of mud snails (Hydrobiidae), *J. Mar. Res.,* 39, 531, 1981.

515. **Levinton, J. S. and Lopez, G. R.,** A model of renewable resources and limitation of deposit-feeding benthic populations, *Oecologia,* 31, 177, 1977.

516. **Levinton, J. S., Lopez, G. R., Lassen, H. H., and Rahu,** U., Feedback and structure of deposit-feeding marine benthic communities, in *11th Eur. Symp. in Mar. Biol.,* Keegan, B. F., Ceidigh, P. O., and Boaden, P. J. S., Eds., Pergamon Press, Oxford, 1977, 409.

517. **Levitt, J.,** *Responses of Plants to Environmental Stresses,* Academic Press, New York, 1972.

518. **Lindall, W. N., Jr. and Soloman, C. H.,** Alteration and destruction of estuaries affecting fishery resources of the Gulf of Mexico, *Mar. Fish. Rev.,* 39, 1, 1977.

519. **Linley, E. A. S. and Field, J. G.,** The nature and ecological significance of bacterial aggregation in a nearshore upwelling ecosystem, *Estuarine Coastal Shelf Sci.,* 14, 1, 1982.

520. **Linthurst, R. A.,** An Evaluation of Biomass, Stem Density, Net Aerial Primary Production (NAPP) and NAPP Methodology for Selected Estuarine Angiosperms in Maine, Delaware and Georgia, M.S. thesis, North Carolina State, Raleigh, 1977.

521. **Linthurst, R. A.,** The effects of aeration on the growth of *Spartina alterniflora* Losiel, *Am. J. Bot.,* 66, 685, 1979.

522. **Linthurst, R. A.,** A growth comparison of *Spartina alterniflora* Loisel ecophenes under aerobic and anaerobic conditions, *Am. J. Bot.,* 67, 883, 1980a.

523. **Linthurst, R. A.,** An evaluation of aeration, nitrogen, pH and salinity as factors affecting *Spartina alterniflora* growth: summary, in *Estuarine Perspectives,* Kennedy, V. S., Eds., Academic Press, New York, 1980b, 235.

524. **Linthurst, R. A. and Reimold, R. J.,** An evaluation of methods for estimating the net aerial primary production of estuarine angiosperms, *J. Appl. Ecol.,* 15, 919, 1978a.

525. **Linthurst, R. A. and Reimold, R. J.,** Estimated net aerial primary productivity for selected marine angiosperms in Maine, Delaware and Georgia, *Ecology,* 59, 945, 1978b.

526. **Linthurst, R. A. and Seneca, E. D.,** Aeration, nitrogen and salinity as determinants of *Spartina alterniflora* Loisel growth response, *Estuaries,* 4, 53, 1981.

527. **Litchfield, C. D. and Seyfried, P. L., Eds.,** Methodology for Biomass Determinations and Microbial Activities in Sediments, ASTM STP 673, American Society for Testing and Materials, Philadelphia, Pa., 1979, 199 pp.

528. **Livingston, R. J.,** Diurnal and seasonal fluctuations of organisms in a north Florida estuary, *Estuarine Coastal Mar. Sci.,* 4, 472, 1976.

529. **Livingston, R. J., Kobylinski, G. J., Lewis, F. G., III, and Sheridan, P. F.,** Long-term fluctuations of epibenthic fish and invertebrate populations in Appachicola Bay, Florida, *Fish. Bull.,* 74, 311, 1976.

530. **Lockwood, S. J.,** An Ecological Study of an O-Group Plaice (*Pleuronectes platessa* L.) Population, Filey Bay, Yorkshire, Ph.D. thesis, University of East Anglia, England, 1972.

531. **Long, S. P. and Mason, C. F.,** *Saltmarsh Ecology,* Blackie and Son, Glasgow, 1983, 160 pp.

532. **Long, S. P. and Woolhouse, H. W.,** Primary production in *Spartina* marshes, in *Ecological Processes in Coastal Environments,* Jeffries, R. J. and Davy, A. J., Eds., Blackwell Scientific, Oxford, 1983, 333.

533. **Longbottom, M. R.,** The distribution of *Arenicola marina* (L.) with particular reference to the effect of particle size and organic matter of the sediments, *J. Exp. Mar. Biol. Ecol.,* 5, 138, 1970.

534. **Loosanoff, V. L., Davis, H. C., and Chanley, P. E.,** Dimensions and shapes of larvae of some marine bivalve molluscs, *Malacologia,* 4, 351, 1966.

535. **Lopez, G. L. and Kofoed, L. H.,** Epipsammic browsing and deposit-feeding in mud snails (Hydrobiidae), *J. Mar. Res.,* 38, 585, 1980.

536. **Lopez, G. R. and Levinton, J. S.,** The availability of microorganisms attached to sediment particles as food for *Hydrobia ventrosa* Montagu (Gastropoda: Prosobranchia), *Oecologia,* 32, 236, 1978.

537. **Lopez, G. R., Levinton, J. S., and Slobodkin, L. B.,** The effect of grazing by the detritivore *Orchestia grillus* on *Spartina* litter and its associated microbial community, *Oecologia,* 30, 111, 1977.

538. **Lord, C. T., III,** The Chemistry and Cycling of Iron Manganese and Sulfur on Salt Marsh Sediments, Ph.D. dissertation, University of Delaware, Newark, 1980.

539. **Lovelund, R. E., Brauner, J. F., Taylor, J. E., and Kennish, M. J.,** Macroflora, in *Ecology of Barnegat Bay, New Jersey,* Kennish, M. J. and Lutz, R. A., Eds., Springer-Verlag, New York, 1984, 78.

540. **Lugo, A. E., Evink, G., Brinson, M. M., Brace, A., and Snedaker, S. C.,** Diurnal rates of photosynthesis, respiration and transpiration in mangrove forests of south Florida, in *Tropical Ecological Systems,* Golley, M. F. B. and Mendina, E., Eds., Springer-Verlag, New York, 1975, 535.

541. **Lugo, A. E., Sell, M., and Snedaker, S. C.,** Mangrove ecosystem analysis, in *Systems Analysis 3,* Pattern, B. C., Ed., Academic Press, New York, 1974, 114.

542. **Lugo, A. E. and Snedaker, S. C.,** The ecology of mangroves, *Annu. Rev. Ecol. Syst.,* 5, 170, 1974.
543. **Lugo, A. E. and Snedaker, S. C.,** Properties of a mangrove forest in southern Florida, in *Proc. Int. Symp. Biol. Manage. Mangroves,* Vol. 1, Walsh, G. E., Snedaker, S. C., and Teas, H. J., Eds., Institute of Food and Agricultural Sciences, University of Florida, Gainesville, 1975, 170.
544. **Lund, E. J.,** Effects of bleedwater, "soluble fraction" and crude oil on the oyster, *Publ. Int. Mar. Sci. Univ. Tex.,* 4, 296, 1957.
545. **MacArthur, R. H.,** *Geographical Ecology,* Harper & Row, New York, 1972, 269 pp.
546. **MacArthur, R. H. and Wilson, E. O.,** *Theory of Island Biogeography,* Princeton University Press, Princeton, N.J., 1967, 203 pp.
547. **McCall, P. L.,** Community patterns and adaptive strategies of the infaunal benthos of Long Island Sound, *J. Mar. Res.,* 35, 221, 1977.
548. **McCarthy, J. J., Taylor, W. R., and Loftus, M. E.,** Significance of nannoplankton in the Chesapeake Bay estuary and problems associated with the measurement of nannoplankton productivity, *Mar. Biol.,* 24, 7, 1974.
549. **McClatchie, S., Juniper, S. K., and Knox, G. A.,** Structure of a mudflat diatom community in the Avon-Heathcote Estuary, New Zealand, *N. Z. J. Mar. Freshwater Res.,* 16, 299, 1982.
550. **McDougall, B. M. and Rovira, A. D.,** Movement of ^{14}C-photosynthate into roots of wheat seedlings and exudation of ^{14}C from intact roots, *New Phytol.,* 69, 37, 1970.
551. **McHugh, J. L.,** Estuarine nekton, in *Estuaries,* Publ. No. 83, Lauff, G., Ed., Am. Assoc. Adv. Sci., Washington, D. C., 1967, 581.
552. **McHugh, J. L.,** Estuarine fisheries: are they doomed?, in *Estuarine Processes,* Vol. 1, Wiley, M., Ed., Academic Press, New York, 1976, 15.
553. **McIntyre, A. D.,** Ecology of the marine meiobenthos, *Biol. Rev.,* 44, 245, 1969.
554. **McIntyre, A. D.,** The range of biomass in intertidal sand with special reference to the bivalve *Tellina tenuis, J. Mar. Biol. Assoc., U.K.,* 50, 561, 1970.
555. **McIntire, C. D.,** Diatom assemblages in Yaquina estuary, Oregon: a multivariate analysis, *J. Phycol.,* 9, 254, 1973.
556. **McIntire, C. D. and Reimer, C. W.,** Some marine and brackish water *Achanthes* from Yaquina estuary, Oregon (U.S.A.), *Bot. Mar.,* 17, 164, 1974.
557. **McLachlan, A., Winter, D., and Botha, L.,** Vertical and horizontal distribution of sublittoral meiofauna in Angola Bay, South Africa, *Mar. Biol.,* 40, 355, 1977.
558. **McLean, R. O., Corrigan, J., and Webster, J.,** Heterotrophic nutrition in *Melosira nummouloides,* a possible role in affecting distribution in the Clyde estuary, *Br. J. Phycol.,* 16, 95, 1981.
559. **McLusky, D. S.,** *The Estuarine Ecosystem,* Blackie and Son, Glasgow, 1981, 150 pp.
560. **McMillan, C.,** Salt tolerance of mangroves and submerged aquatic plants, in *Ecology of Halophytes,* Reimold, R. J. and Queen, W. H., Eds., Academic Press, New York, 1974, 379.
561. **MacNae, W.,** A general account of the fauna and flora of mangrove swamps and forests in the Indo-West Pacific Region, *Adv. Mar. Biol.,* 6, 73, 1968.
562. **McRoy, C. P.,** The Standing Stock and Ecology of Eelgrass, *Zostera marina,* Izembek Lagoon, Alaska, Masters thesis, University of Washington, Seattle, 1966, 138 pp.
563. **McRoy, C. P.,** On the Biology of Eelgrass in Alaska, Ph.D. dissertation, University of Alaska, College, 1970a, 156 pp.
564. **McRoy, C. P.,** Standing stocks and related features of eelgrass populations in Alaska, *J. Fish. Res. Board Can.,* 27, 1811, 1970b.
565. **McRoy, C. P.,** Seagrass productivity: carbon uptake experiments in eelgrass, *Zostera marina, Aquaculture,* 4, 131, 1974.
566. **McRoy, C. P. and Barsdate, R. J.,** Phosphate absorption in eelgrass, *Limnol. Oceanogr.,* 15, 14, 1970.
567. **McRoy, C. P., Barsdate, R. J., and Nebert, M.,** Phosphorus cycling in an eel-grass (*Zostera marina* L.) ecosystem, *Limnol. Oceanogr.,* 17, 58, 1972.
568. **McRoy, C. P. and Goering, J. J.,** Nutrient transfer between the seagrass *Zostera marina* and its epiphytes, *Nature,* 248, 173, 1974.
569. **McRoy, C. P. and Helfferich, C., Eds.,** *Seagrass Ecosystems: A Scientific Perspective,* Marcel Dekker, New York, 1977, 314 pp.
570. **McRoy, C. P. and McMillan, C.,** Production ecology and physiology in seagrasses, in *Seagrass Ecosystems: A Scientific Perspective,* McRoy, C. P. and Helfferich, C., Eds., Marcel Dekker, New York, 1977, 53.
571. **Main, S. P. and McIntire, C. D.,** The distribution of epiphytic diatoms in Yaquina estuary, Oregon, *Bot. Mar.,* 17, 88, 1974.
572. **Majock, W., Cragie, J. S., and McLachlan, J.,** Photosynthesis in algae. I. Accumulation of products in the Rhodophyceae, *Can. J. Bot.,* 44, 541, 1966.
573. **Mangum, C. P.,** Studies on speciation in maldanid polychaetes of the North American Atlantic coast. II. Distribution and competitive interaction of five sympatric species, *Limnol. Oceanogr.,* 9, 12, 1964.

574. **Mangum, C. P.,** Respiratory physiology of annelids, *Am. Sci.,* 58, 641, 1970.

575. **Mann, K. H.,** Macrophyte production and detritus food chains in coastal waters, in *Detritus and its Role in Aquatic Ecosystems,* Melchorri-Santalini, U. and Hopton, J., Eds., Memoir Inst. Italiano Idriobiol., (Suppl. 29), 1972, 325.

576. **Mann, K. H.,** *The Ecology of Coastal Waters,* Blackwell Scientific, Oxford, 1982, 322 pp.

577. **Marchant, C. J.,** Evolution of *Spartina* (Gramineae). I. The history and morphology of the genus in Britain, *J. Linn. Soc. London, Bot.,* 60, 1, 1967.

578. **Margalef, R.,** El ecosistema, in *Ecologica Marina,* Foundation la salle de ciencias naturales, Caracas, 1967, 377.

579. **Marmer, H. A.,** The currents in Barataria Bay, Texas A & M Research Foundation, College Station, Tex., 1948, 30 pp.

580. **Marshall, N.,** Food transfer through the lower trophic levels of the benthic environment, in *Marine Food Chains,* Steele, J. H., Ed., Oliver and Boyd, Edinburgh, 1970, 52.

581. **Marshall, B. E. and Park, R. B.,** The ecotone between *Spartina foliosa* Trin. and *Salicornia virginica* L. in salt marshes in northern San Fransciso Bay. I. Biomass and production, *J. Ecol.,* 64, 421, 1976.

582. **Marshall, N., Durbin, A. G., Gerber, R., and Telek, G.,** Observations on particulate and dissolved organic matter in coral reef areas, *Int. Rev. Gesamten Hydrobiol. Hydrogr.,* 60, 335, 1975.

583. **Marshall, N., Oviatt, C. A., and Skauen, D. M.,** Productivity of the benthic microflora of shoal and estuarine environments in southern New England, *Int. Rev. Gesamten Hydrobiol. Hydrogr.,* 56, 947, 1971.

584. **Mason, C. F. and Bryant, R. J.,** Production, nutrient content and decomposition of *Phragmites communis* Trin. and *Typha angustifolia* L., *J. Ecol.,* 63, 71, 1975.

585. **Massalski, A. and Leppard, G. G.,** Morphological examination of fibrillar colloids associated with algae and bacteria in lakes, *J. Fish. Res. Board Can.,* 36, 922, 1979.

586. **Maturo, F. J. S., Jr.,** A supplementary zooplankton survey at the Crystal River Plant site, in Environmental Status Report, October 1972—March, 1973, Florida Power Corp., St. Petersburg, Fla., 1973, 36.

587. **Mendelssohn, I. A.,** The influence of nitrogen level, form and application method on the growth response of *Spartina alterniflora* in North Carolina, *Estuaries,* 2, 106, 1979.

588. **Mendelssohn, I. A., McKee, K. L., and Patrick, W. H., Jr.,** Oxygen deficiency in *Spartina alterniflora:* metabolic adaptation to anoxia, *Science,* 214, 439, 1981.

589. **Mendelssohn, I. A. and Seneca, E. D.,** The influence of soil drainage on the growth of salt marsh cordgrass *Spartina alterniflora* in North Carolina, *Estuarine Coastal Mar. Sci.,* 11, 27, 1980.

590. **Meyer-Reil, L.-A., Bölter, M., Liebezeit, G., and Schramm, W.,** Short-term variations in microbiological and chemical parameters, *Mar. Ecol. Prog. Ser.,* 1, 1, 1979.

591. **Meyers, A. C.,** Sediment Reworking, Tube Building, and Burrowing in a Shallow Subtidal Marine Bottom Community: Rates and Effects, Ph.D. dissertation, University of Rhode Island, Kingston, 1973.

592. **Meyers, S. P.,** Degradative activities of filamentous marine fungi, in *Biodeterioration of Materials Microbiology and Allied Aspects,* Walter, A. H. and Elphick, J. J., Eds., Proc. 1st Int. Biodeterioration Symp., Elsevier, New York, 1968, 594.

593. **Mihursky, J. A., Heinle, D. R., and Boynton, W. R.,** Ecological Effects of Nuclear Steam Electric Station Operations on Estuarine Systems, (Unpublished Rep.) UMCEES Ref. No. 77-28-CBL, Chesapeake Biological Laboratory, Solomons, Md., 1977.

594. **Miller, C. B.,** The zooplankton of estuaries, in *Estuaries and Enclosed Seas,* Vol. 26, Ketchum, B. H., Ed., Elsevier, Amsterdam, 1983, 103.

595. **Miller, P. C.,** Bioclimate, leaf temperature, and primary production in red mangrove canapies in South Florida, *Ecology,* 53, 22, 1972.

596. **Milliman, J. D., Summerhayes, C. P., and Barrelto, H. T.,** Oceanography and suspended matter off the Amazon River, February — March, 1973, *J. Sediment. Petrol.,* 45, 189, 1975.

597. **Mills, E. L.,** The community in concept in marine ecology with comments on continua and instability in some marine communities: a review, *J. Fish. Res. Board Can.,* 24, 305, 1969.

598. **Milne, H. and Dunnet, G.,** Standing crop, productivity and trophic relationships of the fauna in the Ythan estuary, in *The Estuarine Environment,* Barnes, R. S. K. and Green, J., Eds., Applied Science, London, 1972, 86.

599. **Milner, C. and Hughes, R. E.,** *Methods for the Measurement of the Primary Production of Grasslands,* IBP Handbook No. 6, Blackwell Scientific, Oxford, 1968, 70 pp.

600. **Moebus, K. and Johnson, K. M.,** Exudation of dissolved organic carbon by brown algae, *Mar. Biol.,* 26, 117, 1974.

601. **Montague, C. L.,** A natural history of the temperate western Atlantic fiddler crabs (Genus *Uca*) with reference to their impact on the salt marsh, *Contrib. Mar. Sci. Univ. Tex.,* 23, 25, 1980.

602. **Montague, C. L.,** The influence of fiddler crab burrows and burrowing on metabolic processes in salt marsh sediments, in *Estuarine Comparisons,* Kennedy, V. S., Ed., Academic Press, New York, 1982, 283.

603. **Moore, D., Basher, H. A., and Trent, L.,** Relative abundance, seasonal distribution and species composition of demersal fishes off Louisiana and Texas, 1962—1964, *Contrib. Mar. Sci. Univ. Tex.,* 15, 45, 1970.
604. **Moore, L. B.,** A "loose-lying" form of the brown alga *Hormosira, Trans. R. Soc. N. Z.,* 78, 48, 1950.
605. **Moore, R. T., Miller, P. C., Ehleringer, J., and Lawrence, W.,** Seasonal trends in gas exchange characteristics of three mangrove species, *Photosynthetica,* 7, 387, 1973.
606. **Mooring, M. T., Cooper, A. W., and Seneca, E. D.,** Seed germination response and evidence for height ecophenes in *Spartina alterniflora* from North Carolina, *Am. J. Bot.,* 58, 48, 1971.
607. **Morgan, M. H.,** Annual Angiosperm Production on a Salt Marsh, M.S. thesis, University of Delaware, Newark, 1961.
608. **Morgans, J. F. C.,** The biology of the Heathcote-Avon Estuary, in *The Natural History of Canterbury,* Knox, G. A., Ed., A. H. and A. W. Reed, Wellinton, 1969, 553.
609. **Morris, I. and Glover, H. E.,** Questions of the mechanism of temperature adaptation in marine phytoplankton, *Mar. Biol.,* 24, 147, 1974.
610. **Morris, J. C. and Stumm, W.,** Redox equilibria and measurements of potentials in the aquatic environment, in *Equilibrium Concepts in the Aquatic Environment,* Gould, R. F., Ed., American Chemical Society Publication, Columbus, Ohio, 1967, 270.
611. **Morris, J. T.,** The Nitrogen Uptake Kinetics and Growth Responses of *Spartina alterniflora,* Ph.D. dissertation, Yale University, New Haven, Conn., 1979, 102 pp.
612. **Morris, J. T.,** The nitrogen uptake kinetics of *Spartina alterniflora* in culture, *Ecology,* 61, 1114, 1980.
613. **Morton, J. E. and Miller, M. C.,** *The New Zealand Seashore,* Collins, London, 1968, 638 pp.
614. **Muus, K.,** Settling growth and mortality of the young bivalves in the Øresund, *Ophelia,* 12, 79, 1973.
615. **Myers, A. C.,** Sediment Reworking, Tube Building and Burrowing in a Shallow Subtidal Marine Bottom Community: Rates and Effects, Ph.D. dissertation, University of Rhode Island, Kingston, 1973, 117 pp.
616. **Naimen, R. J. and Sibert, J. R.,** Transport of nutrients and carbon from the Nanaimo River and its estuary, *Limnol. Oceanogr.,* 23, 102, 1978.
617. **Naimen, R. J. and Sibert, J. R.,** Detritus and juvenile salmon production in the Nanaimo Estuary. III. Importance of detrital carbon in the estuarine ecosystem, *J. Fish. Res. Board Can.,* 36, 504, 1979.
618. **Naqvi, S. M.,** Effects of predation on infaunal invertebrates of Alligator Harbor, Florida, *Gulf Res. Rep.,* 2, 313, 1968.
619. **Neihof, R. A. and Loeb, G. I.,** The surface charge of particulate matter in seawater, *Limnol. Oceanogr.,* 17, 7, 1972.
620. **Nelson, B. W.,** Important aspects of estuarine sediment chemistry, in *Symp. on the Environ. Chemistry of Mar. Sediments,* Occas. Publ. No. 1, Marshall, N. Ed., Narragansett Marine Laboratory, University of Rhode Island, Kingston, 1962, 27.
621. **Nestler, J.,** Interstitial salinity as a cause of ecospheric variation in *Spartina alterniflora, Estuarine Coastal Mar. Sci.,* 5, 707, 1977a.
662. **Odum, W. E. and Heald, E. T.,** The detritus-based food web of an estuarine mangrove community, in *Estuarine Research,* Vol. 1, Cronin, L. E., Ed., Academic Press, New York, 1975, 217.
623. **Newell, R. C.,** The role of detritus in the nutrition of two marine deposit feeders, the prosobranch *Hydrobia ulvae* and the bivalve *Macoma balthica, Proc. Zool. Soc. London,* 144, 25, 1965.
624. **Newell, R. C.,** *The Biology of Intertidal Animals,* 3rd ed., Marine Ecological Surveys, Faversham, Kent, 1979, 781 pp.
625. **Nicholaisen, W. and Kanneworf, E.,** On the burrowing and feeding habits of the amphipods *Bathyporeia pilosa* Lindstrom and *Bathyporeia sarsi* Watkin, *Ophelia,* 6, 231, 1969.
626. **Nixon, S. W.,** Between coastal marshes and coastal waters — a review of twenty years of speculation and research on the role of salt marshes in estuarine productivity and water chemisty, in *Estuarine and Wetland Processes,* Hamilton, P. and MacDonald, K., Eds., Plenum Press, New York, 1980, 437.
627. **Nixon, S. W.,** Remineralization and nutrient cycling in coastal marine ecosystems, in *Nutrient Enrichment in Estuaries,* Neilson, B. and Cronin, L. E., Eds., Humana Press, Clifton, N.J., 1981a, 111.
628. **Nixon, S. W.,** Freshwater inputs and estuarine productivity, in *Proc. Natl. Symp. of Freshwater Inflow to Estuaries,* FWS/OBS-81/04 Cross, R. and Williams, D., Eds., Office of Biological Services, U.S. Fish and Wildlife Service, 1981b, 31.
629. **Nixon, S. W. and Oviatt, C. A.,** Preliminary measurements of midsummer metabolism in beds of eelgrass *Zostera marina, Ecology,* 53, 102, 1972.
630. **Nixon, S. W. and Oviatt, C. A.,** Ecology of a New England salt marsh, *Ecol. Monogr.,* 43, 463, 1973.
631. **Nixon, S. W., Oviatt, C., and Hale, S.,** Nitrogen regeneration and metabolism of coastal marine bottom communities, in *The Role of Terrestrial and Aquatic Organisms in Decomposition Processes,* Anderson, J. and Macfadyen, J., Eds., Blackwell Scientific, London, 1976, 269.
632. **Norton-Griffiths, M.,** Some ecological aspects of the feeding behaviour of the oystercatcher *(Haematopus ostralegus)* on the edible mussel, *(Mytilus edulis), Ibis,* 109, 412, 1967.

633. **Novitsky, J. A. and Kepkay, P. E.,** Patterns of microbial heterotrophy through changing environments in a marine sediment, *Mar. Ecol. Prog. Ser.,* 4, 1, 1981.

634. **Odum, E. P.,** The strategy of ecosystem development, *Science,* 164, 262, 1969.

635. **Odum, E. P.,** *Fundamentals of Ecology,* W. B. Saunders, Philadelphia, 1971, 547 pp.

636. **Odum, E. P.,** Halophytes, energetics and ecosystems, in *Ecology of Halophytes,* Reimold, R. J. and Queen, W. H., Eds., Academic Press, New York, 1974, 599.

637. **Odum, E. P.,** *Ecology,* 2nd ed., Holt, Reinhart & Winston, New York, 1975, 243 pp.

638. **Odum, E. P.,** *Basic Ecology,* Saunders College Publishing, Philadelphia, 1983, 613 pp.

639. **Odum, E. P. and de la Cruz, A. A.,** Particulate organic detritus in a Georgia saltmarsh ecosystem, in *Estuaries,* Publ. No. 83, Lauff, G. H., Ed., Am. Assoc. Adv. Sci., Washington, D.C., 1967, 383.

640. **Odum, E. P. and Fanning, M. E.,** Comparison of the productivity of *Spartina alterniflora* and *Spartina cynosuroides* in Georgia coastal marshes, *Bull. G. Acad. Sci.,* 31, 1, 1973.

641. **Odum, H. T.,** Primary production in flowing waters, *Limnol. Oceanogr.,* 1, 102, 1956.

642. **Odum, H. T.,** Productivity measurements in Texas turtle grass and the effects of dredging on intracoastal channel, *Publ. Inst. Mar. Sci. Univ. Tex.,* 9, 45, 1963.

643. **Odum, H. T.,** Energetics of world food production, in *The World Food Problem,* Vol. 3, U.S. Government Printing Office, Washington, D.C., 1967a, 55.

644. **Odum, H. T.,** Biological circuits and the marine ecosystems of Texas, in *Pollution and Marine Ecology,* Olsen, T. A. and Burgess, F. J., Eds., Interscience, New York, 1967b, 99.

645. **Odum, H. T.,** Work circuits and systems stress, in *Cycling and Productivity in Forests,* Young, H., Ed., University of Maine, Orono, 1968, 81.

646. **Odum, H. T.,** *Environment Power and Society,* John Wiley & Sons, New York, 1971, 336.

647. **Odum, H. T.,** An energy circuit language for ecological and social systems: its physical basis, in *Systems Analysis and Simulation,* Patten, B., Ed., Academic Press, New York, 1972, 139.

648. **Odum, H. T.,** Macroscopic minimodels of man and nature, in *Systems Analysis and Simulation in Ecology,* Vol. 4B, Patten, B., Ed., Academic Press, New York, 1976, 249.

649. **Odum, H. T.,** Energy analysis, energy quality and environment, in *Energy Analysis, a New Public Policy Tool,* Gilliland, M., Ed., AAAS Selected Symp., Westview Press, Boulder, Colo., 1978, 55.

650. **Odum, H. T.,** *Ecological and General Systems,* John Wiley & Sons, New York, 1983, 644 pp.

651. **Odum, H. T., Burkholder, P. R., and Rivero, J.,** Measurements of productivity of turtle grass flats, reefs and the Bahia Fosforescente of Southern Puerto Rico, *Publ. Inst. Mar. Sci. Univ. Tex.,* 6, 159, 1959.

652. **Odum, H. T. and Hoskin, C. M.,** Comparative studies of the metabolism of marine waters, *Publ. Inst. Mar. Sci. Univ. Tex.,* 5, 16, 1958.

653. **Odum, H. T., Knox, G. A., and Campbell, D. E.,** Organization of a new ecosystem, exotic *Spartina* salt marsh in New Zealand, Rep. Natl. Sci. Found. Int. Exchange Prog. N.Z.-U.S., Center for Wetlands, University of Florida, Gainesville, 1983, 106 pp.

654. **Odum, H. T. and Odum, E. C.,** *Energy Basis for Man and Nature,* 2nd ed., McGraw-Hill, New York, 1980, 297 pp.

655. **Odum, H. T. and Pinkerton, R. C.,** Time's speed regulator: the optimum efficiency for maximum power output in physical and biological systems, *Am. Sci.,* 43, 331, 1955.

656. **Odum, W. E.,** The ecological significance of fine particle selection by the striped mullet, *Mugil cephalus, Limnol. Oceanogr.,* 13, 92, 1968.

657. **Odum, W. E.,** Pathways of Energy Flow in a South Florida Estuary, Doctoral dissertation, University of Miami, Coral Gables, 1970a.

658. **Odum, W. E.,** Utilization of direct grazing and plant detritus food chains by the striped mullet *Mugil cephalus,* in *Marine Food Chains,* Steele, J. H., Ed., University of California Press, Berkeley, 1970b, 222.

659. **Odum, W. E.,** Insidious alteration of the estuarine environment, *Trans. Am. Fish. Soc.,* 99, 836, 1970c.

660. **Odum, W. E.,** Pathways of Energy Flow in a South Florida Estuary, Sea Grant Tech. Bull. No. 7, University of Miami Sea Grant Program, Coral Gables, 1971, 162 pp.

661. **Odum, W. E. and Heald, E. T.,** Trophic analyses of an estuarine mangrove community, *Bull. Mar. Sci.,* 22, 671, 1972.

662. **Odum, W. E. and Heald, E. T.,** The detritus-based food web of an estuarine mangrove community, in *Estuarine Research,* Vol. 1, Cronin, L. E., Ed., Academic Press, New York, 1975, 17.

663. **Odum, W. E., Woodwell, G. M., and Wurster, C. F.,** DDT residue adsorbed from organic detritus by fiddler crabs, *Science,* 164, 576, 1969.

664. **Officer, C. B.,** *Physical Oceanography of Estuaries (and Associated Coastal Waters),* John Wiley & Sons, New York, 1976, 465 pp.

665. **Oliff, W. D., Ed.,** South African National Marine Pollution Surveys, Annu. Rep. 1, 2nd Annu. Rep., Sect. C, Estuarine Surveys, NIWR, Durban, 1976, 509 pp.

666. **Oliver, J.,** The geographic and environmental aspects of mangrove communities, in *Mangrove Ecosystems in Australia,* Clough, B. F., Ed., Australian National University Press, Canberra, 1982, 19.

667. **Oliver, J. S.,** Processes Affecting the Organization of Soft-Bottom Communities in Monterey Bay, California and McMurdo Sound, Ph.D. thesis, University of California, San Diego, 1979.

668. **Oliver, J. S., Oakden, J. M., and Slattery, P. N.,** Phoxocephalid amphipod crustaceans as predators on larvae and juveniles in marine soft-bottom communities, *Mar. Ecol. Prog. Ser.,* 7, 179, 1982.

669. **Oppenheimer, C. and Wood, E. J. F.,** Note on the effect of contamination of a marine slough and the vertical distribution of unicellular plants in the sediment, *Z. Allg. Mikrobiol.,* 2, 45, 1962.

670. **Oppenheimer, C. H. and Ward, R. A.,** Release and capillary movement of phosphorus in exposed intertidal sediments, *Symp. on Marine Microbiology,* Oppenheimer, C. H., Ed., Charles C Thomas, Springfield, Ill., 1963, 664.

671. **Orians, G. H.,** Diversity, stability and maturity in natural ecosystems, in *Unifying Concepts in Ecology,* van Dobben, W. H. and Lowe-McConnel, R. H., Eds., B. V. Junk, The Hague, 1975, 139.

672. **Orth, R. J.,** The importance of sediment stability in seagrass communities, in *Ecology of Marine Benthos,* Coull, B. C., Ed., University of South Carolina Press, Columbia, 1977, 281.

673. **Oshrain, R. L.,** Aspects of Anaerobic Sulfur Metabolism in Salt Marsh Soils, M.Sc. thesis, University of Georgia, Athens, 1977.

674. **Ott, J.,** Determination of fauna boundaries of nematodes in an intertidal sand flat, *Int. Rev. Gesamten Hydrobiol. Hydrogr.,* 57, 645, 1972.

675. **Ott, J. and Schiemer, F.,** Respiration and anaerobis of free living nematodes in marine and limnetic sediments, *Neth. J. Sea Res.,* 7, 233, 1973.

676. **Otte, G. and Levings, C. D.,** Distribution of macroinvertebrate communities on a mudflat influenced by sewage, Fraser River Estuary, British Columbia, Tech. Rep. No. 476, Fisheries Marine Service, Research Division, Ottawa, Canada, 1975, 88 pp.

677. **Oviatt, C. A. and Kremer, P.,** Predation on the ctenophore *Mnemiopsis leidyi* by butterfish *Peprilus triacanthus* in Narragansett Bay, Rhode Island, *Chesapeake Sci.,* 18, 236, 1977.

678. **Oviatt, C. A. and Nixon, S. W.,** The demersal fish of Narragansett Bay: an analysis of community structure, distribution and abundance, *Estuarine Coastal Mar. Sci.,* 1, 361, 1973.

679. **Pace, M. L., Glasser, J. E., and Pomeroy, L. R.,** A simulation analysis of continental shelf food webs, *Mar. Biol.,* 82, 47, 1984.

680. **Pace, M. L., Shimmel, S., and Darley, W. M.,** The effect of grazing by the gastropod, *Nassarius obsoletus,* on the benthic microbial community of a salt marsh mudflat, *Estuarine Coastal Mar. Sci.,* 9, 121, 1979.

681. **Packham, J. R. and Liddle, M. J.,** The Cefni salt-marsh, Anglesey, and its recent development, *Field Stud.,* 3, 311, 1970.

682. **Paerl, H. W.,** Bacterial uptake of dissolved organic matter in relation to detrital aggregation in marine and freshwater systems, *Limnol. Oceanogr.,* 19, 966, 1974.

683. **Paerl, H. W.,** Microbial organic carbon recovery in aquatic ecosystems, *Limnol. Oceanogr.,* 23, 927, 1978.

684. **Paine, R. T.,** Food web complexity and species diversity, *Am. Nat.,* 100, 65, 1966.

685. **Palmer, J. D. and Round, F. E.,** Persistent, vertical migration rhythms in benthic microflora. I. The effect of light and temperature on rhythmic behavior of *Euglena obtusa, J. Mar. Biol. Assoc. U.K.,* 45, 567, 1965.

686. **Palmer, J. D. and Round, F. E.,** Persistent vertical-migration rhythms in benthic microflora. VI. The tidal and diurnal nature of this rhythm in the diatom *Hantzchia virgata, Biol. Bull.,* 132, 45, 1967.

687. **Palumbo, A. V. and Ferguson, R. L.,** Distribution of suspended bacteria in the Newport River Estuary, North Carolina, *Estuarine Coastal Mar. Sci.,* 7, 521, 1978.

688. **Pamatmat, M. M.,** Ecology and metabolism of a benthic community on an intertidal sandflat, *Int. Rev. Gesamten Hydrobiol. Hydrogr.,* 53, 211, 1968.

689. **Pamatmat, M. M.,** Oxygen consumption by the sea-bed. IV. Shipboard and laboratory experiments, *Limnol. Oceanogr.,* 16, 536, 1971a.

690. **Pamatmat, M. M.,** Oxygen consumption by the sea-bed. V. Seasonal cycle of chemical oxidation and respiration in Puget Sound, *Int. Rev. Gesamten Hydrobiol. Hydrogr.,* 56, 769, 1971b.

691. **Pannier, P. and Pannier, R. F.,** Physiology of viviparity in *Rhizophora mangle,* in *Proc. Symp. Biol. Manage. Mangroves,* Vol. 2, Walsh, G. E., Snedaker, S. C., and Teas, H. J., Eds., Institute of Food and Agricultural Sciences, University of Florida, Gainesville, 1975, 632.

692. **Park, K., Hood, D. W., and Odum, H. T.,** Diurnal pH variation in Texas Bays and its applications to primary production estimation, *Publ. Inst. Mar. Sci. Univ. Tex.,* 5, 47, 1958.

693. **Parrondo, R. T., Gosselink, J. G., and Hopkinson, C. S.,** Effects of salinity and drainage on the growth of three salt marsh grasses, *Bot. Gaz.,* 139, 102, 1978.

694. **Parsons, T. R. and Seki, H.,** Importance and general implications of organic matter in aquatic environments, in *Organic Matter in Natural Waters,* Occas. Publ. 1, Hood, D. W., Ed., Institute of Marine Science, Fairbanks, Alaska, 1970, 1.

695. **Parsons, T. R., Stephens, K., and Strickland, J. D. H.,** On the chemical composition of eleven species of marine phytoplankton, *J. Fish. Res. Board Can.,* 18, 1001, 1961.

258 *Estuarine Ecosystems: A Systems Approach*

696. **Parsons, T. R., Takahashi, M., and Hargrave, B. T.,** *Biological Oceanographic Processes,* 2nd ed., Pergamon Press, Oxford, 1977, 322 pp.
697. **Patrick, W. H. and DeLaune, R. D.,** Nitrogen and phosphorus utilization by *Spartina alterniflora* in a salt marsh in Barataria Bay, Louisiana, *Estuarine Coastal Mar. Sci.,* 4, 59, 1976.
698. **Patriquin, D. G.,** Estimation of growth rate, production and age of the marine angiosperm *Thalassia testudinum* Konig, *Caribb. J. Sci.,* 13, 111, 1973.
699. **Patten, B. C.,** A primer for ecological modeling and simulation with analog and digital computers, in *Systems Analysis and Simulation in Ecology,* Vol. 1, Patten, B. C., Ed., Academic Press, New York, 1971, 4.
700. **Patten, B. C., Bosserman, R. W., Finn, J. T., and Cole, W. G.,** Propagation of cause in ecosystems, in *Systems Analysis and Simulation in Ecology,* Vol. 4, Patten, B. C., Ed., Academic Press, New York, 1976, 173.
701. **Pearson, T. H.,** Studies on the ecology of the macrobenthic fauna of Lochs Linnhe and Eil, west coast of Scotland. II. Analysis of the macrobenthic fauna by comparison of feeding groups, *Vie Milieu Suppl.,* 22, 53, 1971.
702. **Pearson, T. H.,** The benthic ecology of Loch Linnhe and Loch Eil, a sea-loch system on the west coast of Scotland. IV. Changes in the benthic fauna attributable to organic enrichment, *J. Exp. Mar. Biol. Ecol.,* 20, 1, 1975.
703. **Pearson, T. H. and Rosenberg, R.,** A comparative study of the effects on the marine environment of wastes from cellulose industries in Scotland and Sweden, *Ambio,* 5, 77, 1976.
704. **Pearson, T. H. and Rosenberg, R.,** Macrobenthic succession in relation to organic enrichment and pollution of the marine environment, *Oceanogr. Mar. Biol. Annu. Rev.,* 16, 229, 1978.
705. **Penhale, P. A.,** Macro-epiphyte biomass and productivity in an eelgrass (*Zostera marina* L.) community, *J. Exp. Mar. Biol. Ecol.,* 26, 211, 1977.
706. **Penhale, P. A. and Smith, W. O., Jr.,** Excretion of dissolved organic carbon by eelgrass (*Zostera marina*) and its epiphytes, *Limnol. Oceanogr.,* 22, 400, 1977.
707. **Percival, M. and Womersley, J. S.,** *Floristics and Ecology of the Mangrove Vegetation of Papua New Guinea,* Department of Forests, Lae, New Guinea, 1975, 96 pp.
708. **Perkins, E. J.,** The food relationships of the microbenthos with particular reference to that found at Whitstable, Kent, *Ann. Mag. Nat. Hist.,* (Ser. 13), 1, 64, 1958.
709. **Perkins, E. J.,** Penetration of light into littoral soils, *J. Ecol.,* 51, 687, 1963.
710. **Perkins, E. J.,** *The Biology of Estuaries and Coastal Waters,* Academic Press, London, 1974, 678 pp.
711. **Petersen, C. G. J.,** *Om Baendeltangens (Zostera marina) Aarsproduktion: de Dankse Farvande,* Mindeskrift Hapetus Steenstamp, Copenhagen, 1913.
712. **Petersen, C. G. J.,** The sea bottom and its production of food, *Rep. Dan. Biol. Stn.,* 25, 1, 1918.
713. **Peterson, C. H.,** Predation, competitive exclusion, and diversity in the soft sediment benthic communities of estuaries and lagoons, in *Ecological Processes in Coastal and Marine Systems,* Livingston, R. J., Ed., Plenum Press, New York, 1979, 223.
714. **Peterson, C. H.,** Approaches to the study of competition in benthic soft sediments, in *Estuarine Perspectives,* Kennedy, V. S., Ed., Academic Press, New York, 1980, 291.
715. **Peterson, C. H. and Andre, S. V.,** An experimental analysis of interspecific competition among marine filter feeders in a soft-sediment environment, *Ecology,* 61, 129, 1980.
716. **Phillips, R. C.,** Observations on the ecology and distribution of Florida seagrasses, *Prof. Pap. Fla. Board Conserv.,* 2, 1, 1960.
717. **Phillips, R. C.,** Ecological Life History of *Zostera marina* L. (Eelgrass) in Puget Sound, Washington, Ph.D. dissertation, University of Washington, Seattle, 1972.
718. **Phillips, R. C.,** Temperate grass plots, in *Coastal Ecological Systems of the United States,* Odum, H. T., Copeland, B. J., and McMahan, E. A., Eds., Conservation Foundation, Washington, D.C., 1974, 244.
719. **Phillips, R. C. and McRoy, C. P.,** *Handbook of Seagrass Biology: an Ecosystem Perspective,* Garland STPM Press, New York, 1980, 353 pp.
720. **Phleger, C. F.,** Effect of salinity on growth of a salt marsh grass, *Ecology,* 52, 908, 1971.
721. **Phleger, F. B. and Bradshaw, J. S.,** Sedimentary environments in a marine marsh, *Science,* 154, 1551, 1966.
722. **Pianka, E. R.,** Niche overlap and diffuse competition, *Proc. Natl. Acad. Sci.,* 71, 2141, 1974.
723. **Pienkowski, M. W.,** Differences in habitat requirements and distribution patterns of plovers and sandpipers as investigated by studies of feeding behaviour, in Proc. of IWRB Feeding Ecol. Symp., Gwatt, Switzerland, 1980.
724. **Pienkowski, M. W.,** How foraging plovers cope with environmental effects of invertebrate behaviour and availability, in *Feeding and Survival Stategies of Estuarine Organisms,* Jones, N. V. and Wolff, W. J., Eds., Plenum Press, New York, 1981, 179.
725. **Pitts, G., Allam, A. I., and Hollis, J. P.,** *Beggiatoa:* occurrence in the rice rhizosphere, *Science,* 178, 990, 1972.

726. **Platt, H. M. and Warwick, R. M.,** The significance of free-living nematodes to the littoral ecosystem, in *The Shore Environment,* Vol. 2, Price, J. H., Irvine, D. E. G., and Farnham, W. F., Eds., Academic Press, London, 1980, 727.

727. **Pollard, D. A.,** Estuaries must be protected, *Aust. Fish.,* June, p.6, 1976.

728. **Pomeroy, L. R.,** Algal productivity in salt marshes in Georgia, *Limnol. Oceanogr.,* 4, 386, 1959.

729. **Pomeroy, L. R.,** Isotope and other techniques for measuring primary production, in *Proc. Conf. on Primary Productivity Measurements in Marine and Freshwater,* U.S. Atomic Energy Commission TID-7653, 1961, 97.

730. **Pomeroy, L. R.,** The ocean's food web, a changing paradigm, *Bioscience,* 24, 499, 1974.

731. **Pomeroy, L. R.,** Secondary production mechanisms of continental shelf communities, in *Ecological Processes in Coastal and Marine Systems,* Livingston, P. J., Ed., Plenum Press, New York, 1979, 163.

732. **Pomeroy, L. R., Darley, W. M., Dunn, E. L., Gallagher, J. L., Haines, E. B., and Whitney, D. M.,** Primary production, in *The Ecology of a Salt Marsh,* Pomeroy, L. R. and Wiegert, R. G., Eds., Springer-Verlag, New York, 1981, 39.

733. **Pomeroy, L. R. and Imberger, J.,** The physical and chemical environment, in *The Ecology of a Salt Marsh,* Pomeroy, L. R. and Wiegert, R. G., Eds., Springer-Verlag, New York, 1981, 22.

734. **Pomeroy, L. R., Shenton, L. R., Jones, R. D. H., and Reimold, R. J.,** Nutrient flux in estuaries, in *Nutrients and Eutrophication Special Symposium,* Vol. 1, Likens, G. E., Ed., Am. Soc. Limnol. Oceanogr., Allen Press, Lawrence, Kan., 1972, 272.

735. **Pomeroy, L. R. and Wiegert, R. G.,** *The Ecology of a Salt Marsh,* Springer-Verlag, New York, 1981, 271 pp.

736. **Pool, D. J., Lugo, A. E., and Snedaker, C. S.,** Litter production in mangrove forests of Southern Florida and Puerto Rico, *Proc. Int. Symp. Biol. Manage. Mangroves,* 1, 213, 1975.

737. **Postma, H.,** Sediment transport and sedimentation in estuaries, in *Estuaries,* Publ. No. 83, Lauff, G., Ed., Am. Assoc. Adv. Sci., Washington, D.C., 1967, 158.

738. **Prater, A. J.,** The ecology of Morecombe Bay. III. The food and feeding habits of Knot, *J. Appl. Ecol.,* 9, 179, 1972.

739. **Prater, A. J.,** *Estuary Birds of Britain and Ireland,* T. and A.D. Poyser, Calton, England, 1981, 439 pp.

740. **Pravdić, V.,** Surface change characterization of sea sediments, *Limnol. Oceanogr.,* 15, 230, 1970.

741. **Pritchard, D. W.,** Salinity distribution and circulation in the Chesapeake Bay estuarine system, *J. Mar. Res.,* 11, 106, 1952.

742. **Pritchard, D. W.,** Estuarine circulation patterns, *Proc. Am. Soc. Civ. Eng.,* 81, 1, 1955.

743. **Pritchard, D. W.,** What is an estuary, physical viewpoint?, in *Estuaries,* Publ. No. 85, Lauff, G., Ed., Am. Assoc. Adv. Sci., Washington, D.C., 1967a, 37.

744. **Pritchard, D. W.,** Observations of circulation in coastal plain estuaries, in *Estuaries,* Publ. No. 85, Lauff, G., Ed., Am. Assoc. Adv. Sci., Washington, D.C., 1967b, 37.

745. **Puttick, G. M.,** Energy budgets of Curlew Sandpipers at Langebaan Lagoon, South Africa, *Estuarine Coastal Mar. Sci.,* 11, 207, 1980.

746. **Pütter, A.,** *Die Ernährung der Wassetiere und der Stoffhaushault der Gewasser,* Fisher, Jena, 1909, 168 pp.

747. **Quasim, S. Z.,** Some problems related to the food chain in a tropical estuary, in *Marine Food Chains,* Steele, J. H., Ed., Oliver and Boyd, Edinburgh, 1970, 45.

748. **Quasim, S. Z.,** Productivity of backwaters and estuaries, in *The Biology of the Indian Ocean,* Vol. 3, Zeitschel, B., Ed., Springer-Verlag, Berlin, 1973, 143.

749. **Quasim, S. Z. and Sankaranarayanan, V. N.,** Organic detritus in a tropical estuary, *Mar. Biol.,* 15, 193, 1972.

750. **Quasim, S. Z., Bhattathiri, P. M. A., and Devassy, V. P.,** Organic production in a tropical estuary, *Proc. Indian Acad. Sci., Sect. B,* 59, 51, 1969.

751. **Rabinowitz, D.,** Dispersal properties of mangrove propagules, *Biotropica,* 10, 47, 1978.

752. **Rae, D.,** Life History and Predator-Prey Interactions in the Nemertean *Paranemertes peregrina* Cae, Ph.D. thesis, University of Washington, Seattle, 1971, 129 pp.

753. **Ragotzkie, R. A.,** Plankton productivity in estuarine waters of Georgia, *Publ. Inst. Mar. Sci. Univ. Tex.,* 6, 146, 1959.

754. **Randall, J. E.,** Food habits of reef fishes of the West Indies, *Stud. Trop. Oceanogr.,* 5, 665, 1967.

755. **Ranwell, D. S.,** World resources of *Spartina townsendii (sensu lato)* and economic uses of *Spartina* marshland, *J. Appl. Ecol.,* 4, 239, 1967.

756. **Ranwell, D. S.,** *Ecology of Salt Marshes and Sand Dunes,* Chapman and Hall, London, 1972, 258 pp.

757. **Repp, G.,** Ökologische Unterschungen im Halophyten-gebiet an Neusiedlersee, *Jahrb. Wiss. Bot.,* 88, 545, 1939.

758. **Ratcliffe, P. J., Jones, N. V., and Walters, N. J.,** The survival of *Macoma balthica* (L.) in mobile sediments, in *Feeding and Survival Strategies in Estuarine Organisms,* Wolff, W., Ed., Plenum Press, New York, 1981, 91.

759. **Rattray, M.,** Some aspects of the dynamics of circulation in fjords, in *Estuaries,* Publ. No. 83, Lauff, G., Ed., Am. Assoc. Adv. Sci., Washington, D.C., 1967, 52.

760. **Redfield, A. C.,** On the proportions of organic derivatives in seawater and their relation to the composition of plankton, in *James Johnston Memorial Volume,* Liverpool, 1934, 171.

761. **Relevante, N. and Gillmartin, M.,** Characteristics of microplankton and nannoplankton communities of an Australian coastal plain estuary, *Aust. Mar. Freshwater Res.,* 29, 9, 1978.

762. **Rees, C. B.,** A preliminary study of the ecology of a mudflat, *J. Mar. Biol. Assoc. U.K.,* 21, 185, 1940.

763. **Reeve, M. R.,** The ecological significance of zooplankton in the shallow subtropical waters of South Florida, in *Estuarine Research,* Vol. 1, Cronin, L. E., Ed., Academic Press, New York, 1975, 352.

764. **Reeve, M. R. and Cosper, E.,** The plankton and other seston in Card Sound, South Florida in 1971, Univ. of Miami Tech. Rep. UM-RSMAS-73007, 1973, 24 pp.

765. **Reideburg, C. H.,** Intertidal Pump in a Georgia Salt Marsh, M.S. thesis, University of Georgia, Athens, 1978.

766. **Reimold, R. J., Gallagher, J. L., and Thompson, D. E.,** Remote sensing of tidal marsh, *Photogramm. Eng.,* 39, 477, 1973.

767. **Reimold, R. J. and Linthurst, R. A.,** Primary production of minor marsh plants in Delaware, Georgia and Maine, Rep. No. D-77-36, U.S. Army Corps of Engineers, Vicksburg, Miss., 1977.

768. **Reimold, R. J., Linthurst, R. A., and Wolf, P. L.,** Effects of grazing on salt marsh, *Biol. Conserv.,* 8, 105, 1975.

769. **Reisch, D. J.,** Uptake of organic material by aquatic invertebrates. V The influence of age on the uptake of glycine-C^{14} by the polychaete *Neanthes arenaceodentata, Mar. Biol.,* 3, 352, 1969.

770. **Relse, K.,** Predation pressure and community structure of an intertidal soft-bottom fauna, in *Biology of Benthic Organisms,* Keegan, B. F., Ceidigh, P. O., and Boaden, P. J. S., Eds., Pergamon Press, New York, 1977a, 513.

771. **Reise, K.,** Predator exclusion experiments in an intertidal mud flat, *Helgol. Wiss. Meeresunters.,* 30, 263, 1977b.

772. **Reise, K.,** Experiments on epibenthic predation in the Wadden Sea, *Helgol. Wiss. Meeresunters.,* 31, 55, 1978.

773. **Reise, K.,** Moderate predation on meiofauna by the macrobenthos of the Wadden Sea, *Helgol. Wiss. Meeresunters.,* 32, 453, 1979.

774. **Reise, K.,** High abundances of small zoobenthos around biogenic structures in tidal flat sediments of the Wadden Sea, *Helgol. Wiss. Meeresunters.,* 34, 413, 1981.

775. **Reise, K.,** Biotic enrichment of intertidal sediments by experimental aggregates of the deposit-feeding bivalve *Macoma balthica, Mar. Ecol. Prog. Ser.,* 12, 229, 1983.

776. **Remane, A.,** The ecology of brackish water, *Binnengewaesser,* 25, 1, 1971.

777. **Rheinheimer, G., Ed.,** *Microbial Ecology of a Brackish Water Environment,* Ecol. Stud. 25, Springer-Verlag, Berlin, 1977, 291 pp.

778. **Rhoads, D. C.,** Rates of sediment reworking by *Yoldia limatula* in Buzzards Bay, Massachusetts and Long Island Sound, *J. Sediment. Petrol.,* 33, 727, 1963.

779. **Rhoads, D. C.,** Biogenic working of intertidal and subtidal sediments in Barnsable Harbor and Buzzards Bay, Massachusetts, *J. Geol.,* 75, 461, 1967.

780. **Rhoads, D. C.,** The influence of deposit feeding benthos on water turbidity and nutrient recycling, *Am. J. Sci.,* 273, 1, 1973.

781. **Rhoads, D. C.,** Organism-sediment relations on the muddy sea floor, *Oceanogr. Mar. Biol. Annu. Rev.,* 12, 263, 1974.

782. **Rhoads, D. C., Aller, R. C., and Goldhaber, M. B.,** The influence of colonizing benthos on physical properties and chemical diagenesis of the estuarine sea floor, in *Ecology of Marine Benthos,* Coull, B. C., Ed., University of South Carolina Press, Columbia, 1977, 113.

783. **Rhoads, D. C., McCall, P. L., and Yinst, J. Y.,** Disturbance and production on the estuarine seafloor, *Am. Sci.,* 60, 577, 1978.

784. **Rhoads, D. C. and Young, P. K.,** Animal sediment relationships in Cape Cod Bay, Massachusetts. II. Reworking by *Molpadia oolitica, Mar. Biol.,* 11, 255, 1971.

785. **Riaux, C.,** Structure d'un peuplement estuarien de diatomées épipéliques du Nord-Finistere, *Oceanol. Acta,* 6, 173, 1983.

786. **Ribelin, B. W. and Collier, A. W.,** Ecological considerations of detrital aggregates in the salt marsh, in *Ecological Processes in Coastal and Marine Systems,* Livingston, R. J., Ed., Plenum Press, New York, 1979, 47.

787. **Rice, D. L.,** The Trace Metal Chemistry of Aging Detritus Derived From Coastal Macrophytes, Ph.D. dissertation, Georgia Institute of Technology, Atlanta, 1979.

788. **Richards, L.,** Tanaidacea (Crustacea: Peracarida) of San Juan Island, Zool. 533 Rep., Friday Harbor Laboratories, University of Washington, Seattle, 1969, 129 pp.

789. **Richey, J. E., Likens, G. E., Eaton, J. S., Wetzel, R. G., Odum, W. E., Johnson, N. M., Loucks, O. L., Prentki, R. T., and Rich, P. H.,** Carbon flow in four lake ecosystems: a structural approach, *Science,* 171, 1003, 1978.

790. **Rieman, F. and Schrage, M.,** The mucas-trap hypothesis on feeding of aquatic nematodes and implications for biodegradation and sediment textures, *Oecologia,* 34, 75, 1978.

791. **Riley, G. A.,** Patterns of production in marine ecosystems, in *Ecosystem Structure and Function,* Wiens, J. A., Ed., Proc. 3rd Annu. Biol. Colloq., Oregon St. Univ., Oregon State University Press, Corvallis, 1972, 91.

792. **Riser, N. W.,** *Ophelia* as a member of the meiofauna, *Proc. 3rd Int. Meiofauna Conf.,* (Abstract), Hamburg, August, 1977.

793. **Ritchie, L. D.,** Fish and fisheries aspects of mangrove wetlands, in *Environment 77, Proc. 7, Coastal Zone Workshop,* Knox, G. A., Ed., Environment Centre (Canterbury), Christchurch, New Zealand, 1979, 25.

794. **Riznyk, R. Z. and Phinney, H. K.,** Macrometric assessment of interstitial microalgae production in two estuarine sediments, *Oecologia,* 10, 193, 1972.

795. **Robertson, J. J.,** Predation by estuarine zooplankton on tintinnid ciliates, *Estuarine Coastal Shelf Sci.,* 16, 27, 1983.

796. **Roe, P.,** Life history and predator-prey interactions of the nemertean *Paranemertes peregrina* Coe, *Biol. Bull.,* Marine Biological Laboratory, Woods Hole, Mass., 150, 80, 1976.

797. **Rosenberg, R.,** Benthic faunal recovery in a Swedish fjord following the closure of a sulphite mill, *Oikos,* 23, 92, 1972.

798. **Rosenberg, R.,** Succession in benthic macrofauna in a Swedish fjord subsequent to the closure of a sulphite mill, *Oikos,* 24, 244, 1973.

799. **Rublee, P. A.,** Bacteria and microbial distribution in estuarine sediments, in *Estuarine Comparisons,* Kennedy, V. C., Ed., Academic Press, New York, 1982a, 159.

800. **Rublee, P. A.,** Seasonal distribution of bacteria in salt marsh sediments in North Carolina, *Estuarine Coastal Shelf Sci.,* 15, 67, 1982b.

801. **Rublee, P. A. and Dornseif, B. E.,** Direct counts of bacteria in the sediments of a North Carolina salt marsh, *Estuaries,* 1, 188, 1978.

802. **Rusnak, C. A.,** Rates of sediment accumulation in modern estuaries, in *Estuaries,* Publ. 83, Lauff, G. H., Ed., Am. Assoc. Adv. Sci., Washington, D.C., 1967, 180.

803. **Ryther, J. H.,** Geographical variations in productivity, in *The Sea,* Vol. 2, Hill, M. N., Ed., Wiley Interscience, New York, 1963, 347.

804. **Ryther, J. H.,** Photosynthesis and fish production in the sea, *Science,* 166, 72, 1969.

805. **Saenger, P.,** Morphological, anatomical and reproductive adaptations of Australian mangroves, in *Mangrove Ecosystems in Australia,* Clough, B. F., Ed., Australian National University Press, Canberra, 1982, 153.

806. **Saenger, P., Specht, M. M., Specht, R. L., and Chapman, V. J.,** Ecosystems of the World 1, in *Wet Coastal Ecosystems,* Chapman, V. J., Ed., Elsevier, Amsterdam, 1977, 293.

807. **Saelen, O. H.,** Some features of the hydrography of Norwegian fjords, in *Estuaries,* Publ. No. 83, Lauff, G., Ed., Am. Assoc. Adv. Sci., Washington, D.C., 1967, 63.

808. **Sakamito, W.,** Study on the process of river suspension from flocculation to accumulation in estuary, *Bull. Ocean Res. Inst. Univ. Tokyo,* 5, 1, 1972.

809. **Sanders, H. L.,** Benthic studies in Buzzards Bay. I. Animal-sediment relationships, *Limnol. Oceanogr.,* 3, 245, 1958.

810. **Sanders, H. L.,** Benthic studies in Buzzards Bay. III. Structure of the soft-bottom community, *Limnol. Oceanogr.,* 4, 138, 1960.

811. **Sanders, H. L.,** Marine diversity: a comparative study, *Am. Nat.,* 102, 243, 1968.

812. **Sanders, H. L., Gouldsmit, E. M., Mills, E. L., and Hampston, G. E.,** A study of the intertidal fauna of Barnstable Harbor, Massachusetts, *Limnol. Oceanogr.,* 7, 63, 1962.

813. **Sand-Jensen, K.,** Biomass, net production and growth dynamics in an eel grass (*Zostera marina* L.) population in Vellerup Vig, Denmark, *Ophelia,* 14, 185, 1975.

814. **Santos, S. L. and Simon, J. L.,** Response of soft-bottom benthos to annual catastrophic disturbance in a south Florida estuary, *Mar. Ecol. Prog. Ser.,* 3, 347, 1980.

815. **Scheleske, C. L. and Odum, E. P.,** Mechanisms for maintaining high productivity in Georgian estuaries, *Proc. Gulf Caribb. Fish. Inst.,* 14, 75, 1961.

816. **Schlieper, C.,** Physiology of brackish water, *Binnengewaesser,* 25, 211, 1971.

817. **Schneider, D.,** Equalization of prey numbers by migratory shore birds, *Nature,* 271, 353, 1978.

818. **Scholander, P. F.,** How mangroves desalinate water, *Physiol. Plant.,* 21, 722, 1968.

819. **Scholander, P. F., Hammel, H. T., Hemmingsen, E., and Carey, W.,** Salt balance in mangroves, *Plant Physiol.,* 37, 722, 1962.

820. **Scholander, P. F., van Dame, L., and Scholander, S. I.,** Gas exchange in the roots of mangroves, *Am. J. Bot.,* 42, 92, 1955.

821. **Schultz, D. J. and Calder, J. A.,** Organic carbon $^{13}C/^{12}C$ variations in estuarine sediments, *Geochim. Cosmochim. Acta,* 40, 381, 1976.

822. **Sedell, J. R., Naimen, R. J., Cummins, K. W., Minshall, G. W., and Vannate, R. L.,** Transport of particulate organic material in streams as a function of physical processes, *Int. Ver. Theor. Angew. Limnol. Verh.,* 20, 1366, 1979.

823. **Segerstraale, S. G.,** Brackish water classification, a historical survey, *Arch. Oceanogr. Limnol.,* 7 (Suppl. 11), 7, 1959.

824. **Seki, H.,** The role of microorganisms in the marine food chain with reference to organic aggregates, in *Detritus and its Role in Aquatic Ecosystems,* Melchorri-Santalini, U. and Hopton, J., Eds., Memoir Inst. Italiano Idrobiol., (Suppl. 29), 1972, 245.

825. **Sellner, K. G. and Zingmark, R. G.,** Intepretations of the ^{14}C method of measurement of the total annual production of phytoplankton in a South Carolina estuary, *Bot. Mar.,* 19, 119, 1976.

826. **Shamoot, S., McDonald, L., and Bartholomew, W. V.,** Rhizodeposition of organic debris in soil, *Proc. Am. Soc. Soil Sci.,* 32, 817, 1968.

827. **Shaver, G. R. and Billings, W. D.,** Root production and root turnover in a wet tundra ecosystem, Barrow, Alaska, *Ecology,* 56, 401, 1975.

828. **Sheith, M-S. J.,** Nutrients in Narragansett Bay Sediments, M.S. thesis, University of Rhode Island, Kingston, 1974.

829. **Shelton, E. R.,** Bacterial Distribution and Biomass in Sediments and Water in the Newport River Estuary, North Carolina, M.S. thesis, North Carolina State University, Raleigh, 1979.

830. **Sheridan, P. F. and Livingston, R. J.,** Cyclic trophic relationships of fishes in an unpolluted river-dominated estuary in North Florida, in *Ecological Processes in Coastal and Marine Systems,* Livingston, R. J., Ed., Plenum Press, New York, 1979, 143.

831. **Sherk, J. A., Jr.,** The effects of suspended and deposited sediments on estuarine organisms, Contrib. No. 443, National Research Institute, Chesapeake Biological Laboratory, 1971.

832. **Shew, D. M., Linthurst, R. A., and Seneca, E. D.,** Comparison of production methods in a southeastern North Carolina *Spartina alterniflora* salt marsh, *Estuaries,* 4, 97, 1981.

833. **Sidhu, S. S.,** Structure of epidemis and stomatal apparatus of some mangrove species, in *Proc. Int. Symp. Mangroves,* Vol. 2, Walsh, G. E., Snedaker, S. C., and Teas, H. J., Eds., Institute of Food and Agricultural Sciences, University of Florida, Gainesville, 1975, 569.

834. **Siebruth, J. McN.,** Studies on algal substances in the Sea. III. Glebstoff (humic material) in terrestrial and marine waters, *J. Exp. Mar. Biol. Ecol.,* 2, 174, 1969.

835. **Siebruth, J. McN.,** Bacterial substrates and productivity in marine ecosystems, *Annu. Rev. Ecol. Syst.,* 7, 259, 1976.

836. **Siebruth, J. McN.,** Convenor's report on the informal session on biomass and productivity of microorganisms in planktonic systems, *Helgol. Wiss. Meeresunters.,* 30, 697, 1977.

837. **Siebruth, J. McN.,** Grazing of bacteria by protozooplankton in pelagic marine waters, in *Heterotrophic Activity in the Sea,* Hobbie, J. E. and Williams, A. J. Le B., Eds., Plenum Press, New York, 1982.

838. **Siebruth, J. McN. and Jensen, A.,** Production and transformation of extracellular organic matter from littoral marine algae: a resume, in *Organic Matter in Natural Waters,* Hood, D. W., Ed., Inst. Mar. Sci., Occas. Publ. No. 1, 1970, 203.

839. **Siebruth, J. McN., Johnson, K. M., Burney, K. M., and Lavoie, D. M.,** Estimation of *in situ* rates of heterotrophy using diurnal changes in dissolved organic matter and growth rates of picoplankton in diffusion cultures, *Helgol. Wiss. Meeresunters.,* 30, 565, 1977.

840. **Siebruth, J. McN. and Thomas, C. D.,** Fouling on eelgrass (*Zostera marina* L.), *J. Phycol.,* 9, 46, 1973.

841. **Siegfried, W. R.,** The estuarine avifauna of South Africa, in *Estuarine Ecology with Special Reference to Southern Africa,* Day, J. H., Ed., A.A. Balkema, Rotterdam, 1981, 223.

842. **Sikora, W. B.,** The Ecology of *Palaemonetes pugio* in a Southeastern Salt Marsh Ecosystem with Particular Emphasis on Production and Trophic Relationships, Ph.D. thesis, University of South Carolina, Columbia, 1977.

843. **Sikora, W. B. and Sikora, J. P.,** Ecological implications of the vertical distribution of meiofauna in salt marsh sediments, in *Estuarine Comparisons,* Kennedy, V. S., Ed., Academic Press, New York, 1982, 269.

844. **Simon, J. L. and Dauer, D. M.,** A quantitative evaluation of red-tide induced mass mortalities of benthic invertebrates in Tampa Bay, Florida, *Environ. Lett.,* 3, 229, 1972.

845. **Simon, J. L. and Dauer, D. M.,** Reestablishment of a benthic community following natural defaunation, in *Ecology of Marine Benthos,* Coull, B. C., Ed., University of South Carolina Press, Columbia, 1977, 139.

846. **Simons, R. H.,** The algal flora of Saldanha Bay, *Trans. R. Soc. S. Afr.,* 42, 461, 1977.

847. **Skyring, G. W., Oshrain, R. L., and Wiebe, W. J.,** Assessment of sulfate reduction rates in Georgia marshland soils, *J. Geomicrobiol.,* 1, 389, 1979.

848. **Smalley, A. E.,** The Role of Two Invertebrate Populations, *Littorina irrorata* and *Orchelimum fidicinum* in the Energy Flow of a Salt Marsh Ecosystem, Doctoral dissertation, University of Georgia, Athens, 1958.

849. **Smalley, A. E.,** The growth cycle of *Spartina* and its relation to insect populations in the marsh, in *Proc. Salt Marsh Conf.,* Ragotzkie, R. A., Teal, J. M., Pomeroy, L. R., and Scott, D. C., Eds., University of Georgia, Athens, 1959, 96.

850. **Smart, R. M. and Barko, J. W.,** Influence of sediment salinity and nutrients on the physiological ecology of selected marsh plants, *Estuarine Coastal Mar. Sci.,* 7, 487, 1978a.

851. **Smart, R. M. and Barko, J. W.,** Nitrogen nutrition and salinity tolerance of *Distichlis spicata* and *Spartina alterniflora, Ecology,* 61, 630, 1978b.

852. **Smart, R. M. and Barko, J. W.,** Nitrogen nutrition and salinity tolerance of *Distichlis spicata* and *Spartina alterniflora, Ecology,* 61, 620, 1980.

853. **Smayda, T. J.,** The phytoplankton of estuaries, in *Estuaries and Enclosed Seas,* Ketchum, B. H., Ed., Elsevier, Amsterdam, 1983, 65.

854. **Smith, K. K., Good, R. E., and Good, N. F.,** Production dynamics for above and below-ground components of a New Jersey *Spartina alterniflora* tidal marsh, *Estuarine Coastal Mar. Sci.,* 9, 189, 1979.

855. **Smith, K. L., Jr., Burns, H. A., and Teal, J. M.,** *In situ* respiration of benthic communities in Castle Harbour, Bermuda, *Mar. Biol.,* 12, 196, 1972.

856. **Smith, K. L., Jr., Rowe, G. T., and Nichols, J. A.,** Benthic community respiration near the Woods Hole sewage outfall, *Estuarine Coastal Mar. Sci.,* 1, 65, 1973.

857. **Smith, T. J., III and Odum, W. E.,** The effects of grazing by Snow Geese on coastal marshes, *Ecology,* 62, 98, 1981.

858. **Snedaker, S. C.,** Mangroves: their value and perpetration, *Nat. Resour.,* 14, 6, 1978.

859. **Snedaker, S. and Lugo, A.,** The Role of Mangrove Ecosystems in the Maintenance of Environmental Quality and a High Productivity of Desirable Species, U.S. Bureau of Sport Fisheries and Wildlife, Atlanta, Georgia, 1973.

860. **Snelling, B.,** The distribution of intertidal crabs in the Brisbane River, *Aust. J. Mar. Freshwater Res.,* 10, 67, 1959.

861. **Sogard, S. M.,** Feeding Ecology, Population Structure, and Community Relationships of a Grassbed Fish, *Callionymus pauciradiatus* in Southern Florida, M.S. thesis, University of Miami, Coral Gables, 1982.

862. **Sogard, S. M.,** Utilization of meiofauna as a food source by a grassbed fish, the spotted dragonet, *Callionymus pauciradiatus, Mar. Ecol. Prog. Ser.,* 17, 183, 1984.

863. **Sørenson, J., Jorgensen, B. B., and Revsbeck, N. P.,** A comparison of oxygen, nitrate and sulfate respiration in coastal marine sediments, *Microb. Ecol.,* 5, 105, 1979.

864. **Sorokin, Yu. I.,** On the trophic role of chemosynthesis and bacterial biosynthesis in water bodies, in *Primary Productivity in Aquatic Environments,* Goldman, C. R., Ed., University of California Press, Berkeley, 1969, 187.

865. **Sorokin, Yu. I.,** On the role of bacteria in the productivity of tropical oceanic waters, *Int. Rev. Gesamten Hydrobiol. Hydrogr.,* Suppl. 18, 187, 1971a.

866. **Sorokin, Yu. I.,** Population, activity and production of bacteria in bottom sediments of the central Pacific, *Oceanology,* 10, 853, 1971b.

867. **Sorokin, Yu. I.,** Decomposition of organic matter and nutrient regeneration, in *Marine Ecology,* Vol. 4, Kinne, O., Ed., John Wiley & Sons, New York, 1978, 501.

868. **Sorokin, Yu. I.,** Microheterotrophic organisms in marine ecosystems, in *Analysis of Marine Ecosystems,* Longhurst, A. R., Ed., Academic Press, London, 1981, 293.

869. **Soule, D. F. and Soule, J. D.,** The importance of non-toxic urban wastes in estuarine detrital food webs, *Bull. Mar. Sci.,* 31, 786, 1981.

870. **Southward, A. J. and Southward, E. C.,** Observations on the role of dissolved organic compounds in the nutrition of benthic invertebrates. I. Experiments on three species of Pagonophora, *Sarsia,* 45, 69, 1970.

871. **Southward, A. J. and Southward, E. C.,** Observations on the role of dissolved organic compounds in the nutrition of benthic invertebrates. II. Uptake by other animals living in the same habitat as Pagonophora, and by some littoral Polychaeta, *Sarsia,* 50, 29, 1972.

872. **Southward, A. J. and Southward, E. C.,** Observations on the role of dissolved organic compounds in the nutrition of benthic invertebrates. III. Uptake in relation to organic content of the habitat, *Sarsia,* 50, 29, 1974.

873. **Specht, R. L.,** Biogeography of halophytic angiosperms (salt-marsh, mangrove and sea-grass), in *Ecological Biogeography of Australia,* Keast, A., Ed., B.V. Junk, The Hague, 1981, 579.

874. **Stace, C. A.,** The use of epidermal characters in phytogenetic considerations, *New Phytol.,* 65, 304, 1966.

875. **Steers, J. A.,** Physiography, in *Wet Coastal Ecosystems,* Vol. 1, Chapman, V. J., Ed., Elsevier, Amsterdam, 1977, 31.

876. **Steever, Z. E., Warren, R. S., and Niering, W. A.,** Tidal energy subsidy and standing crop production of *Spartina alterniflora, Estuarine Coastal Mar. Sci.,* 4, 473, 1976.

877. **Steffensen, D. A.,** The *Euglena* of the Avon-Heathcote Estuary, B.Sc. (Hons) project in Botany, University of Canterbury, Christchurch, New Zealand, 1969.

878. **Steffensen, D. A.,** An Ecological Study of *Ulva lactuca* L. and other Benthic Algae on the Avon-Heathcote Estuary, Ph.D. thesis, University of Canterbury, Christchurch, New Zealand, 1974a.

879. **Steffensen, D. A.,** Distribution of *Euglena obtusa* Schmitz and *E. salina* Liebetanz on the Avon-Heathcote Estuary, Christchurch, *Mauri Ora*, 2, 85, 1974b.

880. **Steffensen, D. A.,** The effect of nutrient enrichment and temperature on the growth in culture of *Ulva lactuca* L., *Aquat. Bot.*, 2, 337, 1976.

881. **Stephens, G. C.,** Dissolved organic material as a nutritional source for marine and estuarine invertebrates, in *Estuaries*, Publ. No. 83, Lauff, G. H., Ed., Am. Assoc. Adv. Sci., Washington, D.C., 1967, 367.

881a. **Stephens, G. C.,** Dissolved organic matter as a potential source of nutrition for marine organisms, *Am. Zool.*, 8, 95, 1968.

882. **Stephens, G. C.,** Trophic role of dissolved organic material, in *Analysis of Marine Ecosystems*, Longhurst, A. R., Ed., Academic Press, New York, 1981, 271.

883. **Stephens, K. R., Sheldon, W., and Parsons, T. R.,** Seasonal variations in the availability of food for benthos in a coastal environment, *Ecology*, 48, 852, 1967.

884. **Stevenson, F. J. and Cheng, C.-N.,** Organic geochemistry of the Argentine Basin sediments: carbon-nitrogen relationships and Quaternary correlations, *Geochim. Cosmochim. Acta*, 36, 653, 1971.

885. **Stevenson, H. L. and Erkenbrecher, C. W.,** Activity of bacteria in the estuarine environment, in *Estuarine Processes*, Vol. 1, Wiley, M., Ed., Academic Press, New York, 1976, 381.

886. **Stevenson, J. C., Heinle, D. R., Flemer, D. A., Small, R. J., Rowland, R. A., and Ustach, J. F.,** Nutrient exchanges between brackish water marshes and the estuary, in *Estuarine Processes*, Vol. 2, Wiley, M. L., Ed., Academic Press, New York, 1977, 219.

887. **Stewart, R. G., Larhev, F., Ahmad, I., and Lee, J. A.,** Nitrogen metabolism and salt-tolerance in higher plant halophytes in *Ecological Processes in Coastal Environments*, Jeffries, R. L. and Davy, A. J., Eds., Blackwell Scientific, Oxford, 1977, 211.

888. **Stickney, R. R., Taylor, G. L., and Heard, R. W., III,** Food habits of Georgia estuarine fishes. I. Four species of flounders (Pleuronectiformes: Bothidae), *Fish. Bull.*, 72, 515, 1974.

889. **Stickney, R. R., Taylor, G. L., and White, D. B.,** Food habits of five species of young southeastern United States estuarine Sciaenidae, *Chesapeake Sci.*, 16, 104, 1975.

890. **Stout, J. P.,** An Analysis of the Annual Growth and Productivity of *Juncus roemerianus* Scheele and *Spartina alterniflora* Loisel in Coastal Alabama, Ph.D. dissertation, University of Alabama, University, 1978.

891. **Stroud, L. M.,** Net Primary Production of Belowground Material and Carbohydrate Patterns of Two Height Forms of *Spartina alterniflora* in Two North Carolina Marshes, Ph.D. dissertation, North Carolina State University, Raleigh, 1976.

892. **Stroud, L. M. and Cooper, A. W.,** Color-infrared aerial photographic interpretation and net primary productivity of a regularly flooded North Carolina salt marsh, Rep. No. 14, University of North Carolina Water Resource Institute, 1968, 81 pp.

893. **Suberkropp, K., Godshalk, G. L., and Kug, M. J.,** Changes in the chemical composition of leaves during processing in a woodland stream, *Ecology*, 57, 720, 1976.

894. **Sullivan, M. J.,** Diatom communities from a Delaware salt marsh, *J. Phycol.*, 11, 384, 1975.

895. **Sullivan, M. J.,** Edaphic diatom communities associated with *Spartina alterniflora* and *Spartina patens* in New Jersey, *Hydrobiologia*, 52, 207, 1977.

896. **Sullivan, M. J.,** Diatom community structure: taxonomic and statistical analyses of a Mississippi salt marsh, *J. Phycol.*, 14, 468, 1978.

897. **Sullivan, M. L. and Daiber, F. C.,** Responses in production of cord grass, *Spartina alterniflora* to inorganic nitrogen and phosphorus fertilizer, *Chesapeake Sci.*, 15, 121, 1974.

898. **Sullivan, M. L. and Daiber, F. C.,** Light, nitrogen and phosphorus limitation of edaphic algae in a Delaware salt marsh, *J. Exp. Mar. Biol. Ecol.*, 18, 79, 1975.

899. **Summers, R. W.,** Distribution, abundance and energy relationships of waders (Aves: Charadrii) at Langebaan Lagoon, *Trans. R. Soc. S. Afr.*, 42, 483, 1977.

900. **Summerson, H. C.,** The Effects of Predation on the Marine Benthic Invertebrate Community In and Around a Shallow Subtidal Seagrass Bed, M.S. thesis, University of North Carolina, Chapel Hill, 1980.

901. **Sutcliffe, W. H.,** Some relations of land drainage, nutrients, particulate matter and fish catch in two eastern Canadian bays, *J. Fish. Res. Board Can.*, 30, 856, 1972.

902. **Sutcliffe, W. H.,** Correlations between seasonal river discharge and local landings of American lobster *(Homarus americanus)* and Atlantic halibut *(Hippoglossus hippoglossus)* on the Gulf of St. Lawrence, *J. Fish. Res. Board Can.*, 30, 856, 1973.

903. **Sutherland, J. P.,** Multiple stable points in benthic communities, *Am. Nat.*, 118, 859, 1974.

904. **Swedmark, B.,** The interstitial fauna of marine sand, *Biol. Rev.*, 37, 1, 1964.

905. **Anon.,** Symp. Class. Brackish Waters, The Venice System for the classification of marine waters according to salinity, *Oikos,* 9, 311, 1958.
906. **Swennen, C.,** Wadden Seas are rare, hospitable and productive, *Proc. IWPB Conserv. Wetlands and Waterfowl Conf. Heiligenhafen,* Vol. 1974, 184, 1976.
907. **Sykes, J.,** Paper presented to The Marsh and Estuary Manage. Symp., Louisiana, 1967.
908. **Tanner, W. F.,** Florida coastal classification, *Trans. Gulf Coast Assoc. Geol. Soc.,* 10, 259, 1960.
909. **Taylor, A. G.,** The direct uptake of amino acids and other small molecules from seawater by *Nereis virens* Sars, *Comp. Biochem. Physiol.,* 29, 243, 1969.
910. **Taylor, J. E.,** The Ecology and Seasonal Periodicity of Benthic Marine Algae from Barnegat Bay, New Jersey, Ph.D. thesis, Rutgers University, New Brunswick, N.J., 1970.
911. **Taylor, W. R.,** Light and photosynthesis in intertidal benthic diatoms, *Helgol. Wiss. Meeresunters.,* 10, 29, 1964.
912. **Teal, J. M.,** Energy flow in the salt marsh ecosystem of Georgia, *Ecology,* 40, 614, 1962.
913. **Teal, J. M. and Kaniwisher, J.,** Gas exchange in a Georgia salt marsh, *Limnol. Oceanogr.,* 6, 388, 1961.
914. **Teal, J. M. and Kaniwisher, J. W.,** Gas transport in the marsh grass *Spartina alterniflora, J. Exp. Bot.,* 17, 13, 1966.
915. **Teal, J. M. and Teal, M.,** *Life and Death of a Salt Marsh,* Little, Brown, Boston, 1969.
916. **Teal, J. M. and Wieser, W.,** The distribution and ecology of nematodes in a Georgia salt marsh, *Limnol. Oceanogr.,* 11, 217, 1966.
917. **Tenore, K. R.,** Detrital utilization by the polychaete *Capitella capitata, J. Mar. Res.,* 33, 261, 1975.
918. **Tenore, K. R.,** Growth of the polychaete *Capitella capitata* cultured in different levels of detritus derived from various sources, *Limnol. Oceanogr.,* 22, 936, 1977a.
919. **Tenore, K. R.,** Utilization of aged detritus derived from different sources by the polychaete, *Capitella capitata, Mar. Biol.,* 44, 51, 1977b.
920. **Tenore, K. R.,** Food chain pathways in detritus feeding benthic communities: a review, with new observations on resuspension and detrital cycling, in *Ecology of the Marine Benthos,* Coull, B. C., Ed., University of South Carolina Press, Columbia, 1977c, 37.
921. **Tenore, K. R.,** Organic nitrogen and caloric content of detritus. I. Utilization by the polychaete *Capitella capitata, Estuarine Coastal Shelf Sci.,* 12, 39, 1981.
922. **Tenore, K. R.,** Comparison of the ecological energetics of the polychaetes *Capitella capitata* and *Nereis succinea* in experimental systems receiving similar levels of detritus, *Neth. J. Sea Res.,* 16, 46, 1982.
923. **Tenore, K. R., Hanson, R. B., Dornself, B. E., and Wiederhold, C. N.,** The effect of organic nitrogen supplement on the utilization of different sources of detritus, *Limnol. Oceanogr.,* 84, 350, 1979.
924. **Tenore, K. R. and Rice, D. L.,** A review of trophic factors affecting secondary production of deposit-feeders, in *Marine Benthic Dynamics,* Tenore, K. R. and Coull, B. C., Eds., University of South Carolina Press, Columbia, 1980, 325.
925. **Tenore, K. R., Tietjen, J. H., and Lee, J. J.,** Effect of meiofauna on incorporation of aged eelgrass, *Zostera marina* detritus by the polychaete *Nepthys incisa, J. Fish. Res. Board Can.,* 34, 563, 1977.
926. **Thayer, G. W.,** Phytoplankton production and distribution of nutrients in a shallow unstratified estuarine system near Beaufort, North Carolina, *Chesapeake Sci.,* 12, 240, 1971.
927. **Thayer, G. W., Parker, P. L., La Croix, M. W., and Fry, B.,** The stable carbon isotope ratio of some components of an eelgrass, *Zostera marina* bed, *Oecologia,* 35, 1, 1978.
928. **Thayer, G. W., Wolfe, D. A., and Williams, R. B.,** The impact of man on seagrass systems, *Am. Sci.,* 63, 288, 1975.
929. **Thistle, T.,** Natural physical disturbance and communities of marine soft bottoms, *Mar. Ecol. Prog. Ser.,* 8, 223, 1981.
930. **Thom, B. G.,** Mangrove ecology and deltaic geomorphology, Tabasco, Mexico, *J. Ecol.,* 55, 301, 1981.
931. **Thom, B. G.,** Mangrove ecology from a geomorphic viewpoint, in *Proc. Int. Symp. Biol. Manage. Mangroves,* Walsh, G., Snedaker, S., and Teas, H., Eds., Institute of Food and Agricultural Sciences, University of Florida, Gainesville, 1974, 469.
932. **Thom, B. G.,** Mangrove ecology — a geomorphological perspective, in *Mangrove Ecosystems in Australia. Structure, Function and Management,* Clough, B. F., Ed., Australian National University Press, Canberra, 1982, 3.
933. **Thomas, J. P.,** Influence of the Atamaka River on Primary Production Beyond the Mouth of the River, M.Sc. thesis, University of Georgia, Athens, 1966.
934. **Thomas, J. P.,** Release of dissolved organic matter from natural populations of marine phytoplankton, *Mar. Biol.,* 11, 311, 1971.
935. **Thorson, G.,** Bottom communities (sublittoral or shallow shelf), in *Treatise on Marine Ecology and Palaeoecology,* Vol. 1, Hedgpeth, J. W., Ed., Geological Society of America, Boulder, Colo., 1957, 461.

936. **Thorson, G.,** Parallel level bottom communities, their temperature adaptation, and their 'balance' between predators and food animals, in *Perspectives in Marine Biology,* Buzzati-Traverso, A. A., Ed., University of California Press, Berkeley, 1958, 67.

937. **Tietjen, J. H. and Lee, J. J.,** Life history and feeding behaviour of marine nematodes, in *Ecology of Marine Benthos,* Coull, B. C., Ed., University of South Carolina Press, Columbia, 1973, 21.

938. **Trevallion, A., Steele, J., and Edwards, R. R. C.,** Dynamics of a benthic bivalve, in *Marine Food Chains,* Steele, J. H., Ed., Oliver and Boyd, London, 1970, 285.

939. **Tunnicliffe, V. and Risk, M. J.,** Relations between the bivalve *Macoma balthica* and bacteria in intertidal sediments: Miners Basin, Bay of Fundy, *J. Mar. Res.,* 35, 499, 1977.

940. **Turner, R. E.,** Geographic variations in salt marsh macrophyte production: a review, *Contrib. Mar. Sci. Univ. Tex.,* 20, 47, 1976.

941. **Turner, R. E.,** Community plankton respiration in a salt marsh estuary and the importance of macrophyte leachates, *Limnol. Oceanogr.,* 23, 442, 1978.

942. **Turner, R. E. and Gosselink, J. G.,** A note on the standing crops of *Spartina alterniflora* in Texas and Louisiana, *Contrib. Mar. Sci. Univ. Tex.,* 19, 113, 1975.

943. **Udell, H. F., Zarudsky, T. E., Doheny, T. E., and Burkholder, P. R.,** Productivity and nutrient values of plants growing in salt marshes of the town of Hempstead, L. I., *Bull. Torrey Bot. Club,* 90, 42, 1979.

944. **Ulrich, B., Mayer, R., and Heller, H., Eds.,** *Data Analysis and Data Synthesis of Forest Ecosystems,* Göttinger Bodenkundliche Berichte 30, 1974, 459 pp.

945. **University of Miami,** South Florida's Mangrove Bordered Estuaries: their Role in Sport and Commercial Fish Production, Univ. Miami Sea Grant Inf. Bull. No. 4, 1970.

946. **Vaccaro, R. F., Azarn, F., and Hudson, R. E.,** Response of natural marine bacterial populations to copper: controlled ecosystem pollution experiment, *Bull. Mar. Sci.,* 27, 17, 1977.

947. **Valiela, I., Howes, B. L., Howarth, R. W., Goblin, A., Foreman, K., Teal, J. M., and Hobbie, J. E.,** The regulation of primary production and decomposition in a salt marsh ecosystem, in *Proc. 1st Int. Wetlands Conf.,* New Delhi, India, 1983.

948. **Valiela, I. and Teal, J. M.,** Nutrient limitation in a salt marsh vegetation, in *Ecology of Halphytes,* Reimold, R. J. and Queen, W. H., Eds., Academic Press, New York, 1974, 547.

949. **Valiela, I., Teal, J. M., and Persson, N. Y.,** Production and dynamics of experimentally enriched salt marsh vegetation: belowground biomass, *Limnol. Oceanogr.,* 21, 245, 1976.

950. **Valiela, I., Teal, J. M., and Sass, W.,** Production and dynamics of salt marsh vegetation and effect of sewage contamination. Biomass, production and species composition, *J. Appl. Ecol.,* 12, 973, 1975.

951. **Valiela, I., Teal, J. M., Volkman, S., Shorter, D., and Carpenter, E. J.,** Nutrient and particulate fluxes in a salt marsh ecosystem: tidal exchanges and inputs by precipitation and groundwater, *Limnol. Oceanogr.,* 23, 798, 1978.

952. **van der Ben, D.,** Les épiphytes des fueilles de *Posidonia oceanica* sur les côtes francaise de la Méditerranée, *Proc. Int. Seaweed Symp. (Madrid),* 6, 79, 1969.

953. **Vandermeer, J. H.,** Niche theory, *Annu. Rev. Ecol. Syst.,* 3, 107, 1972.

954. **Van Dyne, G. M., Ed.,** *The Ecosystem Concept in Natural Resource Management,* Academic Press, New York, 1969, 383 pp.

955. **Van Raalte, C. D., Valiela, I., and Teal, J. M.,** Production of epibenthic salt marsh algae: light and nutrient limitation, *Limnol. Oceanogr.,* 21, 862, 1976.

956. **Van Valkenberg, S. D. and Flemer, D. A.,** The distribution and productivity of nannoplankton in a temperate estuarine area, *Estuarine Coastal Mar. Sci.,* 2, 311, 1974.

957. **Verwey, J.,** The ecology of the distribution of cockle and mussel in the Dutch Waddensea. Their role in the sedimentation and the source of their food supply with a short review of feeding behaviour in molluscs, *Arch. Neerl. Zool.,* 10, 171, 1952.

958. **Virnstein, R. W.,** The importance of predation by crabs and fishes on the benthic infauna in Chesapeake Bay, *Ecology,* 58, 1199, 1977.

959. **Virnstein, R. W.,** Predator caging experiments in soft sediments: caution advised, in *Estuarine Interactions,* Vol. 1, Wiley, M. L., Ed., Academic Press, New York, 1978, 261.

960. **Virnstein, R. W.,** Measuring effects of predation on benthic communities in soft sediments, in *Estuarine Perspectives,* Kennedy, V. S., Ed., Academic Press, New York, 1980, 281.

961. **Voller, R. W.,** Salinity, Sediment, Exposure, and Invertebrate Macrofaunal Distributions on the Mud Flats of the Avon-Heathcote Estuary, Christchurch, M.Sc. thesis, University of Canterbury, Christchurch, New Zealand, 1975, 61 pp.

962. **Vouglitois, J. J.,** The benthic flora and fauna of Barnegat Bay before and after the onset of thermal addition — a summary analysis of a ten-year study by Rutgers University, Unpublished Tech. Rep., Jersey Central Power and Light Company, 1976.

963. **Waddell, J. E.,** The Effect of Oyster Culture on Eelgrass (*Zostera marina* L.) Growth, M.S. thesis, Humboldt State College, Arcata, Calif., 1964, 48 pp.

964. **Wagner, P. R.,** Seasonal Biomass, Abundance and Distribution of Estuarine Dependent Fishes in the Caminada Bay System of Louisiana, Ph.D dissertation, Louisiana State University, Baton Rouge, 1973.

965. **Waisel, Y.,** *Biology of Halphytes,* Academic Press, New York, 1972, 395 pp.

966. **Waite, T. D. and Mitchell, R.,** The effect of nutrient fertilization on the benthic alga *Ulva lactuca, Bot. Mar.,* 15, 151, 1972.

967. **Waits, E. D.,** Net Primary Productivity of an Irregularly Flooded North Carolina Salt Marsh, Ph.D. thesis, North Carolina State University, Raleigh, 1967, 124 pp.

968. **Walsh, G. E.,** Mangroves: a review, in *Ecology of Halophytes,* Reimold, R. J. and Queen, W. H., Eds., Academic Press, New York, 1974, 51.

969. **Walter, H.,** *Ecology of Tropical and Sub-Tropical Vegetation,* Oliver and Boyd, Edinburgh, 1971, 539 pp.

970. **Walter, M. A.,** The Ecological Significance of the Feeding Behavior of the Ctenophore *Mnemiopsis mccradyi,* M.S. thesis, University of Miami, Coral Gables, 1975.

971. **Walters, C.,** Systems ecology: the systems approach and mathematical models in ecology, in *Fundamentals of Ecology,* Odum, E. P., Ed., W. B. Saunders, Philadelphia, 1971, 276.

972. **Wangersky, D. J.,** The role of particulate matter in the productivity of surface waters, *Helgol. Wiss. Meeresunters.,* 30, 546, 1977.

973. **Warme, J. E.,** Graded bedding in the recent sediments of Mogu Lagoon, California, *J. Sediment. Petrol.,* 37, 540, 1967.

974. **Warme, J. E.,** Paleoecological aspects of a modern coastal lagoon, *Univ. Calif. Publ. Geol. Sci.,* 87, 1, 1971.

975. **Warwick, R. M.,** Population dynamics and secondary production of the benthos, in *Marine Benthic Dynamics,* Tenore, K. R. and Coull, B. C., Eds., University of South Carolina Press, Columbia, 1980, 1.

976. **Warwick, R. M.,** Survival strategies of meiofauna, in *Feeding and Survival Strategies of Estuarine Organisms,* Jones, N. V. and Wolff, W. J., Eds., Plenum Press, New York, 1981, 39.

977. **Warwick, R. M., Joint, I. R., and Radford, P. J.,** Secondary production of the benthos in an estuarine environment, in *Ecological Processes in Coastal Environments,* Jeffries, R. L. and Davy, A. J., Eds., Blackwell Scientific, Oxford, 1979, 429.

978. **Warwick, R. M. and Price, R.,** Macrofauna production on an estuarine mudflat, *J. Mar. Biol. Assoc. U.K.,* 55, 1, 1975.

979. **Watling, L.,** Analysis of the structural variations in a shallow estuarine deposit-feeding community, *J. Exp. Mar. Biol. Ecol.,* 19, 275, 1975.

980. **Watson, J. G.,** Mangrove forests of the Malay Peninsula, *Malay. For. Rec.,* 6, 1, 1928.

981. **Watson, W. S.,** The role of bacteria in an upwelling system, in *Upwelling Ecosystems,* Boje, R. and Tomczak, M., Eds., Springer-Verlag, Berlin, 1978, 139.

982. **Wass, M. C. and Wright, T. D.,** Coastal Wetlands of Virginia, Int. Rep. Gov. Gen. Assemb., Virginia Institute of Marine Science, Spec. Rep., Appl. Mar. Sci. Ocean Eng., 10, 1968, 15 pp.

983. **Wax, C. L., Borengasser, M. J., and Muller, R. A.,** Barataria Basin: synoptic weather types and environmental responses, Sea Grant Publ. No. LSU-T-78-001, Louisiana State University, Center for Wetland Resources, Baton Rouge, 1978.

984. **Wellershaus, S.,** Some aspects of the plankton ecology in the Cochin backwater (a South Indian estuary), *Ger. Schol. India,* 2, 341, 1976.

985. **West, R. C.,** Tidal salt-marsh and mangal formations of Middle and South America, in *Wet Coastal Ecosystems,* Vol. 1, Chapman, V. J., Ed., Elsevier, Amsterdam, 1977, 193.

986. **Westlake, D. F.,** Comparisons of plant productivity, *Biol. Rev.,* 38, 385, 1963.

987. **Wetzel, R. L.,** An Experimental-Radiotracer Study of Detrital Carbon Utilization in a Georgia Salt Marsh, Ph.D. dissertation, University of Georgia, Athens, 1975, 103 pp.

988. **Wetzel, R. L.,** Carbon resources of a benthic salt marsh invertebrate *Nassarius obsoletus,* Say (Mollusca: Nassariidae), in *Estuarine Processes,* Vol. 2, Wiley, M., Ed., Academic Press, New York, 1977, 293.

989. **Wetzel, R. G. and Manny, B. A.,** Decomposition of dissolved organic carbon and nitrogen compounds from leaves in an experimental hardwood stream, *Limnol. Oceanogr.,* 17, 927, 1972.

990. **Wetzel, R. G., Rich, P. H., Miller, M. C., and Allen, H. L.,** Metabolism of dissolved and particulate detrital carbon in a temperate hard-water lake, in *Detritus and its Role in Aquatic Ecosystems,* Melchorri-Santalini, U. and Hopton, J., Eds., Memoir Inst. Italiano Idriobiol., (Suppl. 29), 1972, 185.

991. **White, D. A., Weiss, T. E., Trapani, J. M., and Thien, L. B.,** Production and decomposition of the dominant salt marsh plants in Louisiana, *Ecology,* 59, 751, 1978.

992. **White, D. D., Findlay, R. H., Fazio, S. D., Bobbie, R. J., Nickels, J. S., Davis, M. W., Smith, G. A., and Martz, R. F.,** Effects of bioturbation and predation by *Melita quinquiesperforata* on sedimentary microbial community structure, in *Estuarine Perspectives,* Kennedy, V. S., Ed., Academic Press, New York, 1980, 163.

993. **Whitlach, R. B.,** Food resource partitioning in the deposit-feeding polychaete *Pectinaria gouldii, Biol. Bull.,* 147, 227, 1974.

994. **Whitlach, R. B.,** Methods of resource allocation in marine deposit-feeding communities, *Am. Zool.,* 16, 195, 1976.

995. **Whitlach, R. B.,** Patterns of resource utilization and coexistence in marine intertidal deposit-feeding communities, *J. Mar. Res.,* 38, 743, 1980.

996. **Whitney, D. E., Chalmers, A. G., Haines, E. B., Hanson, R. B., Pomeroy, L. R., and Sherr, B.,** The cycles of nitrogen and phosphorus, in *The Ecology of a Salt Marsh,* Pomeroy, L. R. and Wiegert, R. G., Eds., Springer-Verlag, New York, 1981, 163.

997. **Whitney, D. E. and Darley, W. M.,** A method for the determination of chlorophyll *a* in samples containing degradation products, *Limnol. Oceanogr.,* 24, 183, 1979.

998. **Whitney, D. E., Woodwell, G. M., and Howarth, R. W.,** Nitrogen fixation in Flax Pond: a Long Island salt marsh, *Limnol. Oceanogr.,* 20, 640, 1975.

999. **Whittaker, R. H.,** Gradient analysis of vegetation, *Biol. Rev.,* 42, 207, 1967.

1000. **Wiebe, W. J.,** Anaerobic benthic microbial processes: changes from the estuary to the continental shelf, in *Ecological Processes in Coastal and Marine Systems,* Livingston, R. J., Ed., Plenum Press, New York, 1979, 469.

1001. **Wiebe, W. J., Christian, R. R., Hanson, J. A., King, G., Sherr, B., and Skyring, G.,** Anaerobic respiration and fermentation, in *The Ecology of a Salt Marsh,* Pomeroy, L. R. and Wiegert, R. G., Eds., Springer-Verlag, New York, 1981, 138.

1002. **Wiebe, W. J. and Pomeroy, L. R.,** Microorganisms and their association with aggregates and detritus in the sea: a microscopic study, in *Detritus and its Role in Aquatic Ecosystems,* Melchorri-Santalini, U. and Hopton, J., Eds., Memoir Inst. Italiano Idriobiol., (Suppl. 29), 1972, 325.

1003. **Wiebe, W. J. and Smith, D. F.,** Direct measurement of dissolved organic carbon release by phytoplankton and incorporation by microheterotrophs, *Mar. Biol.,* 42, 213, 1977.

1004. **Wiegert, R. G.,** Ecological processes characteristic of coastal *Spartina* marshes of the south-eastern U.S.A., in *Ecological Processes in Coastal Environments,* Jeffries, R. L. and Davy, A. J., Eds., Blackwell Scientific, Oxford, 1979, 467.

1005. **Wiegert, R. G., Chalmers, A. G., and Randerson, P. F.,** Primary Production of Salt Marshes: Regulation by Soil-Water Movement, Unpublished manuscript, University of Georgia, Athens, 1982.

1006. **Wiegert, R. G., Christian, R. R., and Wetzel, R. L.,** A model view of the marsh, in *The Ecology of a Salt Marsh,* Pomeroy, L. R. and Wiegert, R. G., Eds., Springer-Verlag, New York, 1981, 183.

1007. **Wiegert, R. G. and Evans, F. C.,** Investigations of secondary productivity in grasslands, in *Secondary Productivity of Terrestrial Ecosystems,* Petrusewicz, K., Ed., Polish Academy of Sciences, Krakow, 1964, 499.

1008. **Weigert, R. G. and Wetzel R. L.,** Simulation experiments with a fourteen-compartment model of a *Spartina* salt marsh, in *Marsh-Estuarine Systems Simulation,* Dame, R. F., Ed., University of South Carolina Press, Columbia, 1979, 7.

1009. **Wieser, W.,** Die Beziehung zwischen Mundhöllengengestalt, Ernähnungsweisse und Vorkommen bei frei-lebenden marinen Nematoden. Eine ökologischmorphologische Studie, *Ark. Zool.,* 4, 439, 1953.

1010. **Wieser, W. and Kaniwisher, J.,** Ecological and physiological studies on marine nematodes from a small salt marsh near Woods Hole, Massachusetts, *Limnol. Oceanogr.,* 6, 262, 1961.

1011. **Wildish, D. J.,** Factors controlling marine and estuarine sublittoral macrofauna, *Helgol. Wiss. Meere-sunters.,* 30, 445, 1977.

1012. **Wildish, D. J. and Kristmanson, P. D.,** Tidal energy and sublittoral benthic animals in estuaries, *J. Fish. Res. Board Can.,* 36, 1197, 1979.

1013. **Wilkinson, M.,** Estuarine benthic algae and their environment: a review, in *The Shore Environment,* Vol. 2, Price, J. H., Irvine, D. E. G., and Farnham, W. F., Eds., Academic Press, London, 1980, 425.

1014. **Wilkinson, M.,** Survival strategies of attached algae in estuaries, in *Feeding and Survival Strategies of Estuarine Organisms,* Jones, N. V. and Wolff, W. J., Ed., Plenum Press, New York, 1981, 29.

1015. **Williams, R. B.,** The Ecology of Diatom Populations in a Georgia Salt Marsh, Ph.D. thesis, Harvard University, Cambridge, Mass., 1962.

1016. **Williams, R. B.,** Division rates of salt marsh diatoms in relation to salinity and cell size, *Ecology,* 43, 877, 1964.

1017. **Williams, R. B.,** Annual phytoplankton production in a system of shallow temperate estuaries, in *Some Contemporary Studies in Marine Sciences,* Barnes, H., Ed., George Allen and Unwin, London, 1966, 699.

1018. **Williams, R. B.,** Nutrient levels and phytoplankton productivity in the estuary, in *Proc. 2nd Symp. Coastal Estuary Manage.,* Louisiana State University, Baton Rouge, 1972, 59.

1019. **Williams, R. B. and Murdoch, M. B.,** Phytoplankton production and chlorophyll concentration in the Beaufort Channel, North Carolina, *Limnol. Oceanogr.,* 11, 73, 1966.

1020. **Williams, R. B. and Murdoch, M. B.,** The potential importance of *Spartina alterniflora* in conveying zinc, manganese, and iron into estuarine food chains, in *Proc. 2nd Int. Symp. Radiol.,* Nelson, D. J. and Evans, F. C., Eds., U.S.A.E.C., Springfield, Va., 1969, 431.

1021. **Williams, R. B. and Murdoch, M. B.,** Compartmental analysis of the production of *Juncus roemerianus* in a North Carolina saltmarsh, *Chesapeake Sci.,* 13, 69, 1972.

1022. **Williams, W. T., Bunt, J. S., and Duke, N. C.,** Mangrove litterfall in NE Australia. II. Periodicity, *Aust. J. Bot.,* 29, 555, 1981.

1023. **Wilson, C. A. and Stevenson, L. H.,** The dynamics of the bacterial population associated with a salt marsh, *J. Exp. Mar. Biol. Ecol.,* 48, 123, 1980.

1024. **Witzig, A. S. and Day, J. W., Jr.,** Trophic state index for the Louisiana coastal zone, Louisiana State University Center for Wetland Resources, Baton Rouge, 1982.

1025. **Wolf, P. L., Shanholtzer, S. F., and Reimold, R. H.,** Population estimates for *Uca pugnax* (Smith 1870) on the Duplin estuary marsh, Georgia, U.S.A. (Decapoda, Brachyura, Ocypodidae), *Crustaceana,* 29, 79, 1975.

1026. **Wolfe, D. A.,** The estuarine ecosystem(s) at Beaufort, North Carolina, in *Estuarine Research,* Vol. 1, Cronin, L. E., Ed., Academic Press, New York, 1975, 645.

1027. **Wolff, W. J.,** The estuary as a habitat. An analysis of data on the soft-bottom macrofauna of the estuarine area of the Rivers Rhine, Meuse and Scheldt, *Zool. Verh.,* 126, 1, 1973.

1028. **Wolff, W. J.,** Changes in intertidal benthos after an increase in salinity, *Thalassia Jugosl.,* 7, 429, 1974.

1029. **Wolff, W. J., Ed.,** *Ecology of the Wadden Sea,* A.A. Balkema, Rotterdam, 1980, 1300 pp. in 3 vol.

1030. **Wolff, W. J.,** Estuarine benthos, in *Estuaries and Enclosed Seas,* Vol. 26, Ketchum, B. H., Ed., Elsevier, Amsterdam, 1983, 151.

1031. **Wolff, W. J., van Haperen, A. M. M., Sandee, A. J. J., Bapist, H. J. M., and Sneijs, H. L. F.,** The trophic role of birds in the Grevelingen estuary, the Netherlands, as compared to their role in the saline Lake Grevelingen, in *Proc. 10th Eur. Symp. Mar. Biol.,* Persoone, G. and Jaspers, W., Eds., Universa Press, Wetteren, Belgium, 1976, 673.

1032. **Wood, J. E. F.,** Some aspects of the ecology of Lake Macquarie, N.S.W. with regard to alleged depletion of fish. VI. Plant communities and their significance, *Aust. J. Mar. Freshwater Res.,* 10, 322, 1959.

1033. **Wood, J. E. F.,** *Marine Microbial Ecology,* Chapman and Hall, London, 1965, 243 pp.

1034. **Wood, J. E. F., Odum, W. E., and Zieman, J. C.,** Influence of seagrasses on the productivity of coastal lagoons, in Lagunas Costeras, un Symposia, Symp. Int. Lagunas Costeras, UNAM-UNESCO, Mexico, 1967, 495.

1035. **Woodhouse, W. W., Jr., Seneca, E. D., and Broome, S. W.,** Propagation of *Spartina alterniflora* for substrate stabilization and salt marsh development, Tech. Memo 46, Coastal Engineering Research Center, U.S. Army Corps of Engineers, Fort Belvoir, Va., 1974.

1036. **Woodin, S. A.,** Polychaete abundance patterns in a marine soft-bottom sediment environment: the importance of biological interactions, *Ecol. Monogr.,* 44, 171, 1974.

1037. **Woodin, S. A.,** Adult-larval interactions in dense infaunal assemblages: patterns of abundance, *J. Mar. Res.,* 34, 25, 1976.

1038. **Woodmansee, A.,** The seasonal distribution of zooplankton of Cheekan Key in Biscayne Bay, Florida, *Ecology,* 39, 247, 1958.

1039. **Woodroffe, C. D.,** Mangroves of the Upper Waitemata Harbour. Biomass, Productivity and Geomorphology, Upper Waitemata Harbour Catchment Study Work. Rep. No. 36, Auckland Regional Authority, Auckland, 1982a, 92 pp.

1040. **Woodroffe, C. D.,** Litter production and decomposition in New Zealand mangrove *Avicennia marina* var. *resinifera, N. Z. J. Mar. Freshwater Res.,* 16, 179, 1982b.

1041. **Woodwell, G. M., Hall, C. A. S., Whitney, D. E., and Houghton, R. A.,** The Flax Pond ecosystem study: exchanges of inorganic nitrogen between an estuarine marsh and Long Island Sound, *Ecology,* 60, 695, 1979.

1042. **Woodwell, G. M., Wurster, C. F., and Isaacson, P. A.,** DDT residues in an east coast estuary: a case of biological concentration of a persistent pesticide, *Science,* 156, 821, 1967.

1043. **Woolridge, T.,** The zooplankton of Mngazana, a mangrove estuary in Transkei, southern Africa, *Zool. Afr.,* 12, 307, 1977.

1044. **Wright D.,** Pollen Morphology of Australian Mangroves, Unpublished Honours thesis, University of Adelaide, Adelaide, Australia, 1977.

1045. **Wright, R. T.,** Measurement and significance of specific activity in the heterotrophic activity of natural waters, *Appl. Environ. Microbiol.,* 36, 297, 1978.

1046. **Wright, R. T. and Coffin, R. B.,** Planktonic bacteria in estuaries and coastal waters of northern Massachusetts: spatial and temporal distribution, *Mar. Ecol. Prog. Ser.,* 11, 205, 1983.

1047. **Wright, R. T., Coffin, R. B., Ersing, C. P., and Pearson, D.,** Field and laboratory measurements of bivalve filtration of natural marine bacterioplankton, *Limnol. Oceanogr.,* 27, 91, 1982.

1048. **Wulff, B. L. and McIntire, C. B.**, Laboratory studies of assemblages of attached estuarine diatoms, *Limnol. Oceanogr.*, 17, 200, 1972.

1049. **Young, D. K.**, Effects of infauna on the sediment and seston of a subtidal environment, *Vie Milieu*, Suppl. 22, 557, 1971.

1050. **Young, D. K., Buzas, M. A., and Young, M. W.**, Species densities of macrobenthos associated with seagrasses: a field experiment study of predation, *J. Mar. Res.*, 34, 577, 1976.

1051. **Young, D. K. and Rhoads, D. C.**, Animal-sediment relations in Cape Cod Bay, Massachusetts. I. A transect study, *Mar. Biol.*, 11, 242, 1971.

1052. **Young, D. K. and Young, M. W.**, Community structure of the macrobenthos associated with seagrass of the Indian River estuary, Florida, in *Ecology of Marine Benthos*, Coull, B. C., Ed., University of South Carolina Press, Columbia, 1977, 359.

1053. **Zajac, R. N. and Whitlach, R. B.**, Responses of estuarine infauna to distrubance. I. Spatial and temporal variation of initial recolonization, *Mar. Ecol. Prog. Ser.*, 10, 1, 1982a.

1054. **Zajac, R. N. and Whitlach, R. B.**, Responses of estuarine fauna to disturbance. II. Spatial and temporal variation of succession, *Mar. Ecol. Prog. Ser.*, 10, 15, 1982b.

1055. **Ziegelmeier, E.**, Das Makrobenthos im Ostteil der Deutschen Bucht nach qualitativen Bodengreiferunter-suchungen in der Ziet von 1949—60, *Veroff. Inst. Meeresforsch. Bremerhaven*, 1, 101, 1963.

1056. **Ziegelmeier, E.**, Uber Massenvorkommen verscheidener markrobenthaler Wirbeloser während wiederbes-iedlungsphase nach Schadingungen durch "katastrophale" Umwelteinflusse, *Helgol. Wiss. Meeresunters.*, 21, 9, 1970.

1057. **Zieman, J. C.**, A Study of the Growth and Decomposition of the Seagrass *Thalassia testudinum* Masters thesis, University of Miami, Coral Gables, 1968.

1058. **Zieman, J. C.**, Origin of circular beds of *Thalassia* (Spermatophyta: Hydrocharitaceae) in South Biscayne Bay, Florida, and their relationship to mangrove hummocks, *Bull. Mar. Sci.*, 22, 559, 1972.

1059. **Zieman, J. C.**, Quantitative and dynamic aspects of the ecology of turtle grass, *Thalssia testudinum*, in *Estuarine Research*, Vol. 1, Cronin, L. E., Ed., Academic Press, New York, 1975, 541.

1060. **Zieman, J. C. and Wetzel, R. G.**, Productivity in seagrasses: methods and rates, in *Handbook of Seagrass Biology: an Ecosystem Perspective*, Phillips, R. C. and McRoy, C. P., Eds., Garland STPM Press, New York, 1980, 87.

1061. **Zingmark, R. G.**, *An Annotated Check List of the Biota of the Coastal Zone of South Carolina*, University of South Carolina Press, Columbia, 1978.

1062. **Zwarts, L. and Drent, R. H.**, Prey depletion and the regulation of predator density: oystercatchers *(Haematopus ostralegus)* feeding on mussels *(Mytilus edulis)*, in *Feeding and Survival Strategies of Estuarine Organisms*, Jones, N. V. and Wolff, W. J., Eds., Plenum Press, New York, 1981, 193.

1063. **Asmus, R.**, Field measurements on seasonal variation of the activity of primary producers on a sandy tidal flat in the Northern Wadden Sea, *Neth. J. Sea Res.*, 16, 389, 1982.

1064. **Baker, L. D.**, The ecology of the ctenophore *Mnemiopsis mccradyi* Meyer in Biscayne Bay, Florida, Univ. Miami Tech. Rep., Unpublished, UM-RSMAS-73016, 1973, 131 pp.

1065. **Bruce, A.**, Rep. Biol. Chemical Invest. Waters in the Estuary of the Avon and Heathcote Rivers, Christchurch Drainage Board, Christchurch, New Zealand, 1953, 58 pp.

1066. **Canoy, M. J.**, Diversity and stability in a Puerto Rican *Rhizophora mangle* L. forest., in *Proc. Int. Symp. Biol. Manage. Mangroves*, Walsh, G. E., Snedaker, S. C., and Teas, H. J., Eds., Institute of Food and Agricultural Science, University of Florida, Gainesville, 1975, 344.

1067. **Carey, A. G., Jr.**, Energetics of the benthos of Long Island Sound. I. Oxygen utilization, *Bull. Bingham Oceanogr. Collect.*, 19, 136, 1976.

1068. **Conover, R. J.**, Feeding interactions in the pelagic zones, *Rapp. P. V. Reun. Cons. Int. Explor. Mer.*, 173, 66, 1978.

1069. **DeLaune, R. D., Patrick, W. H., Jr., and Brannon, J. M.**, Nutrient transformations in Louisiana saltmarsh soils, Sea Grant Publ. No. LSU-T-76-009, Louisiana State University, Baton Rouge, 1976, 38p.

1070. **Elmgren, R.**, Baltic benthos communities and the role of meiofauna, *Contrib. Askö Lab.*, 14, 1, 1977.

1071. **Gallagher, J. L. and Daiber, F. C.**, Primary production of edaphic communities in a Delaware salt marsh, *Ecology*, 54, 1160, 1974.

1072. **Hackney, C. T. and de la Cruz, A. A.**, *In situ* decomposition of roots and rhizomes of two tidal plants, *Ecology*, 61, 266, 1980.

1073. **Haedrick, R. L. and Haedrick, S. O.**, A seasonal survey of the fishes in the Mystic River, a polluted estuary in downtown Boston, Massachusetts, *Estuarine Coastal Mar. Sci.*, 2, 59, 1974.

1074. **Haines, E. B.**, Interactions between Georgia salt marshes and coastal waters: a changing paradigm, in *Ecological Processes in Coastal and Marine Systems*, Livingston, R. J., Ed., Plenum Press, New York, 1979a, 35.

1075. **Haines, E. B., Chalmers, A., Hanson, R., and Sherr, B.**, Nitrogen pools and fluxes in a Georgia salt marsh, in *Estuarine Processes*, Vol. 2, Wiley, M., Ed., Academic Press, New York, 1976, 241.

1076. **Hardisky, M. A. and Reimold, R. J.,** Salt marsh plant geratology, *Science,* 198, 612, 1977.
1077. **Hatcher, B. G., Chapman, A. R. O., and Mann, K. H.,** An annual carbon budget for the kelp *Laminaria longicruris, Mar. Biol.,* 44, 85, 1977.
1078. **Hicks, G. R. F.,** Biomass and production estimates for an estuarine benthic copepod, with an assessment of exploitation by flatfish predators, *N.Z. J. Ecol.,* 8, 125, 1985.
1079. **Hutchinson, G. E.,** Concluding remarks, *Cold Spring Harbor Symp. Quant. Biol.,* 22, 415, 1957.
1080. **Ito, S. and Imai, T.,** Ecology of oyster beds. I. On the decline of productivity due to repeated cultures, *Tohoku J. Agric. Res.,* 5, 251, 1955.
1081. **Joris, C., Billen, G., Lancelot, C., Daro, M. H., Mommaerts, J. P., Bertels, A., Bossicart, M., Nijs, J., and Hecq, J. H.,** A budget of carbon cycling in the Belgian coastal zone: relative rates of zooplankton, bacterioplankton and benthos in the utilization of primary production, *Neth. J. Sea Res.,* 16, 260, 1982.
1082. **Marinucci, A. C.,** Carbon and nitrogen flow during decomposition of *Spartina alterniflora* in a flow-through percolator, *Biol. Bull. Woods Hole, Mass.,* 162, 53, 1982.
1083. **Mills, E. L. and Fournier, R. O.,** Fish production and the marine ecosystems of the Scotian Shelf, Eastern Canada, *Mar. Biol.,* 54, 101, 1979.
1084. **Newell, R. C. and Linley, E. A. S.,** Significance of microheterotrophs in the decompositin of phytoplankton estimates of carbon and nitrogen flow based on the biomass of plankton communities, *Mar. Ecol. Prog. Ser.,* 16, 105, 1984.
1085. **Nixon, S. W. and Pilson, M. E. Q.,** Nitrogen in estuarine and coastal marine ecosystems, in *Nitrogen in the Marine Environment,* Academic Press, New York, 1984.
1086. **Patrick, W. H. and DeLaune, R. D.,** Chemical and biological redox systems affecting nutrient availability in the coastal wetlands, *Geosci. Man,* 18, 131, 1977.
1087. **Rhoads, D. C. and Young, D. K.,** The influence of deposit feeding organisms on sediment stability and community trophic structure, *J. Mar. Res.,* 28, 150, 1970.
1088. **Robertson, M. L., Mills, A. L., and Zieman, J. C.,** Microbial synthesis of detritus-like particles from dissolved organic carbon released by tropical seagrasses, *Mar. Ecol. Prog. Ser.,* 7, 279, 1982.
1089. **Sheldon, R. W., Prakash, A., and Sutcliffe, W. H.,** The size distribution of particles in the ocean, *Limnol. Oceanogr.,* 17, 327, 1972.
1090. **Sibert, J., Brown, T. J., Healey, M. C., Kosh, B. A., and Naimen, R. J.,** Detritus based food webs: exploitation by juvenile chum salmon *(Oncorhynchus keta), Science,* 196, 649, 1977.
1091. **Tenore, K. R. and Hanson, R. B.,** Availability of different detritus with aging to a polychaete macro-consumer *Capitella capitala, Limnol. Oceanogr.,* 25, 553, 1980.
1092. **Stephenson, W., Williams, W. T., and Cook, S. G.,** Computer analyses of Petersen's original data on bottom communities, *Ecol. Monogr.,* 42, 387, 1971.
1093. **Sutcliffe, W. H., Baylor, E. R., and Menzel, D. W.,** Sea surface chemistry and Longmuir circulation, *Deep-Sea Res.,* 10, 233, 1963.
1094. **Thompson, E. F.,** An Introduction to the Natural History of the Heathcote Estuary and Brighton Beach, Canterbury, New Zealand. A Study in Littoral Ecology, M.Sc. thesis, Canterbury University College, Christchurch, New Zealand, 1929.
1095. **Weise, W. and Rheinheimer, G.,** Flouroeszenzmikroskopische Untersuchungen über die Bakterein-besiedlung mariner Sandsedimente, *Bot. Mar.,* 22, 99, 1979.
1096. **Winter, J. E.,** Über den Einfluss der Nahrungskonzentration und anderen Faktoren auf Filtierleistung und Nahvungsausnutzung der Muscheln *Acartia islandica* and *Madiolus modiolus, Mar. Biol.,* 4, 87, 1969.
1097. **Woodwell, G. M., Whitney, D. E., Hall, C. A. S., and Haughton, R. A.,** The Flax Pond ecosystem study: exchanges of carbon in water between a salt marsh and Long Island Sound, *Limnol. Oceanogr.,* 22, 833, 1977.
1098. **Yingst, J. Y. and Rhoads, D. C.,** The role of bioturbation in enhancement of bacterial growth rates in marine sediments, in *Marine Benthic Dynamics,* Tenone, K. R. and Coull, B. C., Eds., University of South Carolina Press, Columbia, 1980, 407.

INDEX

E